Development of Non-Teleost Fishes

Development of Non-Teleost Fishes

Editors

Yvette W. Kunz
*University College Dublin
Maurice Kennedy Research Centre
University of Ireland
Dublin
Ireland*

Carl A. Luer
*Mote Marine Laboratory
Sarasota, Florida
USA*

B.G. Kapoor
*Formerly Professor of Zoology
The University of Jodhpur
Jodhpur
India*

Science Publishers

Enfield (NH)   Jersey   Plymouth

Science Publishers
234 May Street
Post Office Box 699
Enfield, New Hampshire 03748
United States of America

www.scipub.net

General enquiries : *info@scipub.net*
Editorial enquiries : *editor@scipub.net*
Sales enquiries   : *sales@scipub.net*

Published by Science Publishers, Enfield, NH, USA
An imprint of Edenbridge Ltd., British Channel Islands
Printed in India

© 2009 reserved

ISBN: 978-1-57808-500-2

Cover Drawings after
George Brown Goode 1884, George Albert Boulenger 1909 and Peter Bartsch 1997

Cover Photographs from Jennifer Wyffels 2009

Library of Congress Cataloging-in-Publication Data

Development of non-teleost fishes/editors, Yvette W. Kunz, Carl A. Luer, B.G. Kapoor.
       p. cm.
   Includes bibliographical references and index.
   ISBN 978-1-57808-500-2 (hardcover)
   1. Relict bony fishes--Development. 2. Chondrichthyes--Development. I. Kunz, Yvette W. II. Luer, Carl A. III. Kapoor, B.G.

QL639.25.D484 2009
597.13'8--dc22

2009003771

All rights reserved. No part of this publication may be reproduced, stored in a retrieval system, or transmitted in any form or by any means, electronic, mechanical, photocopying or otherwise, without the prior permission of the publisher, in writing. The exception to this is when a reasonable part of the text is quoted for purpose of book review, abstracting etc.

This book is sold subject to the condition that it shall not, by way of trade or otherwise be lent, re-sold, hired out, or otherwise circulated without the publisher's prior consent in any form of binding or cover other than that in which it is published and without a similar condition including this condition being imposed on the subsequent purchaser.

Contents

Preface		vii
1.	**Embryonic Development of Chondrichthyan Fishes—A Review** *Jennifer T. Wyffels*	1
2.	**Staging of the Early Development of *Polypterus* (Cladistia: Actinopterygii)** *Salif Diedhiou* and *Peter Bartsch*	104
3.	**Early Development of Acipenseriformes (Chondrostei: Actinopterygii)** *Teresa Ostaszewska* and *Konrad Dabrowski*	170
4.	**Early Ontogeny of Semionotiformes and Amiiformes (Neopterygii: Actinopterygii)** *Marta Jaroszewska* and *Konrad Dabrowski*	230
5.	**Early Development in Sarcopterygian Fishes** *Felisa Kershaw, Gregory H. Joss* and *Jean M.P. Joss*	275
Subject Index		290
Species Index		298

Preface

The term **'fishes'** refers to the most diverse group of vertebrates, and of these **teleostei** are the most speciose, containing up to 27,000 species. However, by the 1980s the voluminous studies concerning the ontogenesis of the teleostei were still overshadowed by the literature on the development of amphioxus, amphibians and birds, which was presented even in atlas form. In recent years genome sequencing projects for the teleosts zebrafish and the medaka have taken the limelight as experimental models for developmental and genetic analyses, entering even medical research and an up-to-date account of the development of teleostei, from classical descriptions to modern day cell and molecular analyses, has become available (Kunz, 2004).

An up-to-date compilation of the development of **non-teleost fishes** has so far been unavailable. These fishes include the jawless fishes (hagfish and lampreys), the cartilaginous fishes (sharks, rays, skates and chimaeras), the forerunners of the teleostei: the cladistia (bichirs and reedfish), the chondrostei (sturgeon and paddlefish), the neopterygii (gar pike and bowfin), and, finally, the closest relations to the tetrapods: the coelacanths ('living fossils') and the lungfishes (Protopterus of Africa, Lepidosiren of South America and Neoceratodus of Australia).

Therefore, the present volume has been devoted to closing the gap by an up-to-date scientific review of the early life-history of these non-teleost fishes (agnathi excepted). The editors have commissioned for each chapter recognised authors specialized in the particular fields. We thank them for their time and effort in making possible this book. We trust that it will be a valuable reference for students and researchers in the different aspects of fish development, comparative, phylogenetical, evolutionary and molecular.

Yvette Kunz
July 2009

1 Embryonic Development of Chondrichthyan Fishes—A Review

Jennifer T. Wyffels*
Daemen College, Amherst, NY 14226, USA.
E-mail: jwyffel@mindspring.com

INTRODUCTION

Class Chondrichthyes is a monophyletic group of cartilaginous fishes that evolved from jawless fishes more than 400 million years ago during the Devonian period (Compagno, 1977, 1999; Grogan and Lund, 2004). Subclass Elasmobranchii, sharks and rays, is composed of approximately 1100 species divided among 10 orders and 59 families (Compagno, 2005). Subclass Holocephali, chimaeras, is composed of 33 species divided into three families within a single order, Chimaeriformes (Compagno, 2005). Chondrichthyan fishes include the largest fish in the ocean, the whale shark, *Rhincodon typus*, which may reach 20 m (Stevens, 2007) while its smallest member, the dwarf dogshark, *Etmopterus perryi*, matures at only 19 cm (Springer and Burgess, 1985). Elasmobranchs are globally distributed throughout the oceans and many species inhabit brackish or even fresh water systems for part or all of their lives.

Sharks, rays and chimaeras have a wide range of life history traits but in general grow slowly, mature at a late age and possess long gestation periods with low fecundity (Stevens *et al.*, 2000; Carrier *et al.*, 2004). These characteristics make chondrichthyans particularly susceptible to overfishing and exploitation. For example, *Centrophorus uyato*, Cayman Trench, Jamaica produces only two pups during a three year reproductive cycle (McLaughlin and Morrissey, 2005). *Chlamydoselachus anguineus* has

---

*In grateful recognition of A.B. "Budd" Bodine.

the longest gestation for a viviparous species, estimated to be 3.5 years or longer (Tanaka et al., 1990). *Bathyraja parmifera* has the longest incubation for an oviparous species, estimated to be 1290 days (Hoff, 2007). The most fecund elasmobranch is the whale shark where a single female can carry nearly 300 embryos (Joung et al., 1996). Among other viviparous sharks, large litter sizes are reported for *Prionace glauca* (135 embryos), *Hexanchus griseus* (108 embryos), *Galeocerdo* (formerly *Galeorhinus) cuvier* (82 embryos), *Scymnodalatias albicauda* (59 embryos), *Galeorhinus galeus* (formerly *zyopterus*) (52 embryos), and *Ginglymostoma cirratum* (50 embryos) (Bigelow and Schroeder, 1948; Pratt and Casey, 1990; Nakaya and Nakano, 1995; Castro, 2000). While fecundity is high for some species, reproductive cycles are typically annual or biennial, with the possibility that some are even longer. For example, *G. cuvier* has a three year reproductive cycle with a gestation of 15-16 months, (Whitney and Crow, 2007) and *G. cirratum* has a two year reproductive cycle with a 5-6 month gestation (Castro, 2000).

Chondrichthyans have been subjects for embryological studies for more than 500 years. The reproduction and development of chondrichthyan fishes from its beginnings with Aristotle through the nineteenth century was recently reviewed (Wourms and Demski, 1993; Wourms, 1997). In brief, Aristotle described reproductive modes, egg cases and reproductive anatomy of elasmobranchs. Lorenzini (1678) was the first to illustrate the blastoderm of a chondrichthyan using four species of ray ((Lorenzini, 1678) within (Wourms, 1997)). Monro (1785) described gill filaments and Leuckart (1836) proposed they were plesiomorphic among chondrichthyans ((Monro, 1785; Leuckart, 1836) within (Wourms, 1997)). Home (1810) illustrated the respiratory canals present in oviparous egg cases and attempted a biochemical analysis of egg jelly. Skates were the subject of developmental studies by Wyman (1864) and Beard (1890). Coste (1850) described but did not illustrate cleavage ((Coste, 1850) within (Wourms, 1997)). Gerbe (1872) illustrated surface views of cleavage through blastula stages embryos ((Gerbe, 1872) within (Wourms, 1997)). Balfour's 1878 monograph, *On the Development of Elasmobranch Fishes*, was the first to establish a set of described embryonic stages for a chondrichthyan fish. Ziegler and Ziegler (1892) used reconstructions from serial sections and surface illustrations to characterize early development of the electric ray, *Torpedo marmorata*. Hoffman (1896) used *Squalus acanthias* (formerly *Acanthias vulgaris*) as a subject for developmental studies. Rückert (1899) demonstrated polyspermy and illustrated cleavage through gastrula stage embryos in surface view and sections. Chimaeroid development was illustrated and described by Schauinsland (1903) and Dean (1906). Scammon (1911) prepared a comprehensive analysis of both internal and external development for *S. acanthias*. Clark (1922, 1927) described incubation and developmental biology of a variety of skates. By 1919, chondrichthyans were established members of comparative embryology textbooks (Kerr, 1919; Nelson, 1953; Witschi, 1956; Pasteels, 1958). Unfortunately, our knowledge of developmental biology for chondrichthyans is little changed from these pioneering works especially considering the time that has elapsed since their completion.

Most current reports on the reproductive biology of elasmobranchs include general data about embryos in an effort to describe the gestation portion of the reproductive cycle. Embryos are often weighed, measured and their development divided into broad categories such as early, mid or late gestation (Hisaw and Albert, 1947; Jones and Ugland, 2001; Hernandez et al., 2005; Braccini et al., 2007). Features of development noted include obvious characters such as presence of gill filaments, an external yolk sac (EYS), fin folds, claspers and body pigmentation (Capapé et al., 2006; Castro et al., 1988; Dral, 1981; Verissimo et al., 2003). Descriptions of embryos during early development, before separation from the yolk by a stalk, are infrequent. Many reports of good detail exist for isolated stages of embryos acquired by chance, but they are usually well-developed specimens (Davy, 1834; Parker, 1882; Southwell and Prashad, 1910; von Bonde, 1945; Nair and Appukuttan, 1974; Appukuttan, 1978; Natanson and Cailliet, 1986; Stehmann and Merrett, 2001; Ribeiro et al., 2006). Tewinkel's (1950) descriptions of *Mustelus canis* embryos detailing consecutive developmental stages in sequentially fertilized eggs is an exception.

Few contemporary scientists focus their efforts on developmental biology of chondrichthyans. Melouk (1949, 1957) defined and described in great detail 10 stages of development in *Carcharhinus melanopterus* and 13 stages of development for the rajiforms *Rhinobatos halavi* and *Rhynchobatus djiddensis*. Included are descriptions of comparative body form differentiation between rajiforms and torpediform sharks. Observations on the development of *Hemiscyllium ocellatum* in captivity were described by West and Carter (1990). Several manuscripts, including details of embryonic development, precede a complete description of normal stages of development for *Scyliorhinus canicula* (Ballard et al., 1993; Lechenault and Mellinger, 1993; Lechenault et al., 1993; Mellinger et al., 1984, 1986, 1987). Their research is the first to include a timetable for development and describes a single species from fertilization until hatching. Development of *Rhizoprionodon terraenovae* was described by Castro and Wourms (1993). Chimaeroids were re-investigated using modern techniques by Didier et al. (1998) and batoid development was described for *Raja eglanteria* (Luer et al., 2007). Captive breeding and embryonic development were described for *Chiloscyllium punctatum* (Harahush et al., 2007). Most recently, the functional morphology of embryonic development was described for *Heterodontus portusjacksoni* (Rodda and Seymour, 2008). This chapter summarizes reproductive habits as they relate to embryonic development and describes both commonalities and distinctions in the developmental biology of chondrichthyan species.

Mating, Ova and Fertilization

All elasmobranchs are anamniotes that utilize internal fertilization. Mating is rarely observed either in captivity or the wild and observations are usually serendipitous (Uchida et al., 1990; Masuda, 1997; Pratt and Carrier, 2001; Chapman et al., 2003). For most studies of reproductive cycles, mating is supposed from the presence of bite marks or mating scars on females. In rare cases, semen can be observed in a smear made from cloacal fluid confirming a recent mating event. Mating among wild populations of nurse sharks, G. cirratum, has been observed and documented in the Dry Tortugas archipelago, Florida (Carrier et al., 1994). Polyandry appears normal and genetic analysis estimate 3-6 sires per litter for nurse sharks (Ohta et al., 2000; Saville et al., 2002; Heist, 2005). Multiple paternity has been demonstrated for *Negaprion brevirostris* (Feldheim et al., 2001, 2004), *Carcharhinus altimus* (Daly-Engel et al., 2006) and *Sphyrna tiburo* (81%) (Chapman et al., 2004). Monogamy and polyandry for Hawaiian *Carcharhinus plumbeus* occur in nearly equal proportions (Daly-Engel et al., 2007) but in the western North Atlantic and Gulf of Mexico 85% of litters had more than one sire (Portnoy et al., 2007). Polyandry is suggested for *Dasyatis americana* (Chapman et al., 2003) and proven for *Raja clavata* L. (Chevolot et al., 2007).

For many species, females are known to store viable sperm within the terminal zone of the oviducal gland (Pratt and Tanaka, 1994; Conrath and Musick, 2002; Hamlett et al., 2002a,b; Smith et al., 2004). This sperm is used to fertilize eggs throughout the reproductive season, resulting in a disjunction between ovulation and mating. Sperm is stored for several weeks to a year depending on the species (Pratt, 1993; White et al., 2002b). Sperm storage is especially common among oviparous species with protracted oviposition (Luer and Gilbert, 1985; Castro et al., 1988).

The reproductive tract of female chondrichthyans is derived from the Müllerian duct during embryogenesis at the time of sexual differentiation (Chieffi, 1967). In this regard chondrichthyans distinguish themselves from agnathans and teleosts who do not possess a Müllerian duct. Chondrichthyans possess single or paired ovaries and oviducts with unilateral or bilateral functionality. The tract is composed of segments which are specialized in function according to specific reproductive strategies. Generalities are common regardless of species and include accepting ova, hosting sperm, investing the egg with protective envelopes, retaining (however transient) and protecting developing embryos and finally delivering eggs or neonates into the environment (Hamlett and Hysell, 1998).

The ovary sequesters vitellogenin released into the bloodstream from the liver and ovarian follicle cells (Perez and Callard, 1992; Marina *et al.*, 2004). Vitellogenin is cleaved yielding phosvitin and lipovitellin. It is stored as membrane delimited yolk platelets (Romek and Kilarski, 1993). The shape and size of platelets differ among species (Rückert, 1899). Other constituents of yolk include lipids, proteins, minerals, oligosaccharides and immunoglobulins (Needham, 1931, 1942; Plancke *et al.*, 1996; Haines *et al.*, 2005). Oocytes are ovulated into the body cavity as in all vertebrates. They migrate to the single conjoined ostium or paired ostia at the anterior of the peritoneal cavity with the help of cilia (Metten, 1939). The ova pass into the oviduct and enter the shell gland where they are invested with egg jelly by the club and papillary zones (Hamlett *et al.*, 2005b). The site of fertilization for *S. canicula* is within the shell gland but direct observations for other species are lacking (Lechenault and Mellinger, 1993). The fragile ova with jelly are enclosed in a tertiary egg envelope by the shell gland. The structure and function of the shell gland, also termed the nidamental or oviducal gland, were recently reviewed (Hamlett *et al.*, 2005b).

Shell gland morphology is representative of the capsule it secretes. There are few exceptions to egg encapsulation and in these cases the shell gland is reduced and only secretes egg jelly (Prasad, 1945; Babel, 1967; Thorson *et al.*, 1983). Ovarian follicles and shell gland size grow in concert until ovulation when the gland encapsulates ova and regresses. In oviparous species, the gland remains enlarged throughout the laying season. Capsule formation requires 12-24 hours, necessitating storage of shell precursor proteins for rapid extrusion (Koob *et al.*, 1986). Control of shell gland activity is independent of ova presence and anovular egg cases with only egg jelly (wind cases) are common (Beard, 1890; TeWinkel, 1950; Smale and Goosen, 1999; Garner, 2003; Koop, 2005). The egg continues from the shell gland into the uterus where it will stay until embryo development is complete for viviparous species. In oviparous chondrichthyans the egg may be retained for several days before being deposited in the substratum or onto upright gorgonian coral and seaweed (Castro *et al.*, 1988; Ellis and Shackley, 1997).

Eggs of chondrichthyans are both megalecithal and telolecithal with meroblastic cleavage. Egg sizes range over two orders of magnitude. The smallest egg, 1 mm, is described from *Scoliodon laticaudus* (Teshima *et al.*, 1978). Among chondrichthyans with the largest eggs are *Centrophorus granulosus* with ova weighing 370 g (Guallart and Vicent, 2001) and *G. cirratum* and *C. anguineus* with ova measuring nearly 100 mm (Gudger, 1940). Pigmentation in fresh eggs varies from pale green and yellow to bright shades of yellow with orange, pink or green hues.

Elasmobranch reproduction combines oogenesis and gestation into a single dependent cycle (Carrier *et al.*, 2004). Some species have concurrent development of oocytes and embryos allowing females to be continuously gravid once mature. Others employ a sequential development of oocytes followed by pregnancy. Sequential development does not necessarily require more time than concurrent development as gestation times may be significantly reduced. Females sometimes utilize a rest period after parturition extending their reproductive cycle to two or more years.

Reproduction and Development Strategy

Chondrichthyans utilize a range of reproductive strategies that are defined by the site of embryonic development and source of energy for development (Wourms, 1977, 1981). If any portion of embryonic development is completed outside of the body, the mode of reproduction is oviparity. Internal development until parturition, live-bearing, is defined as viviparity. These two categories are further divided by nutrient quantity. Lecithotrophic development occurs when the female's investment towards embryonic development is contained within the egg or vitelline mass. Matrotrophy occurs when the female adds energy to the fetal compartment during development. There are multiple forms of matrotrophy defined by the content and source of caloric provisions supplied to the embryo in addition to its yolk sac. Yolk sac viviparous species receive limited or no additional nutrients from the mother. Histotrophy is the release of nutrient containing fluid from the uterine mucosa. Oophagy is nutrient in the form of ova for embryo consumption. These ova may or may not be fertile and in one species results in adelphophagy or intrauterine cannibalism. Placental viviparity is the result of the exsanguinated yolk sac intimating with the uterine wall to form a placenta. Yolk sac viviparity, histotrophy, oophagy and adelphophagy are reproductive modes independent of a placental attachment. The ratio of embryo mass at hatching or parturition to initial egg mass is an indication of the degree of matrotrophy. A decrease in embryo mass occurs in oviparous and yolk sac viviparous species. An increase in embryo mass occurs in placental sharks, histotrophic myliobatoids and oophagous species (Wourms, 1981; Koob, 1999).

For all species, incubation is defined as the time from fertilization until hatching, where hatching is emergence from the tertiary egg envelope. For oviparous species, oviposition is deposition of the eggs outside of the body and eclosion defines the opening of the egg to its environment via the respiratory slits in the egg case. Oviposition precedes eclosion with the possible exception of species utilizing multiple oviparity, *G. cirratum* and *R. typus*. Gestation is defined as the period from fertilization until parturition for viviparous species and includes a short incubation interval. For placental viviparous species, hatching is equivalent to gestation for embryos that remain in their tertiary egg case until parturition.

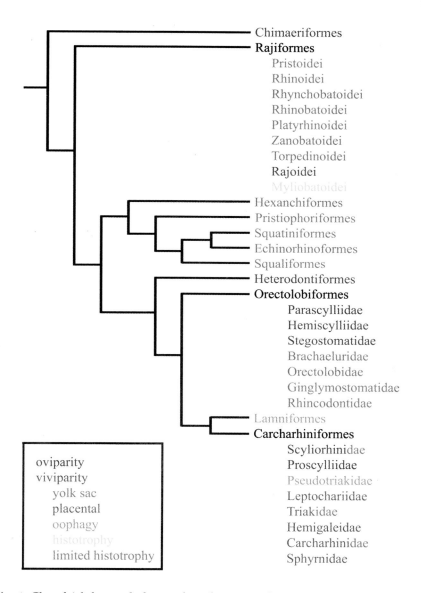

Fig. 1. Chondrichthyan phylogeny based on reproductive mode. The cladogram shows relationships among chondrichthyan orders, with additional detail provided at the suborder level for Order Rajiformes and family level for Order Orectolobiformes and Order Carcharhiniformes. Bi-colored names indicate multiple reproductive modes, with the color of the more frequent mode shown first. Modified from Musick, J.A., and J.K. Ellis (2005) with kind permission of Science Publishers, Enfield, NH.

A recent phylogenetic analysis of reproductive mode for Neoselachii predicts yolk sac viviparity as plesiomorphic with six transitions to oviparity and one reversal (Fig. 1) (Musick and Ellis, 2005). Previous analysis predicted oviparity as plesiomorphic and the prolongation of egg retention a factor for development of yolk sac viviparity (Wourms, 1977; Wourms and Lombardi, 1992; Dulvy and Reynolds, 1997; Carrier et al., 2004). Reproductive habits are uniform for all members of three of the ten chondrichthyan orders, Lamniformes, Chimaeriformes and Heterodontiformes. Regardless of the reproductive mode, neonates are fully formed individuals and no parental care is ever afforded after oviposition in oviparous species or parturition in viviparous species.

Oviparity

Oviparity occurs in approximately 43% of elasmobranchs including all Chimaeriformes, Heterodontiformes, Rajoidea (skates) and Scyliorhinidae as well as a few species of Orectolobiformes (Compagno, 1990). There is no maternal investment except the ova and egg envelopes, representing a pure lecithotrophic mode of nutrition. The surrounding water may be a source for inorganic ions and water after egg case eclosion (Koob, 1991).

Egg cases of skates have traditionally been termed mermaid purses or sailor snuff boxes owing to their shape (Putnam, 1870; Libby, 1959). The tertiary egg envelope or case takes on several interesting forms related to the deposition habitat of the female (Fig. 2). The ability to distinguish species using egg capsule morphology has been understood since the 19th century (Beard, 1890; Hussakof, 1914). In captive environments, egg cases can be attributed to individual females based on shape and pigmentation (Mellinger, 1983; Castro et al., 1988; Howard, 2002). The formation of the egg capsule is similar for all species but its final form is dependent on the activity and arrangement of tubules within the shell gland (Smith et al., 2004). The egg case may be adorned with byssal fibers, hairs and/or tendrils to aid in substrate attachment (Figs. 2-3) (Putnam, 1870; Koob and Straus, 1998; Ebert and Davis, 2007). Heterodontid egg capsules are screw-shaped and while females are believed to lodge eggs into crevices, natural deposition has not been observed (Fig. 2D) (Haswell, 1898; Smith, 1942; McLaughlin and O'Gower, 1971). The egg case of *Sympterygia acuta* has extremely elongated posterior tendrils covered with silky fibers for attachment to surf-aggregated debris because the substrate is devoid of vegetation or other permanent anchoring sites (Oddone and Vooren, 2002). *Callorhinchus milii* has hairs to aid in substrate attachment and camouflage as well as a concave surface that acts to anchor the case to the sea floor (Fig. 2B) (Smith et al., 2004).

Fig. 2A, B. See next page for caption.

Two forms of oviparity are distinguished, single or extended and multiple or retained (Nakaya, 1975; Compagno, 1990). Single or extended oviparity is continual oviposition of individual eggs throughout the reproductive season. Embryonic development occurs exclusively external to the female with the exception of the earliest stages of development. Multiple or retained oviparity is found in females that retain one or more eggs in each oviduct for a significant portion of their incubation but oviposit before hatching (Fig. 4E-F). The retained egg case is tough and leathery, typical of oviparous species, but the tendrils are reduced or absent. Some confusion presents itself when considering the polyovular egg cases of *Raja binoculata* and *Raja pulchra*. Both species are single oviparous but deposit multiple fertilized ova in each egg case (Ishiyama, 1958; Hitz, 1965). The egg case of *R. binoculata* (Fig. 4D) is largest among rays (>300 mm) and

Fig. 2C, D. Chondrichthyan egg case diversity. Egg cases from (A) *Cephaloscyllium ventriosum*, (B) *Callorhinchus milii*, (C) *Raja binoculata*, (D) *Heterodontus francisci*. (Photographs A and D—marinethemes.com/ Mark Conlin; B—marinethemes.com/Kelvin Aitken; C—marinethemes.com/Andy Murch).

contains 2-7 embryos (Hitz, 1965) while *R. pulchra* encapsulates 1-5 embryos (Ishiyama, 1958). The temptation to apply the term multiple oviparity to these two species is palpable but incorrect as defined.

The egg case and associated adornments are products of the shell gland and composed of different proportions of the same collagen-like structural proteins (Koob and Cox, 1993; Koob and Straus, 1998). Confusion and contradictions in the literature exist concerning the designation of anterior and posterior with respect to the case termini (Teshima and Tomonaga, 1986). In chimaerids and skates, the case is secreted by the shell gland beginning with the hatching terminus and finishing with the non-

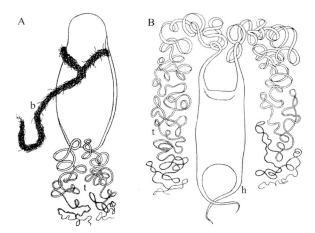

Fig. 3. Egg cases from Patagonia collected and described by Louis Hussakof (1914). (A) Case of an unidentified species but not *Acanthias*, *Centroscyllium* or *Scyllium*. (B) Case of a *Raja*. sp. remarkable for its long tendrils. t—tendrils; h—horn; b—byssal fibers. Reprinted from Hussakof, L. (1914).

hatching terminus (Dean, 1906; Fitz and Daiber, 1963). The completed egg case within the uterus dictates the non-hatching terminus anterior and the hatching terminus posterior in relation to the female (Fig. 4A-B). The embryo within the egg case must orient with its head toward the hatching terminus in order to successfully emerge from the case. The girdle at the non-hatching terminus does not open and prevents escape. The egg case hatching terminus is anterior and the non-hatching terminus posterior with respect to its inhabitant (von Bonde, 1945; Fitz and Daiber, 1963; Luer and Gilbert, 1985). In contrast, heterodontids and scyliorhinids form the non-hatching terminus first (Aiyar and Nalini, 1938; Whitley, 1940; Dodd and Dodd, 1986; Castro *et al.*, 1988; Feng and Knight, 1994). Terminology here is defined with the anterior and posterior of the egg case correlated to function, hatching and non-hatching respectively.

The egg case is composed of a collagen-like matrix in orthogonal layers (Koob and Cox, 1993). Skate capsule formation is synchronous and precedes ovulation. The fertilized egg enters the capsule one third to three fourths through its formation by the baffle zone of the shell gland (Wyman, 1864; Beard, 1890; Hobson, 1930; Metten, 1939; Setna and Sarangdhar, 1948b; Richards *et al.*, 1963; Fitz and Daiber, 1963; Templeman, 1982; Feng and Knight, 1994; Heupel *et al.*, 1999). Oviposition is delayed 2-7 days for *Leucoraja* (formerly *Raja*) *erinacea* while the egg case undergoes sclerotization by a quinine tanning mechanism (Setna and Sarangdhar, 1948b; Callard and Koob, 1993; Koob and Hamlett, 1998). After tanning the egg

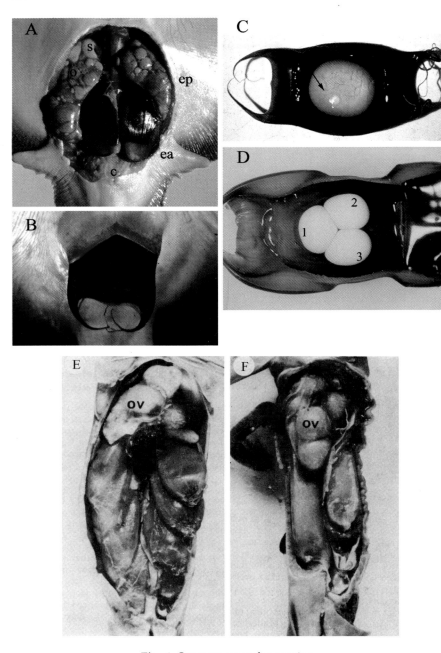

Fig. 4. See next page for caption.

case is structurally and mechanically sound to withstand prolonged incubation conditions in the environment (Hepworth et al., 1994).

Many oviparous species mate and reproduce in captivity (Uchida et al., 1990; Henningsen et al., 2004). Successful artificial insemination in captive oviparous elasmobranchs has been reported for R. eglanteria and Scyliorhinus torazame (Masuda et al., 2003; Luer et al., 2007). Oviparous species either lay eggs throughout the year, S. canicula (Sumpter and Dodd, 1979) and L. erinacea (Fitz and Daiber, 1963; Richards et al., 1963) or have a distinct seasonality, R. eglanteria (Luer and Gilbert, 1985) and H. portusjacksoni (McLaughlin and O'Gower, 1971). Eggs from captive elasmobranchs provide a convenient source of material for developmental studies (West and Carter, 1990; Masuda, 1997).

The egg case contains the yolk, or vitellus, and egg jelly. Initially, the egg jelly was compared to star-shot jelly, regurgitated swollen jelly from frog oviducts deposited in trees by gastro-intestinally distressed birds (Home, 1810). The egg jelly serves as hydrodynamic support for the egg and embryo during early development (von Bonde, 1945; Luer and Gilbert, 1985; Koob and Straus, 1998; Wyffels et al., 2006). If a portion of the egg jelly is removed from the egg case before epiboly is complete, the fragile ova spreads laterally and the plasmalemma soon ruptures (Fig. 5A). Before eclosion, the egg responds to the osmolarity of the environment through dehydration or swelling of the semi-solid egg jelly until equilibrium is achieved. There is no evidence, however, that the egg jelly actively participates in osmoregulation (Read, 1968; Kormanik, 1993; Rodda and Seymour, 2008). Changes in hydration of the egg jelly are probably responsible for observed fluctuations in weight of freshly deposited egg cases (Foulley and Mellinger, 1980; Wyffels et al., 2006). Egg jelly swelling was noted first by Home in S. acanthias, a yolk sac viviparous species. When the eggs within their membranous case were removed from the oviduct and placed in "proof spirit" for preservation the case burst due to swelling of the egg jelly (Home, 1810). The eggs of S. torazame increase or

---

Fig. 4. Egg case orientation. (A) Female *Raja eglanteria* with two egg cases. The egg case is synthesized by the shell gland, s, beginning with the anterior or hatching terminus. o—ovary; ea—egg case anterior; ep—egg case posterior. (B) Cloaca, c, of female with the anterior of an egg case protruding. (C) *R. eglanteria* egg case with a single ovum. The blastodisc is indicated by an arrow. (D) *Raja binoculata* egg case with three ova. (E) Multioviparity in a female *Halaelurus buergeri* with eggs retained in each oviduct. The left oviduct is bisected to show 4 eggs with reduced tendrils. (F) Multioviparity in a female *Proscyllium habereri* with a single retained egg in each oviduct. Images E and F reprinted from Dodd and Dodd (1986) with kind permission of Springer Science and Business Media.
(Photographs A—Jose Castro and Carl Luer; B and C—Carl Luer).

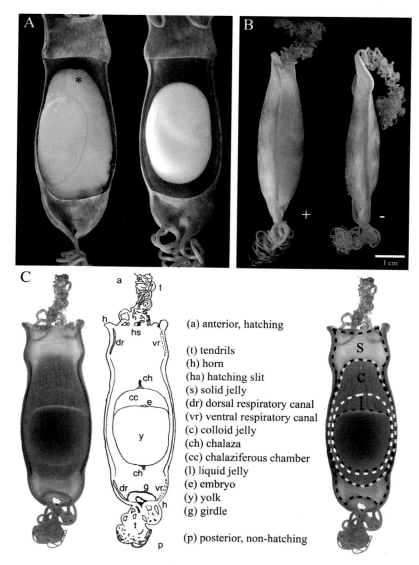

Fig. 5. *Scyliorhinus torazame* egg case terminology and visualization of egg jelly function. (A) Egg jelly provides hydrodynamic support for the fragile vitelline membrane during early development. If the colloid jelly is removed, the yolk expands laterally (*) and will likely cause rupture. (B) The egg jelly responds to the osmolarity of the surrounding water, gaining 11.16 ± 1.99% wet weight after incubation in half strength seawater (+) and losing 10.16 ± 0.79% wet weight when incubated in double strength seawater (-). (C) Terminology for the egg case and its contents.

decrease in weight and volume by approximately 10% when incubated in half strength or double strength seawater respectively, but rarely burst due to the incredible strength of the egg case (Fig. 5B) (Wyffels et al., 2006).

The egg jelly is homologous to the albuminous egg white of chickens, but no evidence for albumen has been found in chondrichthyan eggs. The club and papillary zones of the shell gland secrete egg jelly. Based on staining of the gland in histological preparations with alcian blue and periodic acid-Schiff (PAS), the jelly may contain carbohydrate, neutral mucopolysaccharide, both sulfated and unsulfated glycosaminoglycans, glycoproteins, acidic polysaccharides and sialoglycoproteins (Threadgold, 1957; Rusaouen, 1976; Feng and Knight, 1992; Hamlett et al., 2002a; Smith et al., 2004). Biochemical studies of the liquid jelly have identified proteins and carbohydrates (Koob and Straus, 1998; Wyffels et al., 2006). Egg jelly is also contained within egg cases of viviparous species, although its volume is greatly reduced. The degree of variability in its chemical composition among chondrichthyans is unknown.

Skates are often referred to as chickens of the sea and the terminology used below to describe chondrichthyan eggs is borrowed from avian biology. Within the egg case, the ovum is suspended from its longitudinal equatorial axis by spiral chalazae. There are three layers of egg jelly differing in proximity to the ovum and viscosity (Figs. 5C and 6A). The egg is immediately surrounded by liquid contained within a chalaziferous chamber. The liquid jelly allows the ovum to rotate freely and the embryo maintains an upright position within the egg case as it changes orientation with prevailing seawater currents. As the embryo grows, the liquid enables the embryo to contract its developing muscles. This liquid layer is surrounded by a clear viscous colloid jelly. Finally, the terminal ends of the egg case contain a dense plug of white semi-translucent solid jelly. *S. torazame* has four respiratory canals or fissures in the egg case, two dorsal and two ventral. On each dorsal and ventral surface, the canals are located on the same side of the egg case, one canal on the anterior or hatching terminus and one canal on the posterior or non-hatching terminus. The solid egg jelly occludes the respiratory canals from the environment for the initial portion of incubation (Figs. 5C and 6A). Respiratory canals were illustrated for catsharks by Home (1810) and described by Wyman (1864) for skates.

The layers of egg jelly differ biochemically (Koob and Straus, 1998; Wyffels et al., 2006). The carbohydrate composition of liquid, colloid and solid *S. torazame* egg jelly was measured by HPLC analysis of acid-hydrolyzed samples and six sugars were detected, N-acetylgalactosamine, N-acetylmannosamine, N-acetylglucosamine, fucose, galactose and mannose (Wyffels et al., 2006). The monosaccharide with the highest concentration in the solid and colloid layers was galactose. The monosaccharide with the highest concentration in the liquid jelly (before liquefaction of colloid and solid jelly) was fucose. Solid jelly had the highest total carbohydrate concentration followed by colloid and liquid jelly. The layers of jelly differed in degree of hydration. The solid jelly was 94.2 ± 0.7 % water and the colloid jelly was 84.1 ± 1.9 % water with liquid jelly assumed to be 100 % water. Newly laid egg cases increased in weight 5.9 ± 1.8 % during the first two weeks after laying (Wyffels *et al.*, 2006).

The liquid portion of the egg jelly increases in amount during incubation due to the liquefaction of the colloid and solid layers. Initially, the liquid jelly is confined within the chalaziferous chamber. After 70 ± 9 days of incubation at 14-16°C, the chamber of *S. torazame* is compromised when the colloid layer is liquefied. After 103 ± 6 days of incubation at 14-16°C, liquefaction is complete and the respiratory canals allow passage of the surrounding seawater through the case. The amount of carbohydrate and protein in the liquid layer increases with liquefaction of the colloid and solid layers (Fig. 6A). At eclosion, the carbohydrates and proteins are flushed from the case within 24 hours, precluding the jelly as a source of nutrition (von Bonde, 1945; Koob and Straus, 1998; Wyffels *et al.*, 2006).

Liquefaction is attributed to secretions of the eclosion gland (Ouang, 1931). The presence of the gland has been confirmed in *S. torazame* (pers. obs.), *L. erinacea* (Straus and Koob, 1997), *S. canicula*, *Scyliorhinus stellaris*, *Galeus* (formerly *Pristiurus*) *melastomus*, *Raja asterias*, *T. marmorata* Risso and *Torpedo ocellata* Risso (Ouang, 1931). Liquefaction is complete at 22-49% of incubation (Table 1). The eclosion gland is obvious as a helmet shaped layer of cells covering the head of *S. torazame* embryos (Fig. 6B-C). It is present in rajids under the epithelium and invisible without histology (Ouang, 1931). The secretions of the eclosion gland are not well characterized. Protease activity was identified in the egg jelly of *L. erinacea*. The activity was first observed 33 days after oviposition and increased until eclosion. There were multiple bands of activity ranging in molecular weight from 22 kDa to 125 kDa with the band distribution changing over time (Straus and Koob, 1997). Egg case opening in *H. portusjacksoni* is reported to be independent of an eclosion gland, with no significant difference in time of egg case opening between egg cases containing embryos and egg cases experimentally devoid of yolk and embryos (Rodda and Seymour, 2008).

Ventilation

After eclosion, water flow through the eggs cases occurs by both active and passive mechanisms. If the respiratory canals are blocked, the embryo will die by asphyxiation (Richards *et al.*, 1963). Currents flowing around egg cases induce passive flow through the case (Koob and Summers, 1996).

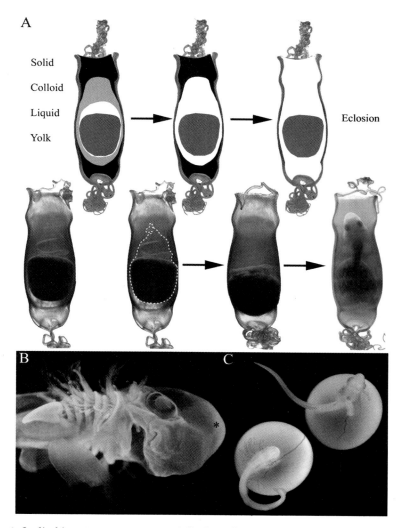

Fig. 6. *Scyliorhinus torazame* egg case jelly liquefaction and eclosion. (A) The ovum is surrounded by three layers of jelly. The innermost layer is liquid supported by a colloid layer, with solid jelly filling the termini and plugging the respiratory canals. The jelly is liquefied during development resulting in pre-hatching or eclosion. The embryo is contained within the chalaziferous chamber until the colloid jelly is liquefied. When the chamber is compromised (dotted line), the embryo drops vertically to the margin of the solid jelly. (B) The eclosion gland (*) is responsible for liquefaction of the jelly layers. (C) The embryos at eclosion (stage 31) have no body pigmentation except around the eye. The yolk sac is completely vascular and the embryo can be removed from its case and cultured in seawater without consequence.

Table 1. Eclosion Among Elasmobranch Species

Species	Incubation temperature, °C	Approximate % of incubation completed[1]	Reference
Hemiscyllium ocellatum	25	22	West and Carter (1990)
Chiloscyllium punctatum	21-25	23	Harahush et al. (2007)
Raja eglanteria	20-22	33	Luer and Gilbert (1985)
Chiloscyllium griseum	25	34	Aiyar and Nalini (1938)
Scyliorhinus canicula	16	36	Ballard et al. (1993)
Heterodontus portusjacksoni	20	37	Rodda and Seymour (2008)
Scyliorhinus torazame	14-16	48	Wyffels et al. (2006)
Scyliorhinus retifer	12-13	49	Castro et al. (1988)

[1]Calculated from information in the original publications.

As the embryo grows and fills the egg case, its respiratory demands exceed the oxygen supply from passive water flow and diffusion across the capsule wall (Diez and Davenport, 1987; Koob and Summers, 1996). Embryos increase water flow through the respiratory canals with the development of specialized and transient tail anatomy (Libby, 1959; Luer and Gilbert, 1985; Mellinger and Wrisez, 1993; Meehan et al., 1997; Eames et al., 2007; Johanson et al., 2007).

Scyliorhinid embryos actively pump water through the respiratory canals after eclosion using embryonic caudal scales (Castro et al., 1988; Mellinger and Wrisez, 1993). The scales are found in a dorsal and ventral row on the both surfaces of the caudal fin, with fewer scales on the ventral row (Figs. 7A and 9C). The scales erupt in sequence, proceeding cephalad after eclosion. *S. torazame* embryos move their tails in a sweeping motion past two respiratory fissures (Fig. 8C). Water enters both proximal fissures, either hatching or non-hatching, in response to the sweeping motion and exits the opposite fissures (pers. obs.). The caudal scales and tail sweeping are used during a period when the gill filaments are at their maximum length. The caudal scales are also present on heterodontid embryos (Johanson et al., 2007).

Cephaloscyllium ventriosum mass specific oxygen consumption rates decrease during development (Meehan *et al.*, 1997). Embryos utilize caudal scales for active ventilation but passive water flow could supply the respiratory demands of the embryo if the egg case is exposed to a 0.2 knot current. *C. ventriosum* embryos increase their oxygen demands until approximately 70% of incubation is complete, paralleling increases in body mass. This is in contrast to *S. canicula* embryos where oxygen demands and body mass increase until hatch (Diez and Davenport, 1987). As the embryo grows and fills the egg case the tail is spatially inhibited, the gill filaments regress and oropharyngeal pumping commences (Fig. 8D). The caudal scales begin to be shed before hatching for most embryos and are completely lost soon after hatching (pers. obs.) (Johanson *et al.*, 2007).

Among rajids, the tail has an elongated whip or filament that oscillates within an egg case horn (Fig. 8A-B) (Clark, 1922; Long and Koob, 1997; Hoff, 2007). The tail filament is composed of notochord and muscle and is 12 mm long in *L. erinacea* (Long and Koob, 1997), 25 mm long in *B. parmifera* (Hoff, 2007) and 7-16 mm long at hatch for *Amblyraja radiata* (Berestovskii, 1994). For *L. erinacea*, the filament movement in the egg case horn creates positive pressure driving water out and results in negative pressure in the three unoccupied horns where water enters the egg case (Fig. 8B) (Long and Koob, 1997). The energetic cost of using the tail filament for pumping water through the case is significant. For *L. erinacea*, a 53-81% increase over basal metabolic rate is required (Leonard *et al.*, 1999). The tail filament is reabsorbed within weeks after hatching (Clark, 1922; Long and Koob, 1997).

Laying Rates

Intervals for egg laying vary widely (Table 2). Oviposition rate for all species is probably influenced by temperature (Holden *et al.*, 1971). Incubation times also vary widely and are highly dependent on temperature with *B. parmifera* requiring 1290 days at a mean developmental temperature of 4.28°C for the longest incubation and *R. eglanteria* requiring 63 days at a mean incubation temperature of 24°C for the shortest incubation (Libby and Gilbert, 1960; Hoff, 2007). For *C. punctatum* an increase in 3°C shortened the incubation period by 12-27% (Garner, 2003; Michael, 2001). *Stegostoma fasciatum* eggs from two aquaria incubated at nearly the same water temperature (24.4 +/- 0.8 and 24.5 +/- 0.8°C) had a mean incubation time that differed by almost 30 days. This indicates the presence of additional undefined factors influencing development and hatching. Egg size is one such factor, with incubation time lengthening with increased vitelline mass.

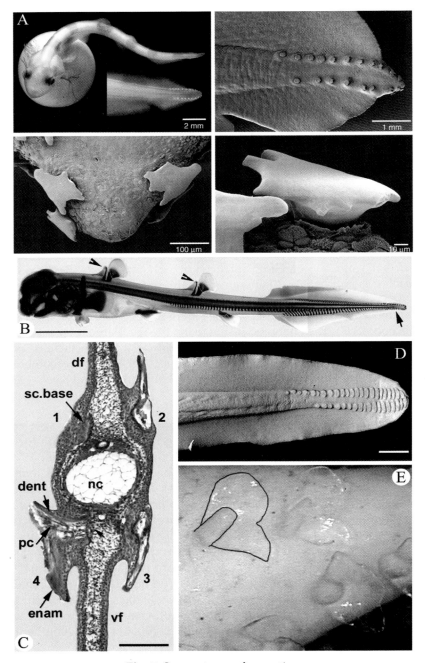

Fig. 7. See next page for caption.

Table 2. Egg Laying Rates Among Oviparous Chondrichthyan Species

Species	Interval between laying of successive egg pairs (days)	Reference
Raja clavata	0-2	Ellis and Shackley (1995)
Raja eglanteria	4.5 ± 2.2	Luer and Gilbert (1985)
Sympterygia bonapartei	4.3 ± 3.2	Janez and Sueiro (2007)
Chiloscyllium griseum	2-9	Jagadis and Ignatius (2003)
Chiloscyllium plagiosum	6-7	Miki (1994); Chen and Liu (2006)
Callorhinchus milii	7-10	Didier *et al.* (1998)
Scyliorhinus retifer	14-17	Castro *et al.* (1988)
Heterodontus francisci	7-22	Farewell (1972)
Parascyllium variolatum	12-20	Caruso and Bor (2007)

Hatching may be facilitated in Scyliorhinids by a set of dorsolateral scales that erupt during incubation (Nakaya, 1975; Ballard *et al.*, 1993; Eames *et al.*, 2007). These scales are proposed to aid hatching by preventing posteriorad movement of the emerging pup (Fig. 9). A thrust of the curled tail followed by alternating lateral flexions is responsible for liberation from the egg case (Grover, 1974). Dorsolateral denticles of similar morphology are not observed among viviparous species and remain undescribed for most oviparous chondrichthyans.

Multiple Oviparity

Multiple oviparity, the retention of one or more eggs for a portion of incubation, occurs in seven species of Carcharhiniformes and may occur in a single species of Orectolobiformes, *S. fasciatum*, where data are sparse.

Fig. 7. Embryonic caudal scale morphology. After eclosion a set of caudal scales erupts on both surfaces of the tail, shown here with *Scyliorhinus torazame* (A) and *Heterodontus* (B-E). (B) Left lateral view of 7.5 cm *Heterodontus portusjacksoni*, black arrowheads indicate calcifying tip of dorsal fin spines and black arrow indicates caudal scales. (C) Transverse section through the posterior caudal fin of *H. portusjacksoni* showing position of four scales (1-4), H&E stain. Opposing, but offset scale rows from left and right sides surround the notochord. dent—dentine; d.f., vf—dorsal and ventral fins; enam—enameloid layer; nc—notochord; pc—pulp cavity; sc. base—scale base (D) Left lateral view of 6 cm *H. portusjacksoni* caudal fin showing opposing rows of scales. (E) Detail showing bilobed and trilobed scales in these rows (one scale being outlined in black) of 8 cm *Heterodontus galeatus*. Scale bars: (B) 1 cm, (C) 0.5 mm, (D) 0.25 cm. Images B-E reproduced from Johanson *et al.* (2007) with kind permission of Blackwell Publishing, Oxford, UK.

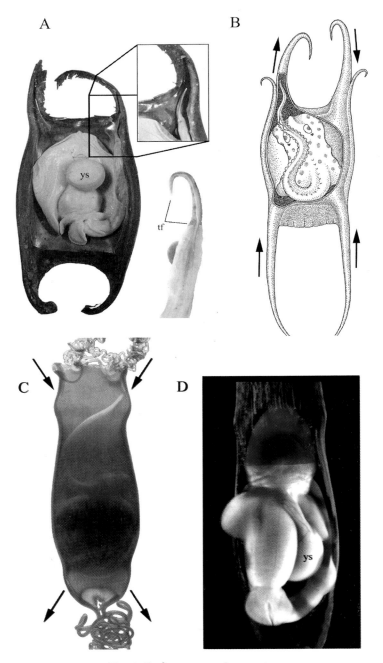

Fig. 8. See next page for caption.

Among Carcharhiniformes, six are scyliorhinids with two species of *Galeus* (*G. melastomus* and *G. atlanticus*) and four species of *Halaelurus* (*H. boesemani, H. buergeri, H. lineatus* and *H. natalensis*). The final member of Genus *Halaelurus, H. quagga*, is predicted to share this reproductive character (Francis, 2006). Unlike the scyliorhinids, *Proscyllium habereri* retains a single egg in each oviduct (Dodd and Dodd, 1986). Proscyllids are rare and biological data are sparse for all three valid species. The reproductive habits of the most recently described congener, *Proscyllium magnificum*, remain undescribed due to a lack of specimens (Last and Vongpanich, 2004).

Within a single uterus, three encapsulated *H. buergeri* embryos differed in total length by as much as 26 mm or as little as 1 mm. The most developed embryos are located caudad and up to six eggs were observed in each uterus. Capsules were retained until embryos were 70 mm total length at most (Kudo, 1959). Multiple oviparity characteristics for *Halaelurus* were recently summarized (Francis, 2006).

S. fasciatum was characterized as multiple oviparous based on the observation of 4 fully formed egg cases within a single oviduct (Compagno, 1984). An internet media report documents 21 eggs laid on November 29, 2004, by a female *S. fasciatum* from the Nanchang aquarium, China, but details about embryos (if present) are not provided. Kunze and Simmons (2004) made observations on a single mature pair of *S. fasciatum* in the Shark Reef exhibit of the Henry Doorly Zoo. Egg laying occurred three successive years with 46 eggs/112 days, 26 eggs/49 days and 18 eggs/24 days during the first, second and third year respectively (Kunze and Simmons, 2004). As many as three pairs of eggs were laid per day and the interval between members of a pair was minutes to several days. Embryos were first visible by egg candling 23-43 days after oviposition, supporting

---

Fig. 8. Egg case aeration. (A) Position of *Bathyraja parmifera* embryo relative to its egg case. The inset is of the tail filament and its position in the horn. Ventral view of the tail of a 210 mm total length specimen with 25 mm tail filament (tf). Specimens preserved in 10% formalin. Reproduced from Hoff, G.R. (2007) with kind permission of G.R. Hoff. (B) Drawing of the dorsal view of an egg capsule of *Leucoraja erinacea*. The filament oscillates creating positive pressure in a single horn to draw seawater through the remaining three horns of the egg case via the respiratory canals as indicated by directional arrows. Modified from Leonard *et al.* (1999) with kind permission of Wiley-Liss, Inc., a subsidiary of John Wiley & Sons, Inc. (C) *Scyliorhinus torazame* embryo after eclosion. The embryo moves its tail in a sweeping motion past two respiratory canals creating negative pressure that draws water inwards. Water is expelled through the two distal canals as indicated by arrows. (D) The caudal scales regress during late gestation when the embryo is too large to utilize the tail for aeration. ys—yolk sac.

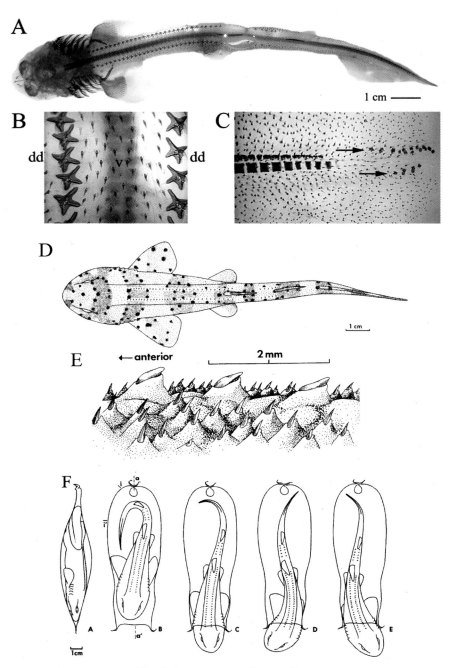

Fig. 9. See next page for caption.

placement of these sharks in the single or extended oviparity reproductive mode. Additional and more detailed observations of embryonic developmental stage at oviposition are needed to categorize *S. fasciatum* oviparity.

Viviparity

Live bearing elasmobranchs use both placental and aplacental methods of reproduction. There are multiple forms of aplacental viviparity among elasmobranchs. Historically these forms were termed ovoviviparous but this term is replaced by the more informative categories, yolk sac viviparity, histotrophy, oophagy and adelphophagy, named for the primary source of nutrition during growth and development (Musick and Ellis, 2005). The term aplacental remains useful only as a descriptor and in no way defines a reproductive mode for fishes.

Yolk Sac Viviparity

Yolk sac viviparous species encapsulate eggs in membranous egg cases. In some species, the embryos hatch but remain within the uterus for the duration of development (Hisaw and Albert, 1947; Verissimo et al., 2003). Other species remain within the egg case throughout gestation (Whitney and Crow, 2007). Uterine compartments are sometimes formed and may function to increase surface area for gas and/or nutrient exchange. Even more complicated is the observation that the same species in geographic

---

Fig. 9. *Cephaloscyllium ventriosum* dorsal scales and hatching. (A) Dorsal view of 15 cm *C. ventriosum* stained with Alcian blue and Alizarin red. (B) Dorsal view of *C. ventriosum* spines, anterior at top. Vertebrae, v, stained with Alcian blue in an embryo that did not undergo acid alcohol treatment. Alizarin red staining shows the development of dermal denticles along the dorsal surface of the spine. (C) Lateral view of *C. ventriosum* caudal fin, dorsal to top, anterior to left. The caudal dermal denticles demonstrate a different morphology than either the initial rows along the dorsal surface or those that eventually cover the body. Images A-C reproduced with modification from Eames *et al.* (2007) with kind permission of Blackwell Publishing, Oxford, UK, and F. Eames. (D) Pigmentation patterns and distribution of juvenile denticles on the dorsum of a newly hatched *C. ventriosum*. (E) Lateral view of three juvenile denticles and the surrounding smaller denticles. (F) Schematic views of *C. ventriosum* in the egg case just before and during hatching. A—Lateral view of sectioned egg case showing enclosed shark. B—Just before emerging. C—Position of the shark after the initial thrust of the tail. D and E—First two lateral flexions, showing the relationship between body movements and the positioning of the two rows of juvenile denticles at the egg case opening. Images D-F reproduced from Grover (1974) with kind permission of NRC Research Press, Ottawa, Ontario, Canada.

continuums differ in gestation strategy. *Mustelus manazo* from four locations circumscribing Japan and a single location in Taiwan were found to have different reproductive habits. Only the population from Aomori, northern Japan, formed uterine compartments, indicating variability of reproductive characters within a species (Yamaguchi *et al.*, 1997).

Some amount of fluid is always present within the uterus of viviparous species and is a potential source of embryonic nutrition. This is indicative of maternal investment of nutrients in the form of histotroph. The term limited or insipient histotrophy is used to describe this system because it is markedly different than that observed for myliobatids in both embryonic organic dry weight gain and necessary uterine specializations (Hamlett *et al.*, 2005c; Musick and Ellis, 2005). Unfortunately, yolk sac viviparous species utilizing limited histotrophy versus strict lecithotrophy are difficult to separate. A comparison of organic dry weight for ova and embryos is the only reliable method to determine the degree of histotrophy, and such studies are uncommon. If nutrition is strictly lecithotrophic, a loss of organic content of approximately 20% or more is expected and accounted for by the energetic cost of tissue synthesis (Hamlett *et al.*, 2005c). Any value less than 20% is attributed to matrotrophic contributions in the form of histotroph. The literature is replete with observations of mucoid or milky fluid within the uterus of yolk sac viviparous species. Few biochemical characterizations of uterine fluid are reported but it is clear that the nutrient type and content vary widely (Lombardi *et al.*, 1993; Amesbury, 1997).

S. acanthias eggs are surrounded by egg jelly and a translucent membranous tertiary egg envelope (Home, 1810). This case is not an ionic barrier but a supportive structure to prevent the delicate ova from rupturing before epiboly and vascularization of the yolk surface is complete (Evans *et al.*, 1982). The encapsulated uterine eggs are often referred to as an egg candle (Fig. 10). Hatching occurs 4-6 months (Atlantic population) or 6-8 months (Eastern Black Sea) into a 22-24 month gestation (Hisaw and Albert, 1947; Demirhan and Seyhan, 2006). Embryos are encapsulated when 7.5 mm TL and hatched by 12 cm TL for the Atlantic population

---

Fig. 10. *Squalus* embryo development. (A) *Squalus acanthias* developmental series with an egg candle, c, containing eggs with blastoderms within the egg capsule. 3.5-7.5 cm TL embryos were removed from their egg candle. Embryos from 12-20 cm TL hatch from the candle and reside inside the uterus until they reach term, 23-29 cm TL. Modification of Figure 1 from Hisaw and Albert (1947) with kind permission of the Marine Biological Laboratory, Woods Hole, MA. (B) *Squalus megalops* egg candle and embryos. (C) *S. acanthias* egg candles, one from each uterus of a single female, with five embryos in each candle. (D) *S. acanthias*, 20 cm embryo. (Photographs B—marinethemes.com/Kelvin Aitken; C and D—Perry Gilbert).

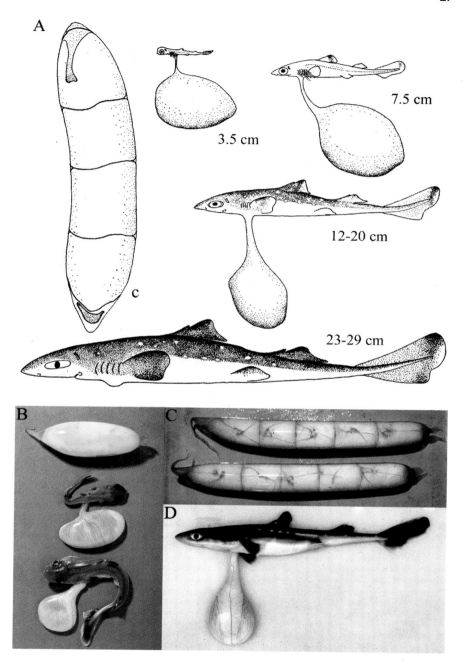

Fig. 10. See previous page for caption.

while hatching occurred at 6-7 cm for Eastern Black Sea embryos. The remaining duration of development is spent within the uterus, without compartmentalization, where the embryos grow to 23-29 cm TL prior to parturition. The EYS is internalized leaving a vitelline scar and limited internal yolk reserves.

Many yolk sac viviparous species have prolonged gestation periods. Gestation is as long as three and a half years for *C. anguineus* (Tanaka *et al.*, 1990), while *Triakis megalopterus* has a 20 month gestation and remains in a membranous egg case until parturition (Smale and Goosen, 1999). *Centroscymnus owstoni* and *Centroscymnus coelolepis* are deep water sharks with two year gestation periods (Yano and Tanaka, 1988; Verissimo *et al.*, 2003).

The egg cases of viviparous species are destroyed when the embryo hatches. Egg case fragments are not commonly reported within later stage uteri and it is unclear if they are consumed, expelled or degraded *in utero*. Even thinner egg envelopes surround embryos that remain within their cases throughout gestation. A few curious exceptions exist. A single *R. typus* aborted egg case recovered by trawl from the Gulf of Mexico contained a viable 36 cm TL embryo (Fig. 11C) (Baughman, 1955). While the egg case was described as amber in color and thin, it is undoubtedly more durable than cases that typify yolk sac viviparous species because it remained intact during bottom trawling and collection (Wolfson, 1983).

The second observation of *R. typus* egg cases was *in utero*. Within a single 10.6 m female harpooned by Taiwanese commercial fishermen, more than 300 embryos were found (Fig. 11A-B) (Joung *et al.*, 1996). The egg cases

---

Fig. 11. *Rhincodon typus*, yolk sac viviparity. (A) Anterior portion of 10.6 m female *R. typus*, harpooned off the coast of Taiwan in July 1995, hanging from a crane before removal of her litter. (B) Most of the dead embryos (307) ranged from 42-64 cm TL with 15 live pups, 58-64 cm TL, placed in aquaria. Reproduced from Joung *et al.* (1996) with kind permission of Springer Science and Business Media. (C) *R. typus* and egg case from which it was removed. The egg case was collected by trawl off the coast of Texas in 1953 after being prematurely expelled by the female. The embryo was alive upon removal from the case. This record resulted in categorizing *Rhincodon* as oviparous until 1995. Reproduced from Baughman (1955) with kind permission of the American Society of Ichthyologists and Herpetologists. (D) *R. typus* empty egg cases and hatched embryo from Taiwanese female described in Fig. 11A. Egg cases in this photo are dehydrated and partially collapsed compared to (C). (E) Egg case of *Ginglymostoma cirratum*. (F) Schematic representation of *R. typus* egg capsule with two respiratory canals (*), hatching terminus, H, and non-hatching terminus, NH. Reproduced with modification from Chang *et al.* (1997) with kind permission of the American Society of Ichthyologists and Herpetologists. (Photographs D—Shoou-Jeng Joung; E—Perry Gilbert).

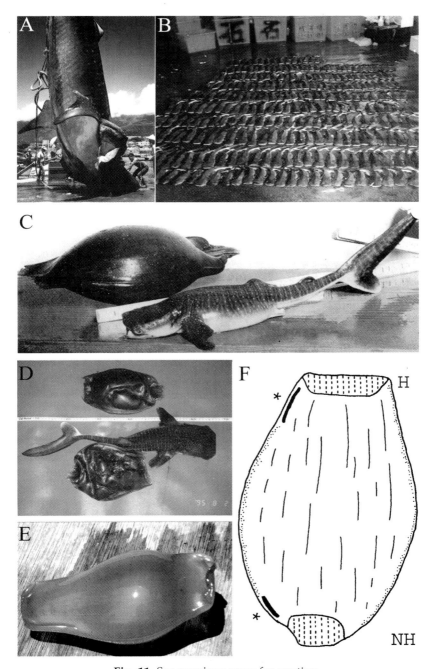

Fig. 11. See previous page for caption.

in this female averaged 21.1 cm in length and 20.1 cm in width, the largest among chondrichthyans (Joung *et al.*, 1996). They lacked tendrils for attachment and were described by the authors as "not as tough as the 'mermaid purses' laid by oviparous species…" Capsule morphology was investigated further and described as amber and smooth with a respiratory fissure on each side (Fig. 11F) (Chang *et al.*, 1997). In photographs from both the trawled Gulf of Mexico specimen and the Taiwanese female, it is evident that the cases have enough structural integrity to support themselves even when empty. More intriguing are the presence of respiratory fissures which had previously only been described for oviparous egg cases.

For 237 embryos, the mean embryo TL was 51.1 cm and weight was 660.2 g. Fifteen term embryos (58-64 cm TL) were released alive. Embryos ranged from 42-64 cm TL and embryos of 58 cm TL and larger had hatched. Among hatched embryos there was no EYS but an obvious vitelline scar. Empty egg cases (approximately 50) were also found within the uterus. Details concerning the number of egg cases versus newly hatched individuals are not presented and it cannot be determined if parturition occurs in waves coincident with each size class or all at once.

G. cirratum has a similar reproductive strategy as *R. typus* (Castro, 2000). Embryos are enclosed in egg capsules for the first 12-14 weeks of a 5-6 month gestation. They hatch when 218-233 mm TL with a large EYS. Nutrition is strictly lecithotrophic and parturition occurs at 280-305 mm TL. Each brood contains embryos in multiple stages of development, reflecting a prolonged period of ovulation. The number of observed term embryos decreases suggesting parturition in waves. A captive female expelled 16-22 empty egg cases during the month before parturition commenced. A total of 21 pups were born during the following two months (Kuenen, 2000). *G. cirratum* egg cases (Fig. 11E) are remarkably similar in organization to *R. typus* (Fig. 11D). Compelling to consider is why the egg cases of *G. cirratum* and *R. typus* are more substantial than most viviparous species and what implications this character may reveal about the evolution of viviparity.

Histotrophy

All myliobatid stingrays employ histotrophy as a matrotrophic reproductive strategy. Fertilized eggs are encapsulated along with a small amount of egg jelly and fluid within a soft and pliable tertiary egg envelope (Fig. 12B). Embryos endure a brief incubation before breaking free of their egg case *in utero*. The uterine mucosa develops glandular trophonemata responsible for the secretion of histotroph or uterine milk (Figs. 11A and 13A) (Wood-Mason and Alcock, 1891). The amount of histotroph increases and its nutrient content changes throughout gestation (Amesbury, 1997).

The trophonemata increase in length and vascularity while developing glandular pits (Lewis, 1982). The trophonemata are at maximum production during mid to late gestation (Fig. 12A) (Amesbury et al., 1998). The epithelium covering peripheral capillaries thins from a 3 µm cuboidal epithelium in rays with encapsulated eggs to 0.3 µm squamous layer during late gestation to facilitate gas exchange (Hamlett et al., 1996).

The histotroph is nutritionally rich with lipids, carbohydrates and proteins enabling rays to increase in size significantly during gestation (Wourms and Bodine, 1983; Luer et al., 1994). The gill filaments and/or vitelline membrane may function to absorb nutrients from the histotroph (Ranzi, 1943; Babel, 1967; Smith and Merriner, 1986). Trophonemata have been observed to invade the spiracles, gills and mouth of mid to late gestation rays (Fig. 13B-C) (Lewis, 1982; Smith and Merriner, 1986; Johnson and Snelson, 1996; White et al., 2001). While this close association may facilitate nutrient transfer it appears not to be limiting. When three or more embryos are present, rays not in close proximity to the trophonemata do not suffer in weight gain or size (Babel, 1967). The trophonemata of *Gymnura* (formerly *Pteroplatea*) *micrura* appear equal in length during early gestation but as the uterus is stretched, the pressure of the embryo on the uterine wall is proposed to cause degeneration of trophonemata, with the exception of trophonemata opposite the spiracles, where they remain elongate (Fig. 13B) (Wood-Mason and Alcock, 1891). This arrangement has not been observed for other myliobatids.

Encapsulated myliobatiform embryos are extraordinarily fragile when compared to oviparous rays at similar stages of development (pers. obs.). This fragility extends through early development and formation of the discoid body. Viviparous rays often abort embryos upon capture or handling stress, destroying precious samples and potentially yielding low estimates of fecundity (Melouk, 1949; Sunye and Vooren, 1997; White et al., 2001; Mollet et al., 2002). Collected and preserved embryos are easily damaged and often detached from their yolk (Thorson et al., 1983; Smith and Merriner, 1986). After the pectoral fins fuse with the body and rostrum the stability is much improved.

Oophagy and Adelphophagy

The synapomorphic character, oophagy, as a reproductive strategy unites 15 species to form the order Lamniformes. The reproductive characteristics for this order were recently reviewed (Mollet et al., 2000; Gilmore, 2005). Female lamnoids, except *Mitsukurina owstoni* and *Odontaspis noronhai*, which remain undescribed, have a vestigial left ovary and a specialized, enlarged right ovary. The right ovary is subdivided into pleats with intervening cavities that connect to a central ellipsoidal lumen (Pratt, 1988).

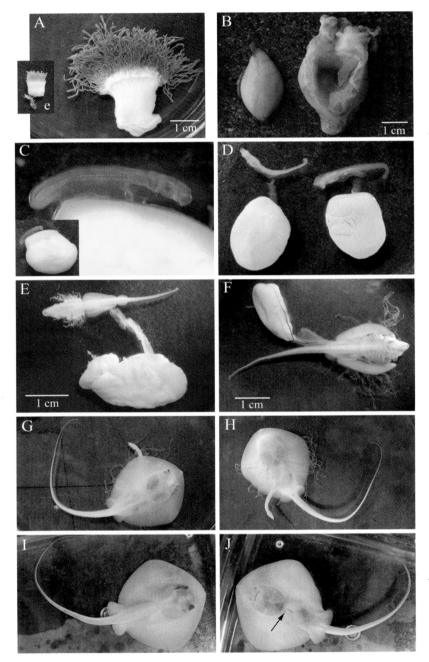

Fig. 12. See next page for caption.

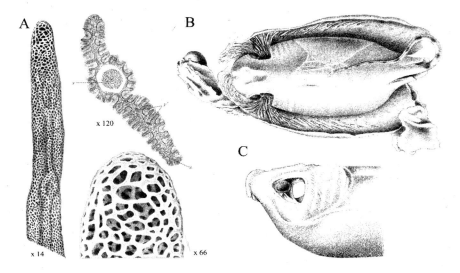

Fig. 13. *Gymnura* (formerly *Pteroplataea*) *micrura*, structure and orientation of trophonemata. (A) Uterine trophonemata taken from within the embryo's spiracle. Distal moiety of a large trophonema from one of the spiracles showing the superficial texture and the outstanding vein. Apex of the same trophonema showing the compound duct openings. Transverse section of a trophonema in its distal half showing the glands in vertical section. a—artery; v—main trunk; v'—branches of vein. (B) Gravid uterus of *G. micrura*, opened by longitudinal incision in its dorsal wall, with the flaps turned back showing the passage of trophonemata through the spiracles into the pharyngeal cavity of a single fetus. (C) Head and shoulders of the same fetus from the left side showing the great size and the lateral position of the spiracles. Reproduced from Wood-Mason and Alcock (1891) with kind permission of the Royal Society.

Fig. 12. Histotrophy and ray development. (A) Section of the uterine wall with trophonemata from *Dasyatis akajei* during early, e, and late gestation at equal scale. (B) Partially dissected uterus of *Dasyatis laevigata* and its egg case. (C) Early embryo of *D. laevigata* with fusiform body. (D) *D. laevigata* embryos with expanding pectoral discs. (E) *D. akajei* embryo with pectoral disc growing laterally and anteriorly to the margin of the pharyngeal region but not yet fusing. (F) *D. akajei* embryo with fusing pectoral disc and long gill filaments. (G and H) *D. laevigata* embryo with completely fused disc and yolk within the yolk stalk. (I and J) *D. akajei* embryos with completely absorbed yolk sac (arrow) and reabsorbed gill filaments. Histotroph becomes the primary source of nutrition until parturition. (Photographs Atsuko Yamaguchi and Keisuke Furumitsu).

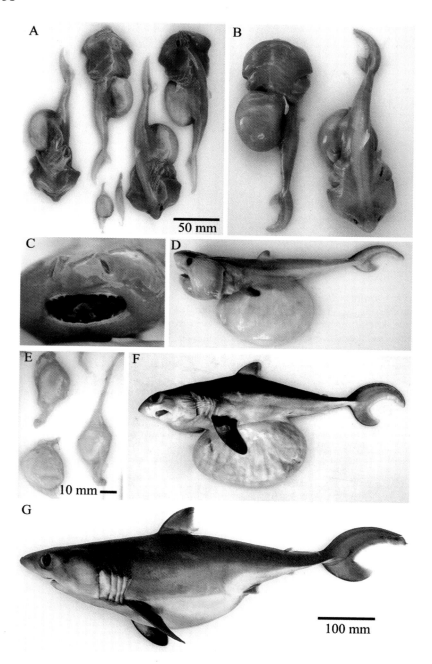

Fig. 14. See next page for caption.

An ovarian pore opens to the abdominal cavity. This arrangement maximizes surface area for ovulation and increases ovarian fecundity to as high as six million for *Cetorhinus maximus*, the highest observed among chondrichthyans (Mathews, 1950). *Carcharias taurus* ova are 10-12 mm in diameter but other lamnoids, including *C. maximus*, *Lamna* spp., *Isurus*, spp. and *Alopias* spp., have 5-6 mm ova (Mathews, 1950; Gilmore, 1983; Gilmore et al., 1983). *Odontaspis ferox* is known from a single observation to contain hundreds of small, 3 mm ova in the ovary (Villavicencio-Garayzar, 1996). Lamnoid oviducts are bilaterally functional and produce large precocious neonates.

All described lamnoid embryos (*Isurus oxyrinchus, Isurus paucus, Lamna nasus, Carcharodon carcharias, Alopias superciliosus, C. taurus, Alopias pelagicus* and *Psuedocarcharias kamoharai*) are oophagous (Springer, 1948; Fujita, 1981; Otake and Mizue, 1981; Gilmore, 1983; Gilmore et al., 1983; Stevens, 1983; Uchida et al., 1996; Francis and Stevens, 2000). The embryos of several lamnoids remain uncharacterized, including *C. maximus, Megachasma pelagios, Odontaspis* spp., and *M. owstoni*. Embryos develop precocious dentition and shed teeth *in utero* that are sometimes ingested (Gilmore, 1993; Uchida et al., 1996; Chen et al., 1997; Francis and Stevens, 2000; Costa et al., 2002; Shimada, 2002). Embryonic teeth useful for hatching and consumption of egg cases and ova are distinct from adult dentition (Fig. 14C).

All lamnoids have a fecundity greater than 2 (1 pup per uterus) except *C. taurus* and alopiids (Moreno and Móron, 1992; Gilmore, 1993; Chen et al., 1997). Gestation time is speculative for most species due to the lack of gravid females examined. *I. oxyrinchus* females carry 2-20 pups throughout a 15-18 month gestation and follow a three year reproductive cycle (Branstetter, 1981; Stevens, 1983; Mollet et al., 2000). Embryos are born at 70 cm total length and the average female is 280 cm when mature, yielding a newborn that is 25% of its mother's size. *C. carcharias* has been observed to carry as many as 14 embryos (Francis, 1996). *C. taurus* has a 9-10 month gestation and reproduces annually (Gilmore, 2005).

Gestation is shortest for *L. nasus*, 8-9 months, with an annual reproductive cycle (Francis and Stevens, 2000). Average fecundity is 3.85 and the smallest embryos examined, 9.6-10.4 cm, already had a distended yolk

---

Fig. 14. *Lamna nasus*, oophagous development. (A) Four embryos ranging in size from 155-162 mm FL (fork length) and two nutritive egg cases, one empty. (B) Male embryo (left) 198 mm FL and female embryo (right) 207 mm FL. (C) Female embryo, 220 mm FL, with precocious dentition for egg consumption. (D) Embryo, 394 mm FL. (E) Nutritive uterine egg cases. (F) Female, 403 mm FL. (G) Female, 580 mm FL. A and B at equal scale; D, F and G at equal scale. Note the expanded pharyngeal region of younger embryos, a characteristic of lamnid embryos. (Photographs Malcolm Francis).

stomach from egg consumption. Lamnid embryos have a unique pharyngeal expansion present during pre-pigmentation stages that gradually diminishes to normal proportions prior to parturition (Fig. 14A-D). Embryonic dentition is present in the smallest embryos and teeth are shed between 34-38 cm (Fig. 14C).

C. taurus utilizes a strategy termed adelphophagy, a specialized form of oophagy where ova and developing conspecifics become a source of nutrition. The result is two embryos, one per uterus, that approach one third the maternal size at parturition or approximately 1 meter. After mating and ovulation, ova are encapsulated and reside in the uterus as in all other viviparous species. The first capsules often contain only egg jelly and sperm (Gilmore *et al.*, 1983). The next capsule usually contains a single fertile ovum. Subsequent capsules contain up to three fertilized ova, thereafter ova increase in number successively up to a maximum of 23 ova per capsule (Gilmore, 2005). Newly fertilized ova are observed during the first 70 days of gestation. Sperm storage has not been observed among lamnoid sharks, due in part to the overt activity of the shell gland (Pratt, 1993; Pratt and Tanaka, 1994; Gilmore, 2005). The activity of the ovary and shell gland of other lamnoids is similar to *C. taurus* (Gruber and Compagno, 1981; Fujita, 1981; Francis, 1996; Uchida *et al.*, 1996).

Embryonic development of *C. taurus* is divided into stages based on the embryo's immediate source of nutrition (Gilmore *et al.*, 1983). The smallest observed sand tiger embryos, about 13 mm, are nourished from endocoelomic yolk within the pharyngeal, pericardial and coelomic cavities. This yolk imparts the embryos with a rotund morphology distinct from other elasmobranchs at similar stages of development (Fig. 15A). The internal yolk and EYS are not connected. The embryonic yolk sac of embryos 18-57 mm, does not decrease in size and no intra-embryonic yolk remains. Potential caloric sources at this time include encapsulated fluid, trans-capsular metabolites and extra-embryonic yolk. After approximately 43 days of incubation, 49-57 mm embryos possess a precocious dentition consisting of 6-7 conical teeth in both the upper and lower jaw that enable

---

Fig. 15. *Carcharias taurus*, adelphophagous development. (A) 13 mm embryo with endocoelomic yolk and a yolk sac (*) and a 31 mm embryo at equal scale. (B) 62 mm embryo, 100 mm embryo with partially cannibalized sibling (51 mm), and 131 mm embryo all drawn at equal scale. The 62 mm embryo hatched from its egg case with a 6 mm yolk sac. The 131 mm male had attacked both a 49 mm and 45 mm sibling. (C) 227 mm female, 271 mm male, and 334 mm female embryo all drawn at equal scale. The 334 mm specimen had four partially consumed siblings within its pharynx. (D) 800 cm late term pup with swollen yolk stomach. Reproduced with modification from Gilmore *et al.* (1983), Scientific Publications Office, National Marine Fisheries Service.

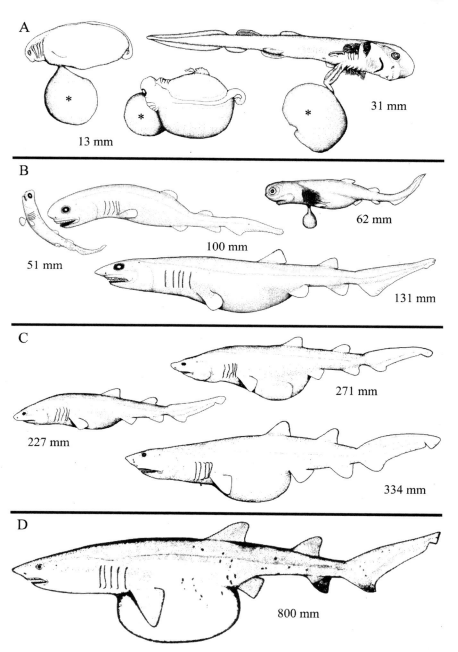

Fig. 15. See previous page for caption.

the embryo to tear free of the egg capsule and swim within the uterus. Embryos, 60-100 mm, utilize their yolk sac and are supplemented with histotroph for 10-12 days before pursuing conspecifics (Fig. 15B). Once embryos grow to 100-335 mm they cannibalize their uterine siblings. Egg capsules with developing embryos are repeatedly attacked while capsules containing infertile ova remain intact. Approximately 100 days of oophagy follow, with freshly deposited egg capsules serving as food for 335-1000 mm embryos (Fig. 15C). Finally, 900-1100 mm embryos digest accumulated yolk and store energy in their enlarged livers in preparation for parturition, 250-270 days after fertilization (Fig. 15D).

The extent to which the endocoelomic yolk of *C. taurus* embryos and the pharyngeal expansion of *lamnids* occur among other lamniforms remains to be discovered. The severely distended yolk stomach of mid-and late gestation is diagnostic for lamniforms from middle to late gestation, albeit diminished for alopiids (Mollet *et al.*, 2000).

A variant form of oophagy is described for carcharhiniform sharks, family Pseudotriakidae, with two described species, *Gollum attenuatus* and *Pseudotriakis microdon*, and one undescribed genus and species (Yano, 1992; 1993; Compagno, 2005). The females of *G. attenuatus* encapsulate 30-80 small ova, 4-8 mm in diameter, in a membranous egg envelope and do not ovulate during gestation (Fig. 16). The number of fertile ova within the egg case is not known but only one embryo develops in each egg case. When it reaches 8-19 mm TL, the remaining ova degenerate and coalesce. The embryo somehow assumes this yolk in its EYS without first consuming it. Hatching occurs from 29-39 mm total length and the embryos remain within a non-compartmentalized uterus bathed in histotroph that increases in quantity throughout gestation. Gill filaments are present on embryos from 40-250 mm total length. Near term embryos with yolk sacs completely reabsorbed were 315-415 mm TL (Yano, 1993). *P. microdon* ovulate infertile ova that are consumed by developing embryos, but the details of ovulation are not known. Unlike lamnoids, the embryos store the ingested yolk in their EYS and do not develop a distended gut (Yano, 1992).

Placental Viviparity

Placentatrophy occurs in family Carcharhinidae, Triakidae, Leptochariidae and Sphyrnidae of order Carcharhiniformes, the ground sharks. Placentas have a plethora of morphologies and when uterine and embryo specializations are included, strategies for nutrient transfer are essentially unique for each species. The uterus forms individual compartments around each developing embryo effectively increasing surface area for histotrophic nutrient transfer and respiration until parturition (Springer, 1960; Otake,

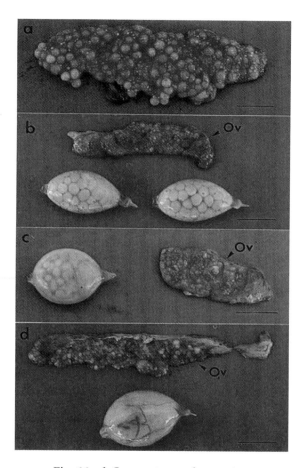

Fig. 16a-d. See next page for caption.

1990; Capapé *et al.*, 2003). The embryos are all initially lecithotrophic. Yolk enters the intestine directly and is never stored in an internal yolk sac (IYS). As yolk diminishes, histotroph supplements nutrition until the placenta is established. Embryos are usually in an advanced stage of development, with body pigmentation, when the placenta forms (Appukuttan, 1978). The yolk sac and stalk are not reabsorbed in placental species. The yolk stalk lengthens and the yolk sac expands as its contents are diminished (Fig. 17). In *M. canis,* the yolk sac increases in surface area by an order of magnitude as the embryos grow from 4 cm to 10 cm total length (TeWinkel, 1963). A portion of the yolk sac intimates with the uterine mucosa to form the fetal portion of the placenta. The yolk stalk, vitelline artery and vitelline vein may now be termed the umbilical stalk, umbilical artery and umbilical

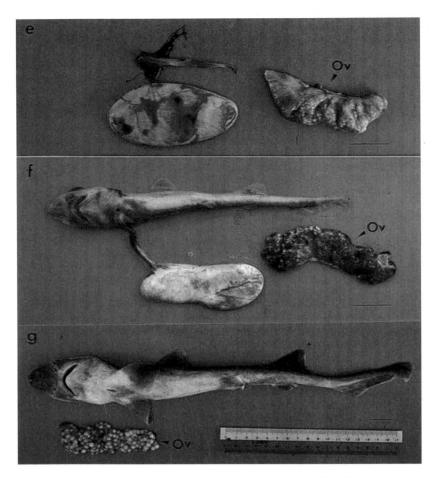

Fig. 16e-g. *Gollum attenuatus*, ovary condition and modified oophagous development. (a) Mature ovary, (b) many ovulated ova with fertilized ovum in rigid egg capsule, and (c) early development of embryo in jelly capsule and non-developing ova in the capsule, consisting of about half intact ova and half mixed yolk material. (d) Embryo with completely formed external yolk sac in jelly capsule. (e) Embryo with external yolk sac hatched from the jelly capsule. (f) Advanced embryo with external yolk sac. (g) Near-term embryo. Ov—ovary; scale bars 20 mm. Reproduced from Yano (1993) with kind permission of Springer Science and Business Media.

vein, respectively, reflecting the newly established placenta. The yolk sac placenta is epitheliochorial and non-invasive (Amoroso, 1952).

With the exception of *S. laticaudus, Iago omanensis* and *P. glauca*, the tertiary egg envelope persists as a semi-permeable barrier or layer between fetal and maternal tissues coupling hatching and parturition (Setna and Sarangdhar, 1948a; Otake and Mizue, 1985; Fishelson and Baranes, 1998). The tertiary egg envelope of placental viviparous species is amber colored and translucent (Fig. 17A). It is reduced in thickness when compared to aplacental viviparous and oviparous species (Heiden *et al.*, 2005). The egg envelope does not stretch but instead is pleated and unfolds on demand, isolating the embryo from direct contact with the uterus (Fig. 18A) (Castro and Wourms, 1993). The egg is encapsulated with a small amount of egg jelly that is liquefied similar to oviparous species, but the mechanism of liquefaction is not understood. The egg case retains fluid throughout gestation. Nutrient transfer through the egg envelope is via diffusion and limited to molecules less than 6 kDa for *S. tiburo* and 5 kDa for *M. canis*. Permeability of the egg case to glucose increases during mid and late gestation in *S. tiburo*, potentially reflecting energetic demands of the growing embryos (Lombardi and Files, 1993; Heiden *et al.*, 2005).

Placentas typically take one of two forms, being composed of both a smooth proximal segment and a vascularized distal segment, or only a vascularized distal segment. The only exception is *Scoliodon*, where the placenta occurs as a trophonematous cup. Yolk sac placental cells stain positively for human chorionic gonadotropin (hCG), human placental lactogen (hPL), alpha-feto protein (AFP), placental alkaline phosphatase (PLAP), pregnancy-specific β_1-glycoprotein (SP_1) and periodic acid-Schiff (PAS) (Carter *et al.*, 2007). The umbilicus may be smooth or develop appendiculae which are superficially reminiscent of trophonemata (Alcock, 1890). The appendiculae may be present as a dense carpet of villous branching extensions or only sparse, long, twisting filaments (Fig. 17D-G) (Southwell and Prashad, 1910; Wourms, 1977). Appendiculae are highly vascular and are composed of two major cell types, one potentially absorptive cell with long microvilli and the other definitively secretory (Hamlett, 1993; Hamlett *et al.*, 2005a). Appendicular secretions are completely uncharacterized.

In *M. canis*, pairs of ova are ovulated in succession from the right ovary (Figs. 18-19). Each egg is fertilized and individually enclosed in a pleated egg envelope (Fig. 18A). The eggs enter both uteri alternatively and fill the uterus from posterior to anterior. A developmental series results from the sequential fertilization of ova with the most developed embryos in the posterior of the uterus (Figs. 18B and 19) (TeWinkel, 1950). Difference in embryonic stage is an indication of time between ovulatory events and the

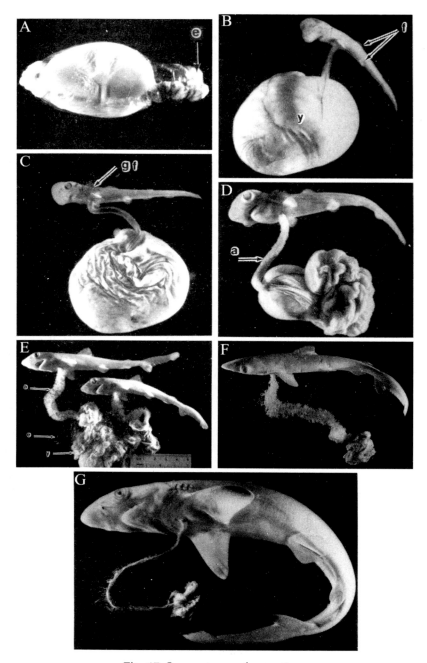

Fig. 17. See next page for caption.

short duration of early developmental stages. During the later portion of embryonic development the disparity between stages would likely be indiscernible.

The fetus is foreign tissue to its host and as a result the maternal immune system must be suppressed to prevent rejection. The pro-inflammatory cytokines, interleukin-1 (IL-1) α and IL-1 β were identified immunohistochemically in the yolk sac endoderm and leukocytes of *M. canis*, while IL-1 α and its receptor were expressed in the uterine epithelium (Paulesu *et al.*, 2000; Cateni *et al.*, 2003). Lymphoid aggregations in the uterine mucosa increase in number and size during gestation in *R. terraenovae* (Haines *et al.*, 2006). These findings are interesting when considering the isolation of the embryo from the mother by the semi-permeable acellular tertiary egg envelope. A recent review of placentation in sharks includes a detailed discussion and summary of the original literature (Hamlett *et al.*, 2005a).

Special Considerations

Scoliodon laticaudus

The ova of *S. laticaudus* are mature at 1 mm, the smallest for any chondrichthyan (Teshima *et al.*, 1978). As development is altered to account for the large yolk mass common to chondrichthyans, it is not unexpected to be different for *S. laticaudus* (Arendt and Nübler-Jung, 1999). The eggs are covered with egg jelly and an egg envelope before delivery into partially

---

Fig. 17. *Rhizoprionodon terraenovae*, placentation and embryo development. (A) Recently fertilized egg. Most of the egg envelope, e, is gathered together at one end of the egg. (B) 25 mm embryo with gill filaments just protruding through the gill slits and well-demarcated fin buds. The yolk sac, y, has begun to lose its turgidity and its distal pole is beginning to become vascular. (C) 40 mm embryo bears external gill filaments, gf, that protrude through the gill slits and spiracle. The yolk sac is very flaccid and its surface is wrinkled. (D) 55 mm embryo, very short appendiculae, a, cover the umbilical cord. Yolk is present only in the turgid, proximal portion of the yolk sac. The distal yolk sac is distended, flaccid, and highly vascularized. (E) 130 mm embryos bear well-developed appendiculae, a, that cover the umbilical cord. The egg envelope, e, is closely associated with the greatly expanded yolk sac, y, and the embryo is firmly attached to the uterine wall forming a placenta (F) 250 mm embryo at midterm. Juvenile pigmentation is visible, the appendiculae have reached their maximum development, and the yolk sac is much reduced. (G) 320 mm embryo at term. The embryo has increased greatly in weight and girth. Appendiculae are sparse. Reproduced from Castro and Wourms (1993) with kind permission of Wiley-Liss, Inc., a subsidiary of John Wiley & Sons, Inc.

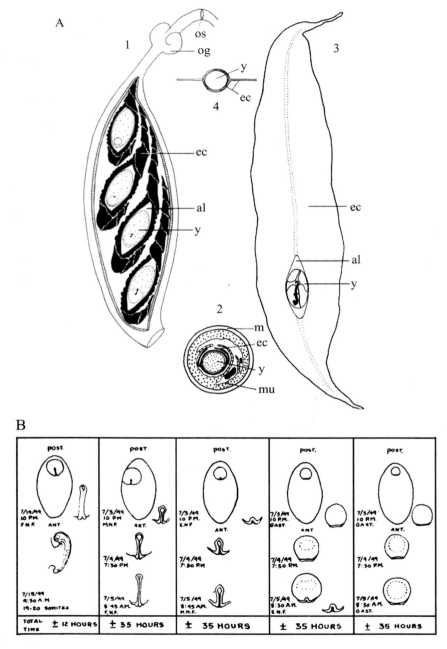

Fig. 18. See next page for caption.

formed uterine compartments (Fig. 20A). The blastoderm expands over the minute vitellus in a modified form of epiboly that needs further characterization (Fig. 20B). Formation of the embryonic axis is described as the result of cell movements through a blastopore at the margin of the blastoderm.

For *S. laticaudus*, the vitelline mass is miniscule and early during development the vitello-intestinal duct, if present, is obliterated (Southwell and Prashad, 1910; Thillayampalam, 1928; Teshima *et al.*, 1978; Wourms, 1993). The embryos hatch from the tertiary egg envelope when they are 1 mm in length and have 8-9 visible somites (Fig. 20C). Placentation occurs by the time embryos reach 2 mm TL, the earliest point in gestation for any elasmobranch. The egg envelope is not present as a barrier to nutrient transfer and the placenta is either epithelio-vitelline or hemovitelline, depending on the status of the maternal epithelium of the trophonematous cup abutting the yolk sac portion of the placenta (Mahadevan, 1940; Setna and Sarangdhar, 1948b; Teshima *et al.*, 1978). The umbilical stalk is adorned with appendiculae that are presumed to aid in nutrient absorption and respiration (Fig. 20D-F) (Southwell and Prashad, 1910; Wourms, 1993). The embryonic portion of the placenta grows from 1 mm to 15 mm during gestation (Teshima *et al.*, 1978).

Diapause

Embryonic diapause, an arrest of development at the blastocyst stage, is a reproductive strategy widely utilized by homeothermic and poikilothermic organisms that extends the time required to complete a reproductive cycle (Mead, 1993; Renfree and Shaw, 2000). Diapause has been documented in a single placental carcharhiniform shark and 11 rajiforms including, six rhinobatids, three dasyatids, and a single urolophid and myliobatid (Table 3). All species are viviparous and in every case except *Rhinobatos rhinobatos*

Fig. 18. (A) 1. Drawing in ventral view of the right oviduct and uterus of *Mustelus canis* showing orientation of egg cases and embryos (embryos are proportionately enlarged for the sake of clearness). 2. Transverse section of uterus showing how folds of uterine mucosa wrap around the egg case. 3. One egg case of *M. canis* spread out to illustrate its dimensions and the position of the yolk. 4. Transverse section of egg case cut through the ovum. al—albumen; ec—egg case; m—muscle; mu—mucosa; og—oviducal gland; os—ostium; y—yolk. (B) Rate of development of five living *M. canis* embryos at 22-23°C. Elapsed time for each embryo is shown at the bottom of each panel. Anterior and posterior labels indicate the orientation of ova *in utero*. Reprinted from TeWinkel (1950) with kind permission of the Marine Biological Laboratory, Woods Hole, MA.

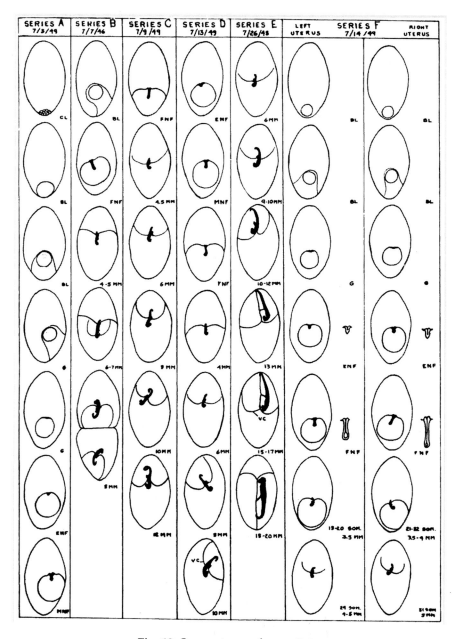

Fig. 19. See next page for caption.

Table 3. Diapause Among Elasmobranch Species

Species	Diapause type	Duration (months)	Reference
Dasyatis sayi	Obligate	9-10	Snelson et al. (1989)
Dasyatis brevis	Obligate	9-10	Melendez (1997)
Pteroplatytrygon violacea	Suspected	6-7	Hemida et al. (2003); Mollet et al. (2002)
Trygonoptera personalis	Obligate	5	White et al. (2002a)
Aetobatus flagellum	Obligate	9	Yamaguchi (2006)
Rhinobatos horkelii	Obligate	8	Lessa et al. (1986)
Rhinobatos productus	Obligate	8	Villavicencio-Garayzar (1993); Marquez-Farias (2007)
Rhinobatos hynnicephalus	Obligate	8-9	Wenbin and Shuyuan (1993)
Rhinobatos cemiculus	Facultative	4	Seck et al. (2004)
Rhinobatos rhinobatos	Facultative	5	Abdel-Aziz et al. (1993); Capapé et al. (1997, 1999)
Trygonorrhina fasciata	Obligate	7-8	Marshall et al. (2007)
Rhizoprionodon taylori	Obligate	7	Simpfendorfer (1992)

and *Rhinobatos cemiculus*, parturition, ovulation and mating occur in rapid succession with an obligatory diapause.

Evidence for diapause is usually limited to the observation of encapsulated uterine eggs without macroscopic embryos for extended periods of time within a reproductively synchronized population. The fertility of these eggs has been questioned but the egg case forms a formidable barrier and

Fig. 19. Vertical columns (Series A-E) represent embryonic development of sequentially fertilized ova contained in individual *Mustelus canis* uteri on the date indicated. Ova are arranged in the sequence and orientation observed *in utero* with anterior ova at the top of each series. Embryos are proportionately enlarged. The most posterior figure in Series B shows a "double-yolked" egg. Circular blastoderms in several figures are surrounded by an area lighter in color than the rest of the yolk. The periphery of the yolk sac is not shown in older stages as it spreads rapidly and is increasingly difficult to distinguish. Early vitelline blood vessels are indicated. The two vertical columns of Series F represent embryonic stages from the left and right uteri of a single specimen to show the similarity in stage of corresponding embryos, indicating that ova are ovulated in pairs. BL—blastodisc; CL—cleavage; ENF—early neural folds; FNF—fusing neural folds; G—gastrulation; MNF—mid-neural folds; VC—vitelline circulation. Reprinted from TeWinkel (1950) kind permission of the Marine Biological Laboratory, Woods Hole, MA.

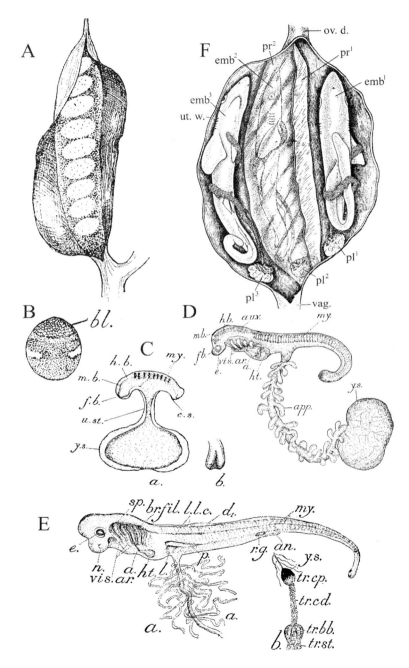

Fig. 20. See next page for caption.

females do not store sperm during gestation (Morris, 1999). The duration of diapause among elasmobranchs varies from five to nine months.

The embryos of *Dasyatis sayi* enter diapause in May and remain arrested as blastula until the following February, however, the degree of arrest for this or any elasmobranch is not known (Fig. 21). Among mammals, blastocysts may be either quiescent or grow at an imperceptibly slow rate (Renfree, 1978). *D. sayi* embryos may maintain a low degree of cell division throughout or during a portion of the diapause period because mitotic figures are observed (Fig. 21E). Development commences in March and parturition follows in May (Snelson et al., 1989). How diapause is maintained in elasmobranchs is uninvestigated but may involve hormones and environmental cues as is common for other vertebrates.

The reasons for embryonic diapause are elusive and likely species specific. Simpfendorfer (1992) proposed that *Rhizoprionodon taylori* neonates enjoy the summer high temperatures and prey abundance that coincide with their parturition. Marshall's (2007) results are partly contradictory because *Trygonorrhina fasciata* parturition occurs during the fall when sea temperatures are dropping. However, a study of diet showed that prey items remain abundant throughout the fall and winter. *Pteroplatytrygon* (formerly *Dasyatis*) *violacea* females with egg capsules were observed during the winter when surface water temperatures were lowest (Hemida et al., 2003). White (2002) postulated that diapause enabled *Trygonoptera personalis* females to thrive during gestation and the embryos to develop quickly because of abundant prey and favorable temperatures in summer. Concomitantly, parturition occurs as water temperatures begin to drop.

Fig. 20. *Scoliodon* development. (A) Uterus in very early stages of pregnancy opened to show more or less horizontal disposition of compartments. (B) Egg with blastodermic cap above and partially consumed yolk below. (C) 1 mm embryo. a—Embryo with yolk sac and umbilical stalk; b—dorsal view of the united caudal region. (D) 3 mm embryo with appendiculated placental cord and yolk sac. (E) 15 mm embryo. a—Embryo; b—yolk sac placenta. *an.*—anus; *br. fil.*—branchial filaments; d_1.—rudiment of first dorsal fin; *l.*—liver; *l.l.c.*—lateral line canal; *n.*—nostril; *p.*—pectoral fin; *r.g.*—rectal gland; *sp.*—spiracle; *tr. cp.*—trophonematous cup; *tr. cd.*—trophonematous cord; *tr. bb.*—trophonematous bulb; *tr. st.*—trophonematous stalk; *bl.*—blastoderm; *c.s.*—caudal swelling; *f.b.*—forebrain; *h.b.*—hind brain; *m.b.*—mid brain; *my.*—myotomes; *u.st.*—umbilical stalk; *y.s.*—yolk sac; *app.*—appendicula; *auv.*—auditory vesicle; *e.*—eye; *ht.*—heart; *vis. ar.*—visceral arches. Reprinted from Setna and Sarangdhar (1948). (F) Right uterus of *Scoliodon laticaudus* (formerly *palasorrah*), cut open to show the intra-uterine embryos. emb^1.—male embryo; emb^2., emb^3.—female embryos; *ov.d.*—anterior end of oviduct; pl^1., pl^2., pl^3.—yolk sac placentas attached to placental cords; pr^1., pr^2.—intra-uterine partitions between embryos; *ut.w.*—uterine wall; *vag.*—vaginal end of uterus. Reprinted from Thillayampalam (1928).

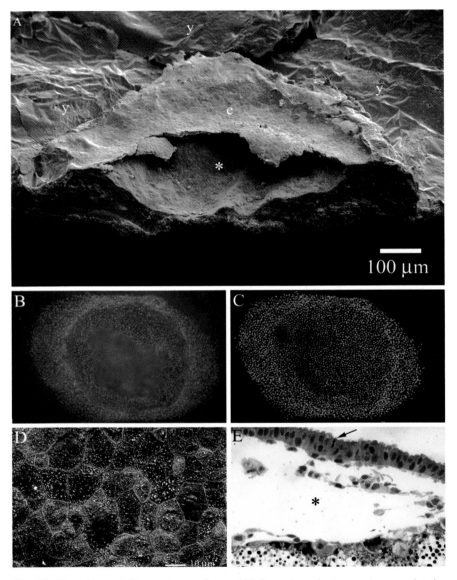

Fig. 21. *Dasyatis sayi* diapausing embryo. (A) Scanning electron micrograph of a blastoderm stage embryo, e, and surrounding vitelline membrane covering the yolk, y. A portion of the embryo has been cut away to reveal a segmentation cavity or blastocoel (*). (B) Fluorescent micrograph and (C) confocal image of a blastoderm. (D) Detail of blastomeres with defined cell borders and sparse microvilli. (E) Section of blastoderm with segmentation cavity (*) separating the epiblast from the yolk. The arrow identifies a mitotic cell.

Diapause may be utilized in species where sexual segregation is observed in order to synchronize parturition, ovulation and mating. While this reason is relevant for *Rhinobatos productus* and *Rhinobatos horkelii*, it is not relevant for non-segregating species like *T. personalis* (Kyne and Bennett, 2002). If for every population a particular event in the reproductive cycle limits the success of all others, then this event is defined as rate-limiting and critical. This event may be as simple as unification of sexes for mating or involve an association of favorable environmental condition with females or neonates. Diapause allows the critical or rate-limiting event in each population's reproductive cycle to occur at a defined point in time. The unifying purpose of diapause is to extend the reproductive cycle during gestation in order to synchronize reproductive parameters among individuals and/or the environment (Renfree, 1978). It also is possible that diapause is vestigial for environmental and social constraints that no longer apply making its derivation difficult to explain.

Parthenogenesis

Reproduction in the absence of male gametes, parthenogenesis, is documented for all jawed vertebrates except mammals. Many observations of reproducing female elasmobranchs in aquaria without male companionship have been noted (Castro *et al.*, 1988; Voss *et al.*, 2001; Heist, 2004). Potential sperm storage and the lack of individual life history details have prevented unequivocal identification of these events as parthenogenetic. Mictic parthenogenesis has been proven genetically for a viviparous placental shark, *Sphyrna tudes* (Chapman *et al.*, 2007). Invalidated reports of parthenogenesis in captive sharks have recently appeared in internet media stories. These include a female blacktip that never had contact with a conspecific mature male during her tenure at the Virginia Aquarium and Marine Science Center, Virginia, but was discovered to be gravid upon necropsy, a female *Chiloscyllium plagiosum* from the Belle Isle Aquarium, Detroit, that laid eggs yielding embryos despite her isolation from a suitable mate for at least six years, and a *C. plagiosum* at Carl Hayden Community High School, Phoenix, AZ, alone for nearly four years, that laid eggs, one of which developed and hatched, producing a healthy offspring.

Developmental Biology

There are several remarkable works concerning chondrichthyan developmental biology including monographs by Dean (1906) describing chimaeras, Scammon (1911) describing *S. acanthias*, Gudger (1940) describing *C. anguineus*, Smith (1942) describing *Heterodontus japonicus*, Ballard *et al.* (1993) describing *S. canicula*, Didier *et al.* (1998) describing *C. milii* and Luer *et al.* (2007) describing *R. eglanteria*. Elasmobranch development begins within the female reproductive tract for all species. It is supposed that chondrichthyan developmental biology is generally uniform among species, especially during the earliest stages. Babel (1967) describes the blastodisc of *Urolophus halleri* but no other early stages, referring instead to Balfour's 1878 monograph and states, "The early stages are similar in all elasmobranchs and will not be reviewed here." Freshly oviposited eggs are usually observed in late segmentation stages. Variability in the amount of time the female retains eggs and environmental temperature influence embryonic stage at oviposition. Elasmobranch ova are generally large and embryonic development necessarily meroblastic.

Fertilization and Segmentation

Images of chondrichthyan blastodiscs are often spoken of in the literature but rarely pictured. Leydig (1852) may have been the first to describe and illustrate a blastodisc for elasmobranchs while examining the eggs of *G. melastomus* (formerly *Pristiurus*) ((Leydig, 1852) within (Wourms, 1997)). Balfour (1878) noted the blastodisc on the eggs of *Pristiurus*, two species of *Scyliorhinus* (formerly *Scyllium*) and *Raja*. Haswell (1898) described the blastodisc of *H. japonicus* (formerly *philippi*) including both surface and sectional views. Rückert (1899) presented a monograph that contains images of blastodiscs from *T. ocellata*, *G. melastomus* and *S. canicula*. Rückert is credited with illustrating polyspermy, the fate of superfluous sperm nuclei and yolk merocytes (yolk nuclei) but his results concerning the sperm nuclei and yolk nuclei have recently been reinterpreted (Lechenault and Mellinger, 1993). Rückert's monograph is beautifully illustrated with surface views and histological sections of blastodiscs that are consistent with discoidal cleavage common to heavily yolked eggs but contradict the corresponding text which largely ignores the nuclear syncytium that forms during early cleavage (Fig. 22A). A comparison of Rückert's drawings with images of *R. eglanteria* and *D. sayi* embryos during cleavage shows many similarities among species (Fig. 22B-D). Babel (1967) described the blastodisc of *U. halleri* as a small, light-colored circular area surrounded by a narrow light colored band. It is thicker in the middle and tapers at the edges.

Fig. 22. Cleavage (A) 1. Egg case of *Galeus* (formerly *Pristiurus*) *melastomus*. 2-11. Early cleavage in surface view for *Torpedo ocellata* (2, 3, 4 and 8), *Scyliorhinus canicula* (7, 9-11), and *Pristiurus* (5 and 6). Reprinted from Rückert (1899). (B and C) Cleavage of *Raja eglanteria* blastodiscs, 24 and 48 hours after oviposition. (D) Scanning electron micrograph of *Dasyatis sayi* early cleavage. (Photographs B and C—Carl Luer).

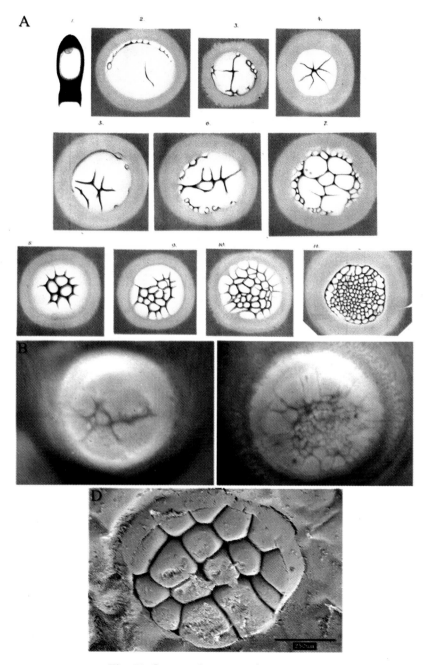

Fig. 22. See previous page for caption.

Bashford Dean made detailed accounts of embryonic development for *H. japonicus* and *C. anguineus*. Nishikawa (1898) collected and described blastodiscs, but they are neither illustrated nor measured. However, Nishikawa shared his histological sections of early and late blastodiscs with Dean who did make notes and illustrations. Nishikawa commented that the gastrula stage *C. anguineus* he obtained was "... oval in form and 3 mm in length. I have nothing special to add about it, as it was like the gastrula of any other shark." Gudger reproduced these drawings despite his statement, "There is nothing very striking about the mode of development portrayed, since it is typically elasmobranch..." In his observations of *Haploblepharus edwardsii,* von Bonde (1945) states, "The segmentation of the germinal disc is very similar to other Selachians... The various later stages of gastrulation with the subsequent origin of the germinal layers is also typical of the class and it is not necessary to elaborate this feature."

Ballard, Mellinger and Lechenault *et al.* (1993) described and defined 34 normal stages of development for *S. canicula*. Their work is the first to describe a single species from fertilization until hatch and the normal stages are referred to by number in the remaining text. Lechenault and Mellinger (1993) further subdivided early stages to describe details of fertilization, cleavage and the origin of yolk nuclei for *S. canicula* (Figs. 23-26). The germinative disc is bright orange and surrounded by a ring of whitish yolk distinguished from the yellow-green vitellus (Fig. 27A). Cleavage is limited to a small cap of cytoplasm that accumulates at one pole and surrounds the female pronucleus (Fig. 27B-H). The yolk granules adjacent to the cytoplasm are 3 µm and gradually increase in size to 8 µm with increased distance from the blastodisc before yielding to yolk platelets in the vitellus (Figs. 23-24). The female pronucleus is larger than the male pronuclei and is 150 µm beneath the vitelline membrane within the smallest yolk granules. Sperm pronuclei (30-50) are observed within the germinative disc of eggs removed from the shell gland within partially formed egg cases. After fertilization by a male pronucleus, the zygote and supernumerary sperm divide successively (stage 1). Supernumerary male pronuclei are displaced by blastomeres to the periphery of the blastodisc. Their division ceases while the diploid nuclei continue mitosis.

Cleavage furrows form from infoldings of the vitelline membrane and result in bulging, open blastomeres. Diploid nuclei continue to divide and supply the syncytium while the supernumerary sperm nuclei degenerate. Blastomeres form from both the surface as well as from division of previously closed blastomeres, eventually reaching a point where there are more than 100 blastomeres (stage 2). The blastodisc assumes a circular shape, thicker in the center, and becomes separated from the yolk by a segmentation cavity (stage 3). Nuclei that remain below the segmentation

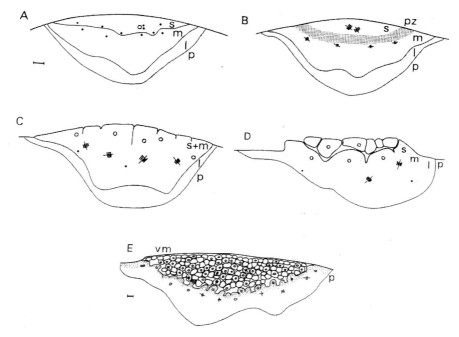

Fig. 23. Five stages in the development of oviductal eggs, based on camera-lucida drawing of representative sections through the middle of germinative discs (A-C) or blastodiscs (D-E). In drawings A-D, more nuclei than are actually visible in a single section have been shown. All nuclei, either at rest or dividing, are indicated by symbols: 0 and larger metaphase symbols indicate the female pronucleus (A) and diploid nuclei (B-E), while dots (A, C, D) and smaller metaphases (B) indicate presumably haploid nuclei, derived from supernumerary sperms. Bars = 50 μm (bar A valid for A-D). (A) Evidence for polyspermy. Two haploid nuclei are found near the female pronucleus, neither of them having already developed into a distinct male pro-nucleus. Several supernumerary sperm nuclei are scattered along the s-m transition zone. (B) Synchronous mitoses, both involving diploid and haploid nuclei, separated by a transient zone of pigmented, dense cytoplasm. (C) Asynchronous, syncytial segmentation. First plasma membranes appearing at the surface of the germinative disc. Small, haploid nuclei at rest remain deep in the s + m zone. (D) Early blastodisc stage. Only one layer of surface blastomeres. Production of deep blastomeres by the syncytial parts of the germinative disc. Some haploid nuclei may still be found in the deepest parts of the s + m + l zone (two of them are shown). (E) End of blastodisc formation: egg-laying stage. Segmentation cavity first delineated above the basal syncytium, where diploid nuclei continue dividing but now only generate giant yolk nuclei. No more haploid nuclei. Reproduced from Lechenault, *et al.* (1993) with kind permission of Wiley-Liss, Inc., a subsidiary of John Wiley & Sons, Inc.

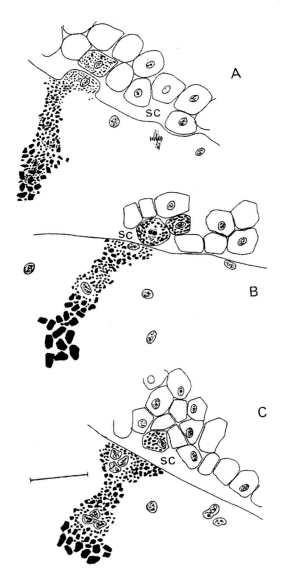

Fig. 24. The first set of yolk nuclei. Camera-lucida drawings. Bar = 50 μm. (A) Detail of Fig. 23E. Size gradient of yolk granules (black) in the s + m + l zone. Unlike yolk platelets, granules have irregular outlines in semi thin sections. (B) Full set of scattered diploid nuclei resulting from final mitoses in the basal syncytium. (C) Late stage, showing groups of 2-4 diploid nuclei, prior to merging. Reproduced from Lechenault, *et al.* (1993) with kind permission of Wiley-Liss, Inc., a subsidiary of John Wiley & Sons, Inc.

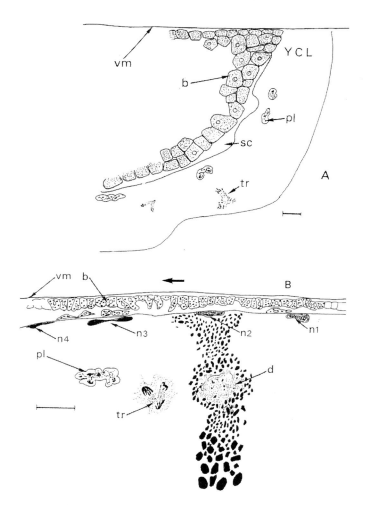

Fig. 25. The first set of yolk nuclei (continued). Semi-diagrammatic sketches. Bars = 50 μm. (A) The stage after that shown in Fig. 24C, showing giant, plurilobed yolk nuclei produced by the merging of diploid nuclei. Abnormal (tripolar) mitoses. The blastodisc has reached its maximal thickness (only its peripheral cells are shown). (B) Epiboly (large arrow). Thin blastoderm, showing a continuous surface layer and sparse deep cells. Peripheral yolk nuclei (n1-n4) show increasing condensation of the chromatin as they migrate over the yolk during epiboly, and they may divide by amitosis (see extended, n3). Deeper nuclei are either plurilobed, dividing, or degenerating. Reproduced from Lechenault et al. (1993) with kind permission of Wiley-Liss, Inc., a subsidiary of John Wiley & Sons, Inc.

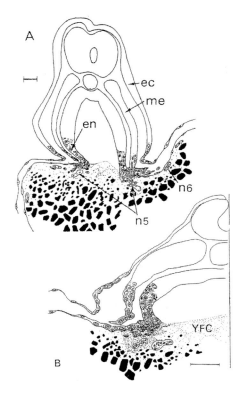

Fig. 26. The second set of yolk nuclei. Camera-lucida drawings of semi thin sections cut transverse through embryos, at stages G-H, attached to their yolk substratum (black granules). Bars = 50 μm. (A) General aspect. n5 = new, giant yolk nuclei resulting from the merging of endoderm nuclei. n6 = new, giant peripheral yolk nucleus participating in epiboly. (B) Detail of another section. Mitoses are not shown, but are common in all tissues and yolk. Reproduced from Lechenault *et al.* (1993) with kind permission of Wiley-Liss, Inc., a subsidiary of John Wiley & Sons, Inc.

cavity form the periblast or syncytium of primary yolk nuclei within the yolk cytoplasmic layer (YCL) of the egg. The yolk nuclei continue to divide, some merging to form giant nuclei (Figs. 24-25). In later stages of development, whole cells and nuclei stream from the embryonic endoderm at its junction with the extra-embryonic endoderm giving rise to a second population of yolk nuclei (Fig. 26). Neither population of yolk nuclei contribute to the embryo proper (Lechenault and Mellinger, 1993).

The accumulation and distribution of deep blastomeres define subsequent stages (Ballard *et al.*, 1993). These stages (4-7) are difficult to identify without invasive techniques such as histology (Figs. 27G-H, 28

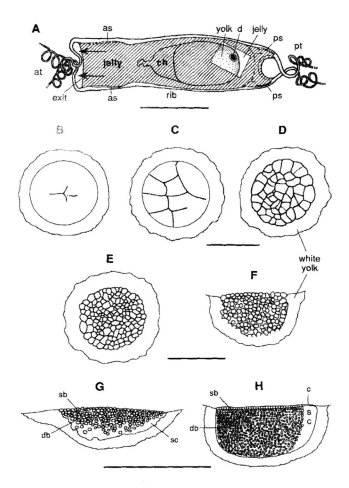

Fig. 27. Egg, eggshell and contents, and earliest cleavages. (A) *Scyliorhinus canicula* eggshell seen from one side, the yolk mass and germinal disc exposed through a window made by shaving its pigmented layer off. Tendrils at either end, shown only in part, can be stretched up to a meter in length. (B-D) Top views of stage 1 germinal discs resting on their white yolk cups, in earliest cleavages. (E-F) Stage 2 in top view and cross section, polarity not yet visible. (G-H) in sagittal sections showing the beginning and the completion of aggregation of the inner blastomeres, with collection of intercellular fluid in a subgerminal cavity and toward the posterior pole under a membranous crescent. as—anterior slits of eggshell; at—anterior tendrils of eggshell; c—posterior crescent; ch—chalazial chamber; d—germinative disc; db—deep blastomeres; ps—posterior slits of eggshell; pt—posterior tendrils of eggshell; sb—surface blastomeres; sc—segmentation cavity. Scale bars 1 mm. Reproduced from Ballard, *et al.* (1993) with kind permission of Wiley-Liss, Inc., a subsidiary of John Wiley & Sons, Inc.

and 29). The segmentation cavity first forms in the caudal aspect of the embryo and the blastomeres accumulate at its margin creating a translucent crescent shaped cavity. The segmentation cavity grows anteriorly pushing the deep blastomeres at its margin forward with the resulting caudal crescent reaching its largest size (stage 5). This stage identifies the orientation of the embryonic axis which eventually bisects the crescent. The deep blastomeres spread, reducing the size of the crescent for stage 6, which is similar in surface view to stage 4. The posterior crescent vanishes by stage 7, which resembles stage 3 in surface view.

Epiboly

The blastodisc spreads to cover the posterior rim of white yolk (stage 8). This marks the onset of epiboly and the morphogenic movements that bring about gastrulation. No white yolk is visible in stage 9 embryos as the blastoderm spreads in all directions on the thin YCL (Fig. 29). The blastoderm develops a posterior thickening on a round or, more commonly, ovoid blastoderm (stage 10) (Fig. 30B). This thickening bulges over the yolk to form a bilamellar epithelial overhang (stage 11) and marks the onset of gastrulation. There is great variability in embryo shape and appearance during stages 8-11, with folds being common in the rapidly expanding blastoderm (Fig. 31). A more uniform appearance emerges during gastrulation and subsequent stages.

Gastrulation

Knowledge of gastrulation (stage 12) in chondrichthyes has benefited from published pictures of specimens in addition to drawings and has been documented in more species than blastulae (Fig. 30C). Kopsch (1950) named the conspicuous triangular bulge that forms as the overhang gathers posteriorly the "notochordal triangle" (Fig. 31). The morphogenic movements that bring about formation of the mesoderm and endoderm remain largely undefined. Many authors illustrate and interpret histological sections of gastrulae, but because they represent rapidly migrating cells at a fixed point in time, differences in interpretation exist. Ziegler and Ziegler (1892) and Swaen (1887) showed the uniform character of elasmobranchs during gastrulation with surface and sectional views of *T. ocellata*. Vanderbrooke (1936) tracked cell movements in the blastoderm of *S. canicula* and prepared a fate map but its validity has been questioned by subsequent researchers (Kopsch, 1950; Ballard *et al.*, 1993). Gastrulation was realized by the movements of notochordal, pre-chordal plate, mesodermal and endodermal cells over the caudal overhang and into the segmentation cavity, eventually obliterating it. Recently, a molecular approach using a

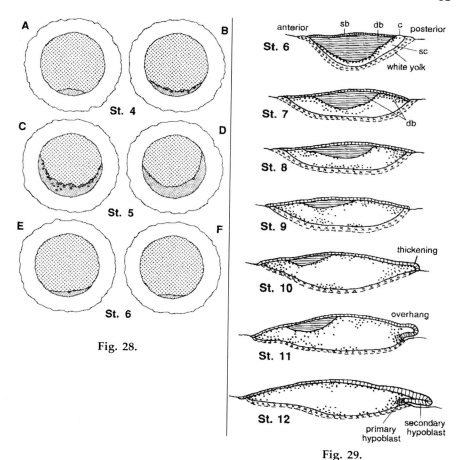

Fig. 28.

Fig. 29.

Fig. 28. Later cleavage and onset of epiboly. Top views of blastodiscs, each surrounded by white yolk. (A-B) The posterior crescent just forming (as in Fig. 27H). (C-D) Maximum development of the crescent and start of scattering of deep mesenchyme. (E-F) Narrowing of the crescent by epibolic spread of the surface epithelium and inner mesenchyme. See Figure 29 also. Much variation in these stages. Reproduced from Ballard *et al.* (1993) with kind permission of Wiley-Liss, Inc., a subsidiary of John Wiley & Sons, Inc.

Fig. 29. Early epiboly stages, diagrammed from serial sections to show the surface epithelium (vertical hatching), white yolk cup (cross hatching), consolidated mass of inner blastomeres (horizontal hatching), mesenchyme (dots) spreading through the subgerminal cavity, and appearance of the embryonic shield (to be compared with Fig. 31). All sections are mid-sagittal except the one for stage 12, which is parasagittal. c—posterior crescent; db—deep blastomeres; sb—surface blastomeres; sc—segmentation cavity. Reproduced from Ballard *et al.* (1993) with kind permission of Wiley-Liss, Inc., a subsidiary of John Wiley & Sons, Inc.

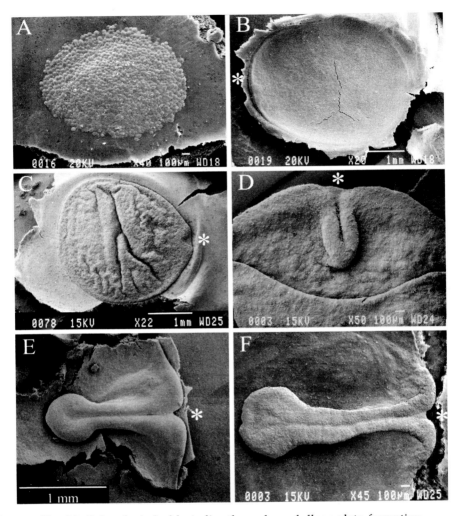

Fig. 30. *Raja eglanteria*, blastodisc through medullary plate formation. (A) More than 1000 cells comprise the blastodisc 24 hours after oviposition. (B) An overhang forms at the posterior margin of the blastoderm marking the orientation of the embryonic axis (*). (C) Gastrulation results in a notochordal triangle and folds on the surface of the rapidly expanding blastoderm are common. (D) The embryonic axis is formed from the convergence of the long sides of the notochordal triangle. (E) The axis elongates from posterior growth and the anterior limit swells into a head enlargement. (F) The head enlargement spreads laterally into a medullary plate, stage 15.

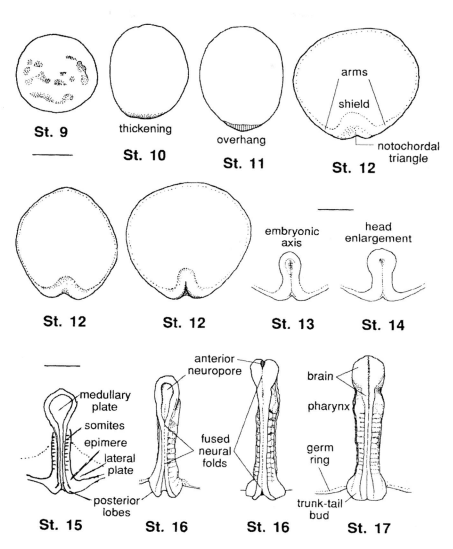

Fig. 31. Principal morphogenetic cell movements and earliest organogenesis shown in top views. Shading in the Stage 9 figure represents depressions in the wrinkled surface epithelium (variably seen in this and several succeeding stages). Compare stages 9-12 with sectional diagrams of Fig. 29. Stage 12 is seen in rapidly succeeding early, middle, and late forms. Early and late stage 16 drawings show rapid closure of neural folds. Scale bar 1 mm. Reproduced from Ballard *et al.* (1993) with kind permission of Wiley-Liss, Inc., a subsidiary of John Wiley & Sons, Inc.

Brachyury-related gene that specifies nascent mesodermal and endodermal cells in *S. canicula* embryos supported Vanderbrooke's earlier research demonstrating that mesoendodermal cells circumscribe the blastoderm and move posteriorly to the overhang and notochordal triangle where they are internalized (Fig. 32) (Sauka-Spengler *et al.*, 2003).

Gastrulation results in the formation of an embryonic axis (stage 13) as the lateral edges of the shield which form the long sides of the notochordal triangle converge (Figs. 30D and 31). The fixed anterior location lengthens only marginally before swelling to form a head enlargement (stage 14) (Figs. 30E and 31). The embryo lengthens with two posterior lobes extending past the expanding blastoderm. Somites are visible posterior to the flared medullary plate (stage 15) (Figs. 30F and 31). The neural folds begin to fuse (Figs. 31 and 33A) in the region of the first somites and fusion proceeds cephalad and caudad simultaneously (stage 16). This marks completion of embryo formation and early embryonic development.

Pharyngeal Stages

Successive stages are identified by the appearance of pharyngeal pouches, their opening into clefts, and adornment with gill filaments (Figs. 33B-F, 34 and 35). External gill filaments are a plesiomorphic character for chondrichthyans. The filaments are sinusoidal loops that extend from the branchial arches of embryos before gills develop. They are composed of a capillary loop covered by a two-layered squamous epithelium (Kryvi, 1976). The formation of gill filaments during early stages of development, when

---

Fig. 32. Expression pattern of *Brachyury*-related gene, *ScT*, in gastrulating *Scyliorhinus canicula* embryos (stages 12-14). (A) Blastodisc at stage 12. At this stage, *ScT* is expressed in the marginal zone of the blastoderm and the notochordal triangle in the posterior midline. Arrowheads point to the posterior limit of the blastocoel. (B, B', C, C') Magnifications of sagittal sections of the embryo shown in (A) at the level of the margin (B, C—posterior margin; B', C'—anterior margin). (D) Transverse section of a stage 12 embryo. (E) Dorsal view of a stage 13 embryo. The marginal signal becomes restricted posteriorly and persists in the axial mesoderm. (F–I) Transverse sections of the embryo shown in (E) at different anteroposterior levels. *ScT* transcripts are found in the prechordal plate (F, G) and the forming notochord (H, I), as well as at the posterior margin (I). (J, K) Magnification of a mid-sagittal section of a stage 13 embryo, (J—anterior margin; K—posterior margin). (L) Dorsal view of a stage 14 embryo. (M-O) Transverse sections of the stage 14 embryo shown in (L) at different anteroposterior levels. Thin lines indicate the planes of sections. Anterior is to the top in (A, E, L). np—neural plate; nt—notochordal triangle; n—notochord; pl—posterior lobes; pm—prechordal mesoderm; so—somite. Scale bars 0.5 mm.
Reproduced from Sauka-Spengler *et al.* (2003) with permission of Elsevier.

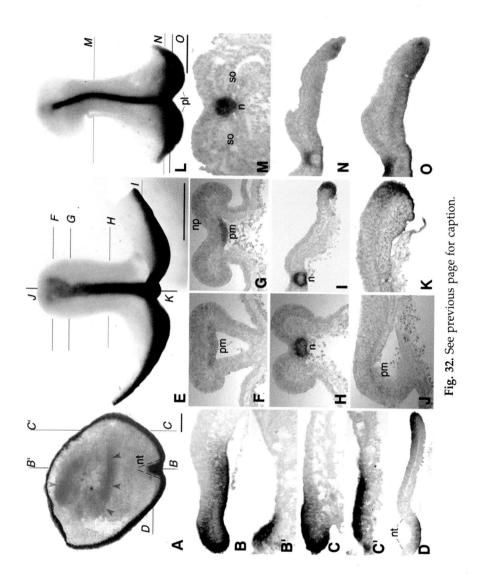

Fig. 32. See previous page for caption.

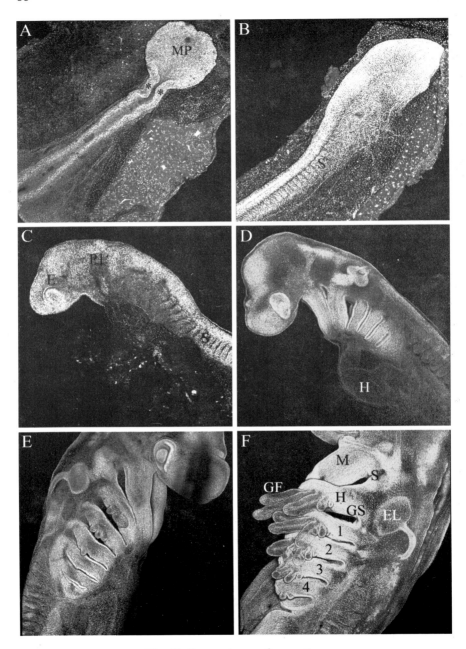

Fig. 33. See next page for caption.

all elasmobranchs are lecithotrophic, supports the importance of increased surface area for respiration at this time (Beard, 1890). For oviparous species, regression of length and diameter of external filaments occurs in concert with the development of internal gills and ram ventilation (Pelster and Bemis, 1992). Among viviparous species the gill filaments also may be absorptive (Davy, 1834; Alcock, 1892; Hamlett et al., 1985a). Before hatching or parturition the gill filaments are completely reabsorbed.

The gill pouches appear sequentially but do not follow in their order of opening (Figs. 33-35). First to open is pouch two (P2) followed by pouch one (P1) then sequentially until all pouches (P3-P6) are open. In *L. erinacea*, five gill filaments extend first from the hyoid arch and then successively from gill arches 1-4 (Pelster and Bemis, 1992). The central of the five filaments appears first, flanked linearly by filaments on adjacent areas of the same arch until five filaments are formed. For *S. canicula*, gill filaments extend first from the hyoid arch, followed in sequence by pharyngeal arches 3-5 and finally, almost simultaneously from the mandibular arch and arch 6 (Ballard et al., 1993). Melouk presents data on gill filaments in *C. melanopterus*, *R. djiddensis* and *R. halavi* developing in all clefts (Melouk, 1949). *C. melanopterus* filaments develop first on the dorsal aspect and later ventral while filaments in batoid species are restricted to the ventral portion of the arches. How much variability exists among chondrichthyans with regard to gill filaments is not known, although the appearance, opening and adornment sequence is consistent among elasmobranchs examined to date. Pharyngeal or gill clefts (C) in the holocephalan *C. milii* open in the sequence C2, C3, C1, C4, C5 and C6 with C1 retarded and opening coincident with C3 or even C4 (Didier et al., 1998). The spiracular cleft, C1, is nearly closed again by the time the last 2 clefts are completely opened. Gill filaments appear on the hyoid arch followed by gill arch 1, 2, 3 and 4.

Potentially confusing is the terminology describing the clefts and

---

Fig. 33. *Raja eglanteria*, neural fold through gill arch formation. (A) Fusion of the neural folds begins posterior to the medullary plate, MP, in the region of the first somite and proceeds cephalad and caudad simultaneously. (B) Stage 16 embryo. (C) Early stage 18 embryo marked by a distinct dorsal curvature of the body, obvious somites, S, and eyes, E. The first pharyngeal pouch, P1, is visible. (D) Early stage 22 embryo with pharyngeal pouch or cleft 2 open and the spiracle or pharyngeal pouch 1 starting to open. Clefts 3-5 are visible as lines perpendicular to the body axis. The heart, H, is present as a bent tube and the endolymphatic sac is well developed. (E) Stage 24 embryo with gill filaments adorning the hyoid arch as well as the first and second gill arches. (F) Stage 27 embryo with all clefts open and all gill arches adorned with gill filaments. M—mandibular arch; S—spiracle or hyomandibular cleft; H—hyoid arch; EL—endolymphatic sac; GS—gill slit or cleft; GF—gill filament; 1, 2, 3, 4—gill arches 1-4.

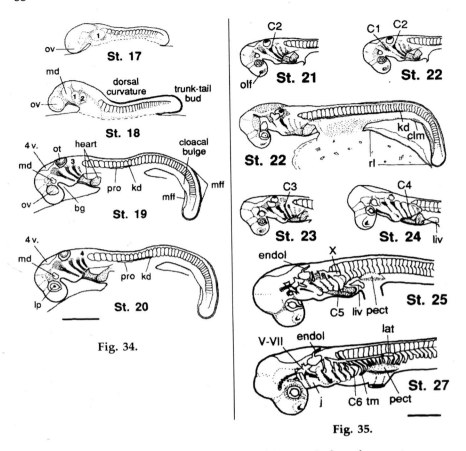

Fig. 34. Side views of lengthening *Scyliorhinus canicula* embryos at same scale during the appearance of successive pharyngeal pouches before any clefts break through. ov—optic vesicle; md—mandibular head cavity (coelomic); 4 v.—fourth brain ventricle; ot—otic placode; pro—pronephros; kd—kidney duct; mff—median (sagittal) finfold; lp—lens placode. Scale bar 1 mm. Reproduced from Ballard *et al.* (1993) with kind permission of Wiley-Liss, Inc., a subsidiary of John Wiley & Sons, Inc.

Fig. 35. Formation of the pharyngeal clefts in *Scyliorhinus canicula* embryos. Stages all shown at the same magnification. C1—hyomandibular cleft; C2-C6—branchial clefts; clm—cloacal membrane; endol—endolymphatic duct; j—jaws or mandibular arch; kd—kidney duct; lat—lateral line canal; liv—liver; olf—olfactory placode; pect—pectoral fin; rl—vitelline vein or "red line"; tm—trunk myotomes; x—vagus ganglion or placode; V-VII—placodes or ganglia of cranial nerves V and VII. Scale bar 1 mm. Reproduced from Ballard, *et al.* (1993) with kind permission of Wiley-Liss, Inc., a subsidiary of John Wiley & Sons, Inc.

arches. The first pharyngeal arch is the mandibular arch and the second pharyngeal arch is the hyoid arch. The third pharyngeal arch is the first gill arch and successive arches follow consecutive numbering. The first pharyngeal cleft is the spiracle or hyomandibular cleft, the second cleft is the hyobranchial. The gill slits correspond to pharyngeal clefts 2-6. In order to keep descriptions of embryos applicable to all taxa, respect for the origin of the cleft rather than its final disposition should dictate its number.

In addition to pharyngeal characters, stage 17 embryos can be identified by their closed neuropore, tail bud and first muscular contractions. A distinct dorsal curvature marks stage 18 embryos. The connection to the yolk remains wide and the embryo's movements are pronounced as it whips to and fro. The heart is present as a straight non-contractile tube. In stage 19 embryos the tail is curved downward with a pronounced cloacal bulge and the heart tube is bent and beating.

Fin development begins with pectoral ridges present in stage 24 embryos that become easily identified along with pelvic fin rudiments in stage 25 embryos. The posterior margins of the anal and dorsal fin folds are marked by a shallow cut in stage 30 embryos. At this stage, the only pigment present on the embryo surrounds the eye. The caudal tail scales appear as buds in preparation for eclosion in scyliorhinids and heterodontids (Figs. 7 and 36A). The rostrum is first recognizable in stage 31 embryos and as it develops in subsequent stages, movement of the mandibular arch changes the shape of the mouth from a diamond (stage 27) to an oval (stage 28) and finally to the adult form of a straight or downward curved line for *S. canicula* (stage 29) (Fig. 37). During fin specification and rostrum development, embryos can be identified by body form, namely batoid or fusiform. For species with remarkable characters like *Pristis*, identification to the genus or even species level is possible (Fig. 38). The job of identification becomes increasingly simplified with advancing stages.

Approximately midway through stage 31, *S. canicula* eggs open to the environment (eclosion). Yolk is visible in the stalk as it moves toward the embryonic gut. Gill filaments are at maximal length. Body pigmentation develops in stage 32 embryos, whose yolk sac is not yet significantly reduced in size. Most of the yolk is consumed or transferred to the IYS by stage 33 embryos. Prehatching, stage 34, is a waiting period where the embryo's IYS is engorged with yolk transferred from the EYS, present as a shriveled stump. The EYS is completely reabsorbed before hatching. A comparison of embryo size at eclosion and prehatching illustrates the great amount of growth that takes place during the latter half of incubation (Fig. 36).

Epiboly begins during stage 8 but because of the size of the yolk mass, the blastopore is split into an embryonic blastopore under the tail bud and

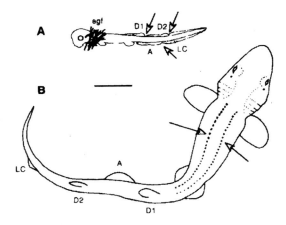

Fig. 36. Early prehatching *Scyliorhinus canicula* fetus and hatchling at same magnification as an indication of growth attained during the prehatching period of about 3 months. External gill filaments regress entirely. Stippled areas of the sagittal finfold are thin membrane (later lost). The plain parts are the thicker rudiments of the two dorsal fins, anal fin, and upper and lower parts of the caudal fin. The newly hatched fish has two rows of special placoid denticles along its back (arrows). These and the four rows of special caudal denticles (not shown) all appear before hatching. egf—external gill filaments; D1 and D2—first and second dorsal fins; A—anal fin; LC—lower half of the caudal fin. Scale bar 1 mm. Reproduced from Ballard *et al.* (1993) with kind permission of Wiley-Liss, Inc., a subsidiary of John Wiley & Sons, Inc.

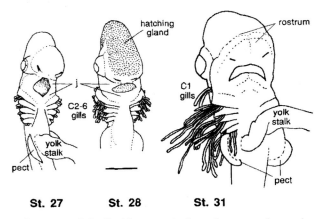

Fig. 37. Ventral views of *Scyliorhinus canicula* embryo to show changes in the mouth, gills, and pharyngeal region during development. pect—pectoral fin; j—jaw or mandibular arch; C2-6—branchial clefts; C1—hyomandibular cleft. Scale bar 1 mm. Reproduced from Ballard *et al.* (1993) with kind permission of Wiley-Liss, Inc., a subsidiary of John Wiley & Sons, Inc.

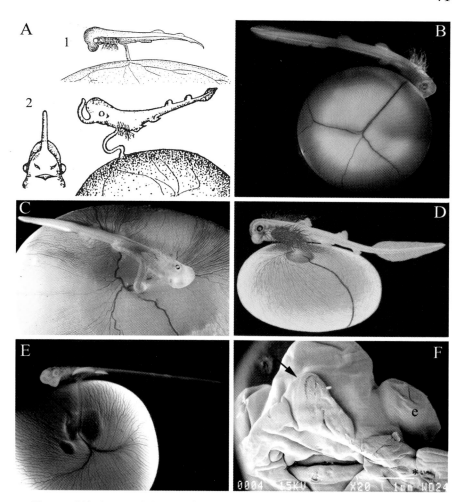

Fig. 38. (A) *Anoxypristis cuspidata* (formerly *Pristis cuspidatus*). 1. A 33 mm embryo with yolk sac only partly depicted. Note the rudimentary gill region. The yolk sac is vascularized only half-way down its walls. 2. A 53 mm embryo with ventral view of rostrum and head. Vascularization of the yolk sac is complete. Reprinted from Setna and Sarangdhar (1948). (B) *Chiloscyllium punctatum*, 55 days post oviposition. (C) *Ginglymostoma cirratum*. (D) *Heterodontus japonicus*, reprinted from Smith (1942) with kind permission of the American Museum of Natural History. (E) *Raja eglanteria*, 24 days post oviposition. (F) *R. eglanteria*, 38 days post oviposition. Ventral view of the mouth, eye, e, and developing rostrum (arrow). Note the movement of the pectoral fin forward at the fusion point with the head (*). (Photographs B—Dr. Nicholas James Cole; C—Carl Luer; E—Jose Castro and Carl Luer).

a yolk blastopore. The embryonic blastopore closes with fusion of the posterior lobes. The yolk blastopore does not close until stage 25 when epiboly is complete. The margin of the blastoderm is conspicuous as it converges under the trunk of the embryo. The edges are pigmented and fuse forming an efferent vitelline vein that is visible as a red line growing in length from the posterior of the embryo in successive stages (Fig. 39). The vitelline artery originates from the dorsal aorta at the pronephros level and is visible during stage 19. It grows forward and bifurcates into wide arches. The lateral plate mesoderm spreads from the embryo to the extraembryonic yolk, converting the bilamellar yolk sac to a trilamellar structure (Mossman, 1937). Vascularization of the mesoderm proceeds radially from the vitelline vessels outward. The vascular splanchnic mesoderm may be partially or completely separated from the non-vascular somatic mesoderm

Fig. 39. Polarity and symmetry of dogfish (*Scyliorhinus canicula*) yolk and embryonic system. Development of external yolk sac (A, D, E), subsequent expansion of vascular area (B, C, E-G), and final pattern of vitelline vessels (H, I). Orientation: A-C—in profile; D-F and I—caudal; G-H—cephalic side views. (A) Polarity of egg cell marked by the presence of a whitish polar yolk cap, PY, around the animal pole, AP, most conspicuous in green-colored egg cells. The blastodisc, B, is thus rejected in the upper corner and generally behind one edge of the egg case. Curves are the successive steps of extraembryonic blastoderm spreading (arrows). Its growth toward the vegetal pole, VP, can be scaled as indicated on the left. The open umbilicus, U, and suture, S, grow from the AP on midline. (B) Yolk sac completely closed, with a red spot on the closed umbilicus, U, which rises during the following days (arrow). Subsequent expansion of vascular area (arrows) proceeds in three steps, 1, 2, 3. Embryo, E, now in polar position (note its orientation: head facing left). (C) Further steps of vascular area expansion, 4 and 5, with arterial lining (note orientation of arterioles toward umbilicus). (D) Caudal view of the open umbilicus, U, with the yolk sac suture above, and vascular area (islets shown as dots) lined in part by branches of presumptive vitelline artery (solid, symmetric lines). (E) Umbilicus just closed, U, vascular area, VA, still unexpanded. Prominent "red streak", RS, marking the suture of the yolk sac. This pigmented streak already existed along blastoderm edge during its spreading over the yolk. (F) Same structures during step 3 of vascular area expansion (arrows). Raised umbilicus, U, crossed by vitelline vein, V. Steps 5 and 6 terminating vascular area expansion: vitelline artery, A, is bifurcated. (H) Full development of vitelline artery, A. Distal bundles of arterial capillaries, ac, are emitted by both arterial branches, while blood is drained by venous capillaries, vc, to the opposite side. (I) Opposite side. Caudal view showing vitelline vein and dense network of venous capillaries, vc. A-I reproduced from Mellinger *et al.* (1986) with kind permission of the Ichthyological Society of Japan. (J) *Raja eglanteria* vitelline vein returning blood to the embryo and vitelline artery (K) carrying blood away from the embryo. Blood flow is indicated by directional arrows. (Photographs J and K—Carl Luer).

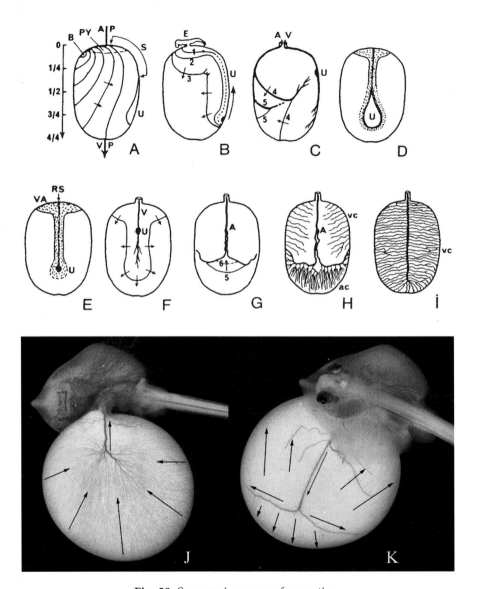

Fig. 39. See previous page for caption.

by an exocoelom. When vascularization of the yolk sac is complete, blood flows over the entire yolk sac surface in a fine branching capillary bed, eventually collected by the vitelline vein (stage 27).

Rays

Melouk (1949) described Rhinobatidae embryos from 5 mm TL to 62 mm TL. He divided their development into four general phases. In the first phase, stages 1-3 (corresponding to Ballard, Mellinger and Lechenault (BML) stages 21-25), the embryo resembles any cartilaginous fish with a notable lack of paired fins. In the second phase, stages 4-8 (BML stages 26-29), general features common to gnathostome fishes develop, including paired pectoral fins, median fins and differentiation of the mandibular arch into the jaw. By stage 8, the position and size of the pectoral fins is already distinct between sharks and rays and in Melouk's words, "The embryonic body-now considerably flattened anteriorly-may be said to have passed over the line separating the rajiform from the squaliform type." The third phase, stages 9-11 (BML stages 30-33), marks a definite distinction between fusiform sharks and batoids with the obvious extension of the pectoral fins forward under the branchial region. The rostrum becomes distinct and grows anteriorly. Gill filaments are at their maximum length and pigmentation circumscribes the eye. In the fourth phase, stage 12 onwards (BML stages 33-34), the embryo is unmistakably rajiform. The pectoral fins begin to fuse with the branchial region and the gill slits have closed dorsally. The rostrum is extended forward and the ray takes on adult characters with a large portion of its gestation still remaining. Similarly, these anatomical features are useful for adapting BML stages to skates (Fig. 40) (Miyake *et al.*, 1999).

Thorson described *Potamotrygon circularis* embryos in early, mid and late gestation (Thorson *et al.*, 1983). Early rays (11-13 mm DW) are already relatively advanced in development with 5 mm gill filaments. The pectoral fins are fusing to the body and the yolk sac is not reduced. Mid gestation rays, 28-30 mm DW, are attached to the yolk sac (still not reduced in size) by a 2 cm stalk. The pectoral fins are fused except at the anterior limit where the rostrum interjects. Gill filaments are 30 mm long and the eye has slight pigmentation. The tail spine is not yet visible. Late gestation embryos, 76-87 mm DW, have reabsorbed their gill filaments. The pectoral fin has commenced fusion and overlaps the pelvic fin as in adults. Pigmentation is evident in the eye and sheathed spine. The yolk stalk is 27 mm long and the yolk is nearly exhausted. Trophonemata have lengthened from 10 mm to 20 mm in preparation for histotroph secretion. The tail of embryos is longest in early embryos and shortens progressively from its terminus throughout development. Size at birth is estimated at 100-125 mm DW and gestation requires no more than three months.

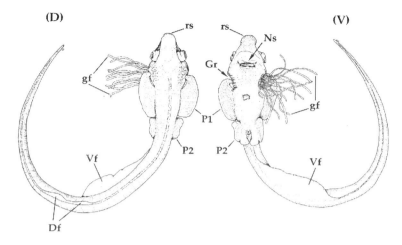

Fig. 40. External embryonic characters of *Leucoraja erinacea*. These characters facilitate staging skate embryos using a staging table for shark embryos (Ballard *et al.*, 1993). Dorsal (D) and ventral (V) view. Df—dorsal fin fold; Gr—gill arches; gf—gill filaments; Ns—nasal areas; P1—pectoral fins; P2—pelvic fins; rs—rostrum; Vf—ventral fin fold. Reproduced from Miyake, *et al.* (1999) with kind permission of Wiley-Liss, Inc., a subsidiary of John Wiley & Sons, Inc.

The embryos of *Dasyatis sabina* have no fin folds and appear "shark-like" when 8.6 mm TL and 1.2 mm DW (Johnson and Snelson, 1996). *D. sabina* is distinguishable as a batoid when the disc width is 3.5 mm due to the presence of a posteriorly elongated pectoral fin and the absence of dorsal and anal fins (Lewis, 1982). Four features of the embryos are described that separate them morphologically from torpediform sharks in subsequent stages of development. First, the dorsal position of the spiracle and its enlarged form are prominent by 7 mm DW. Position and flexure of the olfactory bulb and fusion of the dorsal portion of the gill slits are evident by 11 mm DW. Finally, anterior growth of the pectoral fins and their fusion with the body and rostral cartilage are apparent by 23 mm DW. Claspers are developed at 17 mm DW. Gill filaments are present in the smallest embryo, 1.2 mm DW, and grow until 60 mm DW (Johnson and Snelson, 1996). The embryos have absorbed the EYS after two months of gestation and 75 mm DW. Hatching occurs after 4-6 weeks and gestation requires 3-4 months.

Observations on embryos of *Dasyatis akajei* and *D. laevigata* are similar to *D. sabina* (A. Yamaguchi, pers. com.). The embryo is not distinct from sharks in early stages of development (Fig. 12C). The presence of gill filaments on the ventral portion of the gill arches and lateral extension of pectoral fins are indications of the batoid body plan and become increasingly obvious in subsequent stages (Fig. 12D-F). Pectoral fin fusion with the rostrum occurs late in development but before gill filament reabsorption and yolk consumption (Fig. 12G-H). Body pigmentation is restricted to the eyes until later stages of development when nutrition is histotrophic (Fig. 12I-J).

Rhinoptera bonasus embryos are encapsulated within the uterus for no more than three weeks (Smith and Merriner, 1986). While multiple fertilized eggs may be present at the start of gestation only one ray develops to parturition. The smallest embryos described are already "batoid in appearance" presumably indicating fusion of the pectoral fins to the body. Gill filaments (15-30 mm) are present on 20 mm DW embryos and are resorbed by 89 mm DW. Three-quarter term embryos have 3 mm yolk stalks and almost completely absorbed yolk sacs. The spine is developed but heavily sheathed and trophonemata have a maximum length of 2-3 cm. Full term embryos grow to 405 mm TL and possess pigmentation similar to adults (Hamlett *et al.*, 1985b).

Gestation in *Urolophus lobatus* lasts 10 months and embryos resorb their yolk sac within 5-6 months leaving trophonemata to provide nutrition and oxygen for the latter half of development. Uterine fecundity decreases during gestation similar to *R. bonasus* but remains higher than one. Neonates (105 mm DW) are 44-52% of adult size depending on sex (White *et al.*, 2001).

Babel (1967) describes *U. halleri* embryos as closely resembling sharks until 16 mm TL when "the two forms are separated by a widening gulf". A 9 mm TL embryo is raised off of the vitellus by a short yolk stalk. The eyes and visceral arches are recognizable. Gill filaments extend from the ventral portion of the gill arches in 13 mm TL embryos. Pectoral and pelvic fin buds appear but the lack of medial fins distinguishes rays from sharks. The pectoral fins grow laterally to 3 mm in 16 mm TL embryos. Rays have a conspicuous elongated head and the pectoral fins are posterior to the pharyngeal region. In 20 mm TL embryos, the pectoral fins have extended laterally to 4 mm and backward to the pelvic fins but still remain posterior to the branchial arches (Fig. 41). Gill filaments are 5 mm long and are limited to the ventral portion of each gill arch. The esophagus, stomach, spiral valve and caudal fin are recognizable. The disc width of 24 mm TL embryos has expanded to 7.5 mm and the pectoral fins invade the branchial region but do not fuse. The gill arches with 7 mm filaments are covered dorsally and the entire region flattens. After one month of development the embryo is 40 mm TL and easily recognized as a batoid with a 14 mm DW. Fusion of the pectoral fins is complete except at the rostrum. Within

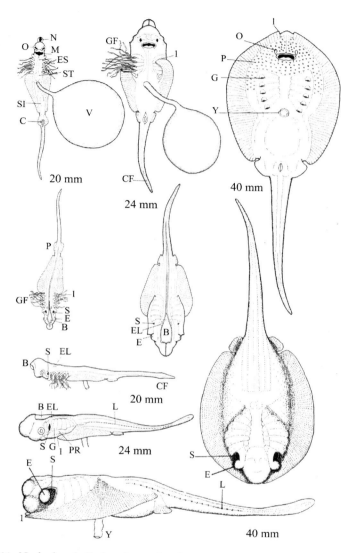

Fig. 41. *Urolophus halleri* embryo development. Smallest embryo is 20 mm TL (total length), medium embryo is 24 mm TL and the largest embryo is one month old and 40 mm TL. All drawn at equal scale. B—unroofed cranium of forebrain; C—cloaca; E—eye; EL—endolymphatic sac; ES—esophagus; G—gill slits; GF—gill filaments; L—lateral line; M—mouth; N—open neuropore; O—olfactory pit; P—papilla; PR—propterygium; S—spiracle; SI—spiral intestine; ST—stomach; V—vitellus; Y—yolk stalk; 1—anterior margin of the unfused pectoral disc. Reproduced with modification from Babel (1967), California Department of Fish and Game.

two weeks, the gill filaments begin to regress and developing trophonemata invade the enlarged spiracles. After two months gestation (64 mm TL, 32 mm DW), the yolk sac is half its original size, but empties quickly and is resorbed before parturition at 75 mm TL. The spiral valve accumulates a green pigmented mass until after birth. The pectoral fins roll around the body and enclose the tail with its sheathed spine to protect the uterus from accidental injury.

For stingray embryos, the tail and its spine pose a significant risk to the female during gestation. In only one case has this potential danger been realized and documented, where a fetal spine punctured the uterine wall and extended into the maternal coelom (Babel, 1967). Protection is afforded through several mechanisms, including flexure of the tail, a protective sheath, and a cartilaginous terminal bulb. The tail flexes anteriorly and is enclosed by the folded pectoral fins of the embryo. The spine is oriented away from the uterine wall and has a sheath or protective knob at its terminus (Babel, 1967; Lewis, 1982; Smith and Merriner, 1986; Johnson and Snelson, 1996; White *et al.*, 2001). The pectoral fins of batoids may be dorsally or ventrally folded, or possess one fin in each orientation (Babel, 1967; Lewis, 1982). A similar situation presents itself for *Pristis* embryos, in which the saw is described to be sheathed (Hussakof, 1912).

Holocephali

The development of holocephalans has been described in great detail but the combined works suffer from a lack of specimens spanning the entire developmental period for any single species (Schauinsland, 1903; Dean, 1906; Didier *et al.*, 1998). Dean noted two peculiarities of chimaera development while observing *Hydrolagus* (formerly *Chimaera*) *colliei*. The first, partitioning of the yolk, has subsequently been discounted by Didier *et al.* (1998) who determined through observations on *C. milii* that yolk fragmentation occurs upon embryo death. Secondly, gastrulation in *H. colliei* was described to occur through an open blastopore connected to the archenteron near the edge of the blastoderm. It is transiently open and bottle cells are pictured migrating through it. Its existence is short and after gastrulation, the appearance of embryos is typical of elasmobranchs until later stages of development. This observation needs re-examination and confirmation.

Didier *et al.* (1998) successfully applied elasmobranch staging (17 to hatching) to *C. milii*. The total number of stages was expanded from 34 to 36, reflecting differences in development between elasmobranchs and holocephalans (Figs. 42-43). Embryonic features that identify chimaerid fishes include a rostral bulb (stages 17-29), an elongated tail, and differences in the sequence of pharyngeal cleft opening as discussed previously.

Later stages of development are easily identified by head morphology and the operculum, a structure never present among elasmobranchs. Cleavage and gastrulation are remarkably similar to elasmobranchs (Fig. 44A-B), but even before stage 17 the embryo is distinctly chimaerid. During development, the yolk sac is at first an ovoid mass, unremarkable in appearance. By stage 17, its form is altered and it appears to be a supportive surface for the head. Two dorsolateral outgrowths extend and cradle the developing rostrum (Fig. 44C-D). This also was observed in *H. colliei* and may be plesiomorphic for order Chimaeriformes (Didier et al., 1998). Among elasmobranchs, the yolk is usually ovoid until after epiboly and vascularization are complete. It then assumes a more or less spherical shape that deflates like a balloon as yolk is consumed.

Fetal Yolk Utilization

All chondrichthyan embryos are initially dependent on yolk for nutrition. Yolk is first consumed intracellularly by blastomeres that bulge with yolk spheres. During epiboly, the yolk is covered by the expanding blastoderm. Because the yolk mass is so large, the embryonic and vitelline blastopores become separate and later, close in sequence. All three germ layers overgrow the yolk and the resulting yolk sac is a trilamellar extension of the embryonic gut. The yolk margin is delimited by a YCL or yolk syncytial layer (YSL). An extra-embryonic coelom may separate splanchnic and somatic mesoderm (TeWinkel, 1943). The yolk sac is a fetal membrane and does not contribute tissue towards the final body of the embryo. The yolk sac does not dissociate from the embryo like most fetal membranes but instead is gradually reabsorbed leaving a vitelline scar. The vitelline scar is unfortunately named because its presence is transient. The role of the EYS in yolk digestion is ill-defined due to a paucity of observations.

In the yolk sac viviparous shark, *S. acanthias*, the endodermal layer of the EYS, IYS, ductus vitellointestinalis and spiral intestine are lined, in part, with ciliated cells that move yolk platelets (TeWinkel, 1943). The yolk sac contains smooth muscle cells and elastic fibers and if prematurely emptied, rapidly shrinks in size. However, there are no observations of yolk sac contractions moving yolk toward the embryo. For *S. acanthias*, glycogen and fat droplets are observed within the yolk endoderm indicating platelet digestion. Vascular ingrowths from the splanchnopleure extend into the vitelline mass increasing the potential surface area for absorption (TeWinkel, 1943). The importance of the EYS in yolk digestion diminishes with development of the spiral intestine, pancreas and liver.

The EYS is active in yolk digestion in the placental sharks, *R. terraenovae, N. brevirostris, Carcharhinus limbatus* and *Carcharhinus acronotus*.

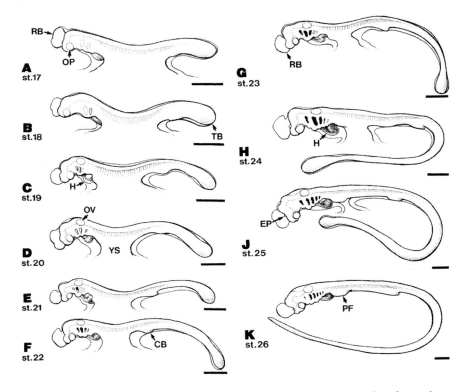

Fig. 42. *Callorhinchus milii*. Line drawing showing details of embryonic stages 17–26. Opening of the gill slits is shown first as a dashed outline indicating presumptive gill slits as opaque regions in the lateral wall of the pharynx. A dark line around the gill slit indicates that the ectoderm and endoderm are in contact. Once opened, the gill slits are shaded black. Tiny gill filaments appear as buds on pharyngeal arches 2 and 3 in stages 25 and 26. CB—cloacal bulge; EP—Epiphysis; H—heart tube; OP—optic cup; OV—otic vesicle; PF—Pectoral Fin; RB—rostral bulb; TB—tail bud; YS—yolk stalk. Scale bar 1 mm. Reproduced from Didier et al. (1998) with kind permission of Wiley-Liss, Inc., a subsidiary of John Wiley & Sons, Inc.

The YCL of pre-implantation yolk sacs contains yolk nuclei (merocytes or naked nuclei) and a variety of cellular organelles, including Golgi bodies, rough endoplasmic reticulum and mitochondria. Yolk platelets are partially digested and transferred to the endoderm. Movement from the endoderm to the vitelline vessels remains obscure because yolk inclusions are not observed within the vascular endothelium (Hamlett and Wourms, 1984). No relationship between yolk nuclei and platelet degradation was observed (Hamlett et al., 1987).

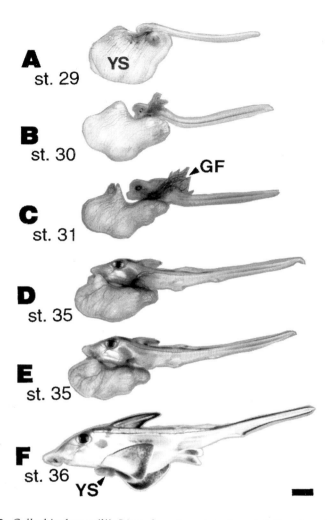

Fig. 43. *Callorhinchus milii*. Line drawing showing details of embryonic stages 29–36. Bulges of the yolk sac, especially those surrounding the head of the embryo are characteristics of *Callorhincus*. A—stage 29, 59.4 mm TL. B—stage 30, 68.0 mm TL. C—stage 31, 73.8 mm TL. D and E—stage 35, 104 mm TL, showing the variation in gill filament length for embryos of the same stage. F—stage 36, 134.6 mm TL. GF—external gill filaments. YS—yolk sac. Scale bar 1 mm. Reproduced from Didier *et al.* (1998) with kind permission of Wiley-Liss, Inc., a subsidiary of John Wiley & Sons, Inc.

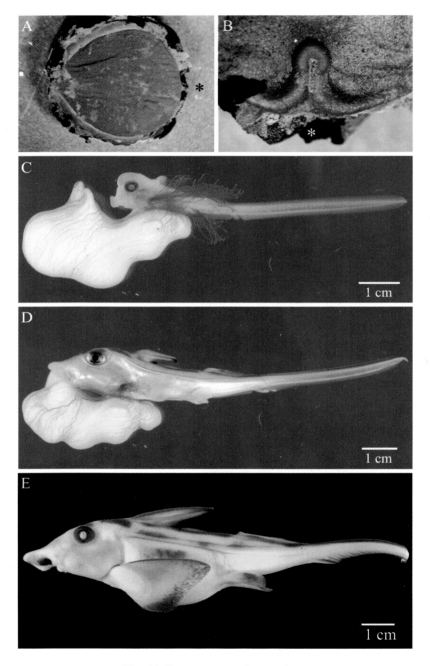

Fig. 44. See next page for caption.

For the oviparous shark, *S. canicula*, the EYS does not decrease in mass until after eclosion (Lechenault and Mellinger, 1993). The EYS is not active in yolk digestion except in the regions surrounding the yolk stalk. The thickness of the EYS membrane differs depending upon the area sampled and is manifested by a fibrous connective tissue layer (FCL). The FCL is thickest immediately surrounding the yolk stalk and vascular ingrowths. Both face a thickened YCL with yolk nuclei and platelets. The endodermal cells are enlarged and contain glycogen, indicating platelet digestion (Lechenault and Mellinger, 1993). The yolk nuclei degenerate at eclosion and are not found within the IYS (Beard, 1896).

The EYS contents are imported via the yolk stalk to the IYS and finally the spiral intestine. The yolk stalk contains a vitelline artery and vein, as well as the ductus vitellointestinalis. Yolk is moved by ciliary action in the vitelline duct of the yolk stalk. Coincident with eclosion in oviparous species, the IYS forms from a local dilation of the ductus distal to the spiral intestine (Lechenault and Mellinger, 1993). The IYS is less vascular than the EYS and occurs in yolk sac viviparous and oviparous species. The IYS functions to store yolk and is limited in size by coelomic space. Eventually, the EYS is completely emptied to the IYS. Yolk is transferred from the IYS to the spiral valve for digestion during the duration of development with only a small reserve of yolk within the IYS remaining at hatch (Wrisez *et al.*, 1993). Torpedo embryos with completely absorbed EYS were removed from the uterus and maintained for five months without eating. After accidental death, necropsy revealed a small accumulation of yolk remaining in the spiral intestine (Davy, 1834). *A. radiata* hatchlings in 2.5-4.5°C water utilize stored yolk for 2-4 months after hatching (Berestovskii, 1994). A rectal occlusion prohibits waste excretion until hatching in the oviparous shark *S. canicula* (Mellinger *et al.*, 1987). *C. plumbeus*, a placental shark, does not defer excretion and feces are found within capsular fluid (Baranes and Wendling, 1981).

Fig. 44. *Callorhinchus milii* embryo development. Chimaeroids are indiscernible from elasmobranchs during early development (A-B) but quickly develop characters that distinguish them from elasmobranchs (C-E). (A) Stage 11, the bilamellar overhang at the posterior of the embryo (*) marks the future axis. (B) Stage 13, the embryonic axis is visible and apex of the notochordal triangle is just beginning to swell marking the future head. The posterior lobes mark the edge of the blastoderm as in (A). (C) Stage 32, 75 mm total length. (D) Stage 35, 105 mm total length. (E) Hatchling stage 36, 133 mm total length. (Photographs A and B—Jen Wyffels and Dominique Didier; C, D and E—Dominique Didier).

Parturition and Umbilical Healing

Parturition is rarely observed but in most cases birth is tail first. *C. taurus* is an exception with birth head first (Gilmore *et al.*, 1983). *Aetobatus narinari* pups are born either head or tail first (Uchida *et al.*, 1990). The extended cephalofoil of *S. tiburo* is folded against the body and unfolds within a day after parturition (Michael, 2001). *Triaenodon obesus* pups are born tail first with abdomens opposed (Schaller, 2006). Umbilical scars are present on neonates and are often used as an indication of recent hatch or birth. The umbilical scars of the placental *C. limbatus* and *Carcharhinus isodon* heal completely in 2-3 weeks (Castro, 1996) and 3-4 weeks (Castro, 1993) following parturition, respectively. The umbilicus of the oviparous *H. edwardsii* is healed within two weeks of hatching (von Bonde, 1945).

SUMMARY

Every description of normal stages of development depends on a defined set of easily recognized characters. The universality of the stages relates to commonalities in developmental mechanisms among similar organisms. Chondrichthyan fishes are diverse in reproductive strategy, body form and size. Despite this, a clade-wise uniform staging system has been described. Chondrichthyan eggs are both megalecithal and telolecithal with meroblastic cleavage. Cell movements during gastrulation are unique among vertebrates. Some chondrichthyan taxa possess features that immediately identify their embryos at the onset of embryonic development. Examples include the rostral bulb of chimaeroids and the endocoelomic yolk of *C. taurus*. *S. laticaudus* embryos almost never conform to the generalized chondrichthyan model due to their reduced vitelline mass. External gill filaments are plesiomorphic. An internal yolk sac is always present in oviparous species and never present in placental or histotrophic species. Lamnoids develop precocious embryonic dentition that is shed and replaced before parturition. Body form for all chondrichthyans is initially fusiform. Fin development and rostral specializations distinguish species relatively early during embryonic development and body pigmentation is always acquired relatively late during development. How well chondrichthyans conform to a shared set of developmental stages remains to be discovered through careful observation of additional species, both primitive and derived, throughout the clade.

Acknowledgements

This work is a result of work with many collaborators who deserve recognition. Unpublished observations of *R. eglanteria* collected with Carl Luer,

A.B. Bodine and J.P. Wourms. *S. torazame* collaborators include Motoyasu Masuda, Yoshiaki Itoh and Junichi Sakai. *D. sayi* observations collected with Julie Sullivan, F.F. Snelson and J.P. Wourms. *Dasyatis* collaborators include Atsuko Yamaguchi, Keisuke Furumitsu and Elena Amesbury. Images were generously provided by Malcolm Francis, Carl Luer, Dominique Didier, Shoou-Jeng Joung, Blake Harahush, Dr. Nicholas James Cole, Elena Amesbury, Zerina Johanson, Brian Eames and Jerry Hoff. I thank Peter Bor and Jose Castro for their effort concerning *G. cirratum* egg cases. Information from Motoyasu Masuda, Alan Henningsen, Colin Simpfendorfer, Henry Mollet and Dave Ebert was timely and appreciated. I thank Carolyn Atherton and Laura Edsberg for the gifts of help and time and Carl Luer for patience, meticulous editing and consultation.

References

Abdel-Aziz, S.H., A.N. Khalil and S.A. Abdel-Maguid. 1993. Reproductive cycle of the common guitarfish, *Rhinobatos rhinobatos* (Linnaeus, 1758), in Alexandria waters, Mediterranean Sea. *Australian Journal of Marine and Freshwater Research* 44: 507-517.

Aiyar, R.G. and K.P. Nalini. 1938. Observations on the reproductive system, egg-case, embryos and breeding habits of *Chiloscyllium griseum* Muller and Henle. *Proceedings of the Indian Academy of Sciences*. B 7: 252-269.

Alcock, A. 1890. Observations on the gestation of some sharks and rays. *Journal of the Asiatic Society of Bengal* 59: 51-56.

Alcock, A. 1892. Embryonic history of *Pteroplatea micrura*. *Annals and Magazine of Natural History* 6: 2.

Amesbury, E. 1997. *Embryo development and nutrition in the Atlantic stingray, Dasyatis sabina (Elasmobranchii, Dasyatidae)*. Ph.D. Thesis. University of Central Florida.

Amesbury, E., J. Wyffels, J.P. Wourms and A.B. Bodine. 1998. Morphology of the annual uterine cycle of an aplacental viviparous elasmobranch, *Dasyatis sabina*. *American Zoologist* 38: 65A.

Amoroso, E.C. 1952. Placentation. In: *Marshall's Physiology of Reproduction*, A.S. Parkes (Ed.). Longmans Green and Co., London, Vol. 2, pp. 127-311.

Appukuttan, K.K. 1978. Studies on the developmental stages of hammerhead shark *Sphyrna-Blochii* from the Gulf of Mannar India. *Indian Journal of Fisheries* 25: 41-51.

Arendt, D. and K. Nübler-Jung. 1999. Rearranging gastrulation in the name of yolk: evolution of gastulation in yolk-rich amniote eggs. *Mechanisms of Development* 81: 3-22.

Babel, J.S. 1967. Reproduction, life history, and ecology of the round stingray, *Urolophus halleri* Cooper. *California Department of Fish and Game Fish Bulletin* 137: 1-104.

Balfour, F.M. 1878. *A Monograph on the Development of Elasmobranch Fishes*. MacMillan and Company, London.

Ballard, W.W., J. Mellinger and H. Lechenault. 1993. A series of normal stages for development of *Scyliorhinus canicula*, the lesser spotted dogfish (Chondrichthyes: Scyliorhinidae). *Journal of Experimental Zoology* 267: 318-336.

Baranes, A. and J. Wendling. 1981. Early stages of development in *Carcharhinus plumbeus*. *Journal of Fish Biology* 18: 159-176.

Baughman, J.L. 1955. The oviparity of the whale shark, *Rhineodon typus*, with records of this and other fishes in Texas waters. *Copeia* 1955: 54-55.

Beard, J. 1890. On the development of the common skate (*Raja batis*). *Eighth Annual Report of the Fishery Board for Scotland. Part III Scientific Investigations*: 300-311, Plates 9-11.

Beard, J. 1896. The yolk-sac, yolk and merocytes in Scyllium and Lepidostes. *Anatomischer Anzeiger* 12: 334-347.

Berestovskii, E.G. 1994. Reproductive biology of skates of the family Rajidae in the seas of the far north. *Journal of Ichthyology* 34: 26-37.

Bigelow, H.B. and W.C. Schroeder. 1948. Lancelets, cyclostomes and sharks. In: *Fishes of the Western North Atlantic*, J. Tee-Van (Ed.). Sears Foundation for Marine Research, Yale University, New Haven, pp. 59-546.

Braccini, J.M., W.C. Hamlett, B.M. Gillanders and T.I. Walker. 2007. Embryo development and maternal-embryo nutritional relationships of piked spurdog (*Squalus megalops*). *Marine Biology* 150: 727-737.

Branstetter, S. 1981. Biological notes on the sharks of the north central Gulf of Mexico. *Contributions in Marine Science* 24: 13-34.

Callard, I.P. and T.J. Koob. 1993. Endocrine regulation of the elasmobranch reproductive tract. *Journal of Experimental Zoology* 266: 368-377.

Capapé, C., R. Ben-Brahim and J. Zaouali. 1997. Aspects de la biologie de la reproduction de la guitare commune (*Rhinobatos rhinobatos*) des eaux tunisiennes (Méditerranée centrale). *Ichtyophysiologica Acta* 20: 113-127.

Capapé, C., Y. Diatta, M. Diop, Y. Vergne and O. Guélorget. 2006. Reproductive biology of the smoothhound, *Mustelus mustelus* (Chondrichthyes: Triakidae) from the coast of Senegal (eastern tropical Atlantic). *Cybium* 30: 273-282.

Capapé, C., A. Gueye-Ndiaye and A.A. Seck. 1999. Observations sur la biologie de la reproduction de la guitare commune, *Rhinobatos rhinobatos* (L. 1758) (Rhinobatidae) de la presqu'île du Cap-Vert (Sénégal, Atlantique oriental tropical). *Ichtyophysiologica Acta* 22: 87-101.

Capapé, C., F. Hemida, A.A. Seck, Y. Diatta, G. Olivier and J. Zaouali. 2003. Distribution and reproductive biology of the spinner shark, *Carcharhinus brevipinna* (Müller and Henle, 1841) (Chondrichthyes: Carcharhinidae). *Israel Journal of Zoology* 49: 269-286.

Carrier, J.C., J.H.L. Pratt and J. Castro. 2004. Reproductive biology of elasmobranchs. In: *Biology of Sharks and Their Relatives*, J. Carrier, J. Musick and M. Heithaus (Eds.). CRC Press, Boca Raton, pp. 269-286.

Carrier, J.C., H.L. Pratt, Jr. and L.K. Martin. 1994. Group reproductive behaviors in free-living nurse sharks, *Ginglymostoma cirratum*. *Copeia* 1994: 646-656.

Carter, A.M., B.A. Croy, V. Dantzer, A.C. Enders, S. Hayakawa, A. Mess and H. Soma. 2007. Comparative aspects of placental evolution: A workshop report. *Placenta* 28: S129-S132.

Caruso, J. and P.H.F. Bor. 2007. Egg capsule morphology of *Parascyllium variolatum* (Duméril, 1853) (Chondrichthyes; Parascylliidae), with notes on oviposition

rate in captivity. *Journal of Fish Biology* 70: 1620-1625.
Castro, J.I., P.M. Bubucis and N.A. Overstrom. 1988. The reproductive biology of the chain dogfish, *Scyliorhinus retifer*. *Copeia* 1988: 740-746.
Castro, J.I. and J.P. Wourms. 1993. Reproduction, placentation, and embryonic development of the Atlantic sharpnose shark, *Rhizoprionodon terraenovae*. *Journal of Morphology* 218: 257-280.
Castro, J.I. 1993. The biology of the finetooth shark, *Carcharhinus isodon*. *Environmental Biology of Fishes* 36: 219-232.
Castro, J.I. 1996. Biology of the blacktip shark, *Carcharhinus limbatus*, off the southeastern United States. *Bulletin of Marine Science* 59: 508-522.
Castro, J.I. 2000. The biology of the nurse shark, *Ginglymostoma cirratum*, off the Florida east coast and the Bahama Islands. *Environmental Biology of Fishes* 58: 1-22.
Cateni, C., L. Paulesu, E. Bigliardi and W.C. Hamlett. 2003. The interleukin 1 (IL-1) system in the uteroplacental complex of a cartilaginous fish, the smoothhound shark, *Mustelus canis*. *Reproductive Biology and Endocrinology* 1: 25.
Chang, W.-B., M.-Y. Leu and L.-S. Fang. 1997. Embryos of the whale shark, *Rhincodon typus*: Early growth and size distribution. *Copeia* 1997: 444-446.
Chapman, D.D., M.J. Corcoran, G.M. Harvey, S. Malan and M.S. Shivji. 2003. Mating behavior of southern stingrays, *Dasyatis americana* (Dasyatidae). *Environmental Biology of Fishes* 68: 241-245.
Chapman, D.D., P.A. Prodohl, J. Gelsleichter, C.A. Manire and M.S. Shivji. 2004. Predominance of genetic monogamy by females in a hammerhead shark, *Sphyrna tiburo*: implications for shark conservation. *Molecular Ecology* 13: 1965-1974.
Chapman, D.D., M.S. Shivji, E. Louis, J. Sommer, H. Fletcher and P.A. Prodohl. 2007. Virgin birth in a hammerhead shark. *Biological Letters* 3: 425-427.
Chen, C.-T., K.-M. Liu and Y.-C. Chang. 1997. Reproductive biology of the bigeye thresher shark, *Alopias superciliosus* (Lowe, 1839) (Chondrichthyes: Alopiidae), in the northwestern Pacific. *Ichthyological Research* 44: 227-235.
Chen, W.-K. and K.-M. Liu. 2006. Reproductive biology of whitespotted bamboo shark *Chiloscyllium plagiosum* in northern waters off Taiwan. *Fisheries Science (Tokyo)* 72: 1215-1224.
Chevolot, M., J.R. Ellis, A.D. Rijnsdorp, W.T. Stam and J.L. Olsen. 2007. Multiple paternity analysis in the thornback ray *Raja clavata* L. *Journal of Heredity* 98: 712-715.
Chieffi, G. 1967. The reproductive system of elasmobranchs: developmental and endocrinological aspects. In: *Sharks, Skates and Rays*, P.W. Gilbert, R.F. Mathewson and D.P. Rall (Eds.). The Johns Hopkins Press, Baltimore, pp. 553-580.
Clark, R.S. 1922. Rays and skates (Raia). No. I. Egg capsule and young. *Journal of the Marine Biological Association of the United Kingdom* 12: 577-643.
Clark, R.S. 1927. Rays and skates. No. 2. Description of embryos. *Journal of the Marine Biological Association of the United Kingdom* 14: 661-683.
Compagno, L.J.V. 1977. Phyletic relationships of living sharks and rays. *American Zoology* 17: 303-322.
Compagno, L.J.V. 1984. *FAO species catalogue. Vol. 4. Sharks of the world. An annotated and illustrated catalog of shark species known to date. Part 1. Hexanchiformes*

to *Lamniformes*. FAO (Food and Agriculture Organization of the United Nations) Fisheries Synopsis.

Compagno, L.J.V. 1990. Alternative life-history styles of cartilaginous fishes in time and space. *Environmental Biology of Fishes* 28: 33-76.

Compagno, L.V.J. 1999. Systematics and body form. In: *Sharks, Skates, and Rays: The Biology of Elasmobranch Fishes*, W.C. Hamlett (Ed.). Johns Hopkins University Press, Baltimore, pp. 1-42.

Compagno, L.V.J. 2005. Checklist of living chondrichthyans. In: *Reproductive Biology and Phylogeny of Chondrichthyes: Sharks, Batoids and Chimaeras*, W.C. Hamlett (Ed.). Science Publishers, Enfield, NH, pp. 503-548.

Conrath, C.L. and J.A. Musick. 2002. Reproductive biology of the smooth dogfish, *Mustelus canis*, in the northwest Atlantic Ocean. *Environmental Biology of Fishes* 64: 367-377.

Costa, F.E.S., F.M.S. Braga, C.A. Arfelli and A.F. Amorim. 2002. Aspects of the reproductive biology of the shortfin mako, *Isurus oxyrinchus* (Elasmobranchii Lamnidae), in the southeastern region of Brazil. *Brazilian Journal of Biology* 62: 239-248.

Coste, M. 1850. Recherches sur la segmentation de la cicatricule chez les Oiseaux, les Reptiles écailleux, les Poissons cartilagineux. *Comptes rendus hebdomadaires des séances de l'Académie des Sciences* 30: 638-642.

Daly-Engel, T.S., R.D. Grubbs, B.W. Bowen and R.J. Toonen. 2007. Frequency of multiple paternity in an unexploited tropical population of sandbar sharks (*Carcharhinus plumbeus*). *Canadian Journal of Fisheries and Aquatic Sciences* 64: 198-204.

Daly-Engel, T.S., R.D. Grubbs, K.N. Holland, R.J. Toonen and B.W. Bowen. 2006. Assessment of multiple paternity in single litters from three species of carcharhinid sharks in Hawaii. *Environmental Biology of Fishes* 76: 419-424.

Davy, J. 1834. Observations on the torpedo, with an account of some additional experiments on its electricity. *Philosophical Transactions of the Royal Society of London* 124: 531-550.

Dean, B. 1906. Chimaeroid fishes and their development. In: *Publication no. 32*, Carnegie Institute, Washington, D.C., pp. 1-172, Plates 1-11.

Demirhan, S.A. and K. Seyhan. 2006. Seasonality of reproduction and embryonic growth of spiny dogfish (*Squalus acanthias* L., 1758) in the eastern Black Sea. *Turkish Journal of Zoology* 30: 433-443.

Didier, D.A., E.E. Leclair and D.R. Vanbuskirk. 1998. Embryonic staging and external features of development of the chimaeroid fish, *Callorhinchus milii* (Holocephali, Callorhinchidae). *Journal of Morphology* 236: 25-47.

Diez, J.M. and J. Davenport. 1987. Embryonic respiration in the dogfish (*Scyliorhinus canicula* L.). *Journal of the Marine Biological Association of the United Kingdom* 67: 249-261.

Dodd, J.M. and M.H.I. Dodd. 1986. Evolutionary aspects of reproduction in cyclostomes and cartilaginous fishes. In: *Evolutionary Biology of Primitive Fishes*, R.E. Foreman, A. Gorbman, J.M. Dodd and R. Olsson (Eds.). Plenum Press, New York, Vol. 103, pp. 295-319.

Dral, A.J. 1981. Reproduction en aquarium de requin de fond tropical *Chiloscyllium griseum* Müller et Henle (Orectolobidés). *Revue Francaise d'Aquariologie* 7: 99-104.

Dulvy, N.K. and J.D. Reynolds. 1997. Evolutionary transitions among egg-laying, live-bearing and maternal inputs in sharks and rays. *Proceedings Royal Society, B* 264: 1309-1315.

Eames, B.F., N. Allen, J. Young, A. Kaplan, J.A. Helms and R.A. Schneider. 2007. Skeletogenesis in the swell shark *Cephaloscyllium ventriosum*. *Journal of Anatomy* 210: 542-554.

Ebert, D.A. and C.D. Davis. 2007. Descriptions of skate egg cases (Chondrichthyes: Rajiformes: Rajoidea) from the eastern north pacific. *Zootaxa* 1393: 1-18.

Ellis, J.R. and S.E. Shackley. 1995. Observations on egg-laying in the thornback ray. *Journal of Fish Biology* 46: 903-904.

Ellis, J.R. and S.E. Shackley. 1997. The reproductive biology of *Scyliorhinus canicula* in the Bristol Channel, U.K. *Journal of Fish Biology* 51: 361-372.

Evans, D.H., A. Oikari, G.A. Kormanik and L. Mansberger. 1982. Osmoregulation by the prenatal spiny dogfish *Squalus acanthias*. *Journal of Experimental Biology* 101: 295-306.

Farewell, C. 1972. Observations on egg laying in the horn shark, *Heterodontus francisci* (Girard). *Drum and Croaker* 13: 20.

Feldheim, K.A., S.H. Gruber and M.V. Ashley. 2001. Multiple paternity of a lemon shark litter (Chondrichthyes: Carcharhinidae). *Copeia* 2001: 781-786.

Feldheim, K.A. 2004. Reconstruction of parental microsatellite genotypes reveals female polyandry and philopatry in the lemon shark, *Negaprion brevirostris*. *Evolution* 58: 2332-2342.

Feng, D. and D.P. Knight. 1992. Secretion and stabilization of the layers of the egg capsule of the dogfish *Scyliorhinus canicula*. *Tissue and Cell* 24: 773-790.

Feng, D. 1994. Structure and formation of the egg capsule tendrils in the dogfish *Scyliorhinus canicula*. *Philosophical Transactions. B* 343: 285-302.

Fishelson, L. and A. Baranes. 1998. Observations on the Oman shark, *Iago omanensis* (Triakidae), with emphasis on the morphological and cytological changes of the oviduct and yolk sac during gestation. *Journal of Morphology* 236: 151-165.

Fitz, E.S. and F.C. Daiber. 1963. An introduction to the biology of *Raja eglanteria* Bosc 1802 and *Raja erinacea* Mitchell 1825 as they occur in Delaware Bay. *Bulletin of the Bingham Oceanographic Collection* 18: 69-97.

Foulley, M.M. and J. Mellinger. 1980. Étude chronologique, structurale et biométrique de l'œuf et de son développement chez a petite roussette (*Scyliorhinus canicula*) élevée en eau de mer artificielle. *Reproduction, nutrition, développement* 20: 1835-1848.

Francis, M.P. 1996. Observations on a pregnant white shark with a review of reproductive biology. In: *Great White Sharks: The Biology of Carcharodon carcharias*, A.P. Klimley and D.G. Ainley (Eds.). Academic Press, San Diego, pp. 157-172.

Francis, M.P. 2006. Distribution and biology of the New Zealand endemic catshark, *Halaelurus dawsoni*. *Environmental Biology of Fishes* 75: 295-306.

Francis, M.P. and J.D. Stevens. 2000. Reproduction, embryonic development, and growth of the porbeagle shark, *Lamna nasus*, in the southwest Pacific Ocean. *Fishery Bulletin (Washington DC)* 98: 41-63.

Fujita, K. 1981. Oviphagous embryos of the pseudocarchariid shark *Pseudo-*

carcharias kamoharai from the Central Pacific. *Japanese Journal of Ichthyology* 28: 37-44.

Garner, R. 2003. Annual fecundity, gestation period and egg survivorship in the brown-banded bamboo shark, *Chiloscyllium punctatum* in captivity. *Thylacinus* 27: 4-9.

Gerbe, Z. 1872. Recherches sur la segmentation de la cicatricule et la formation des produits adventifs de l'oeuf des Plagiostomes ey particuliérement des Raies *Journal de l'anatomie et de la physiologie normales et pathologiques de l'homme et des animaux* 8: 609-676.

Gilmore, R.G. 1983. Observations on the embryos of the longfin mako *Isurus paucus* and the bigeye thresher *Alopias superciliosus*. *Copeia* 1983: 375-382.

Gilmore, R.G. 1993. Reproductive biology of lamnoid sharks. *Environmental Biology of Fishes* 38: 95-114.

Gilmore, R.G. 2005. Oophagy, intrauterine cannibalism and reproductive strategy in lamnoid sharks. In: *Reproductive Biology and Phylogeny of Chondrichthyes: Sharks, Batoids and Chimaeras*, W.C. Hamlett (Ed.). Science Publishers, Enfield, NH, Vol. 3, pp. 435-462.

Gilmore, R.G., J.W. Dodrill and P.A. Linley. 1983. Reproduction and embryonic development of the sand tiger shark *Odontaspis taurus*. *Fishery Bulletin (Washington DC)* 81: 201-226.

Grogan, E.D. and R. Lund. 2004. The origin and relationships of early Chondrichthyes. In: *Biology of Sharks and Their Relatives*, J. Carrier, J. Musick and M. Heithaus (Eds.), CRC Press, Boca Raton, pp. 3-31.

Grover, C.A. 1974. Juvenile denticles of the swell shark *Cephaloscyllium ventriosum* function in hatching. *Canadian Journal of Zoology* 52: 359-363.

Gruber, S.H. and L.J.V. Compagno. 1981. Taxonomic status and biology of the bigeye thresher *Alopias superciliosus*. *Fishery Bulletin (Washington DC)* 79: 617-640.

Guallart, J. and J.J. Vicent. 2001. Changes in composition during embryo development of the gulper shark, *Centrophorus granulosus* (Elasmobranchii, Centrophoridae): An assessment of maternal-embryonic nutritional relationships. *Environmental Biology of Fishes* 61: 135-150.

Gudger, E.W. 1940. The breeding habits, reproductive organs and external embryonic development of *Chlamydoselachus*, based on notes and drawings by Bashford Dean. In: *The Bashford Dean Memorial Volume Archaic Fishes*, E.W. Gudger (Ed.). American Museum of Natural History, Vol. 7, New York, pp. 523-633.

Haines, A.N., M.F. Flajnik, L.L. Rumfelt and J.P. Wourms. 2005. Immunoglobulins in the eggs of the nurse shark, *Ginglymostoma cirratum*. *Developmental and Comparative Immunology* 29: 417-430.

Haines, A.N., M.F. Flajnik and J.P. Wourms. 2006. Histology and immunology of the placenta in the Atlantic sharpnose shark, *Rhizoprionodon terraenovae*. *Placenta* 27: 1114-1123.

Hamlett, W.C. 1993. Ontogeny of the umbilical cord and placenta in the Atlantic sharpnose shark, *Rhizoprionodon terraenovae*. *Environmental Biology of Fishes* 38: 253-267.

Hamlett, W.C. and M.K. Hysell. 1998. Uterine specializations in elasmobranchs. *Journal of Experimental Zoology* 282: 438-459.

Hamlett, W.C. and J.P. Wourms. 1984. Ultrastructure of the preimplantation shark yolk sac placenta. *Tissue and Cell* 16: 613-626.

Hamlett, W.C., C.J.P. Jones and L.R. Paulesu. 2005a. Placentatrophy in sharks. In: *Reproductive Biology and Phylogeny of Chondrichthyes: Sharks, Batoids and Chimaeras*, W.C. Hamlett (Ed.). Science Publishers, Enfield, Vol. 3, pp. 463-502.

Hamlett, W.C., F.J. Schwartz and L.J. DiDio. 1987. Subcellular organization of the yolk syncytial-endoderm complex in the preimplantation yolk sac of the shark, *Rhizoprionodon terraenovae*. *Cell and Tissue Research* 247: 275-85.

Hamlett, W.C., J.P. Wourms and J.W. Smith. 1985b. Stingray placental analogues: structure of trophonemata in *Rhinoptera bonasus*. *Journal of Submicroscopic Cytology* 17: 541-550.

Hamlett, W.C., M. Reardon, J. Clark and T.I. Walker. 2002b. Ultrastructure of sperm storage and male genital ducts in a male holocephalan, the elephant fish, *Callorhynchus milii*. *Journal of Experimental Zoology* 292: 111-128.

Hamlett, W.C., D.J. Allen, M.D. Stribling, F.J. Schwartz and L.J.A. Didio. 1985a. Permeability of external gill filaments in the embryonic shark *Rhizoprionodon terraenovae* electron microscopic observations using horseradish peroxidase as a macromolecular tracer. *Journal of Submicroscopic Cytology* 17: 31-40.

Hamlett, W.C., L. Fishelson, A. Baranes, C.K. Hysell and D.M. Sever. 2002a. Ultrastructural analysis of sperm storage and morphology of the oviducal gland in the Oman shark, *Iago omanensis* (Triakidae). *Marine and Freshwater Research* 53: 601-613.

Hamlett, W.C., D.P. Knight, F.T.V. Pereira, J. Steele and D.M. Server. 2005b. Oviducal glands in Chondrichthyes. In: *Reproductive Biology and Phylogeny of Chondrichthyes: Sharks, Batoids and Chimaeras*, W.C. Hamlett (Ed.), Science Publishers, Enfield, NH, pp. 301-335.

Hamlett, W.C., G. Kormanik, M. Storrie, B. Stevens and T.I. Walker. 2005c. Chondrichthyan parity, lecithrotrophy and matrotrophy. In: *Reproductive Biology and Phylogeny of Chondrichthyes: Sharks, Batoids and Chimaeras*, W.C. Hamlett (Ed.). Vol. 3, Science Publishers, Enfield, NH, pp. 395-434.

Hamlett, W.C., J.A. Musick, A.M. Eulitt, R.L. Jarrell and M.A. Kelly. 1996. Ultrastructure of uterine trophonemata, accommodation for uterolactation, and gas exchange in the southern stingray, *Dasyatis americana*. *Canadian Journal of Zoology* 74: 1417-1430.

Harahush, B.K., A.B.P. Fischer and S.P. Collin. 2007. Captive breeding and embryonic development of *Chiloscyllium punctatum* Müller & Henle, 1838 (Elasmobranchii: Hemiscyllidae). *Journal of Fish Biology* 71: 1007-1022.

Haswell, W.A. 1898. On the development of *Heterodontus philipi*. *Proceedings of the Linnean Society of New South Wales* 22: 96-103.

Heiden, T.C.K., A.N. Haines, C. Manire, J. Lombardi and T.J. Koob. 2005. Structure and permeability of the egg capsule of the bonnethead shark, *Sphyrna tiburo*. *Journal of Experimental Zoology* 303A: 577-589.

Heist, E. 2004. Genetics of sharks, skates and rays. In: *Biology of Sharks and Their Relatives*, J.C. Carrier, J.A. Musick and M. Heithaus (Eds.), CRC Press, Boca Raton, pp. 471-486.

Heist, E.J. 2005. Population and reproductive genetics in Chondrichthyes. In: *Reproductive Biology and Phylogeny of Chondrichthyes: Sharks, Batoids and*

Chimaeras, W.C. Hamlett (Ed.), Science Publishers, Enfield, NH, Vol. 3, pp. 27-44.

Hemida, F., R. Seridji, S. Ennajar, M.N. Bradaï, E. Collier, O. Guélorget and C. Capapé. 2003. New observations on the reproductive biology of the pelagic stingray, *Dasyatis violacea* Bonaparte, 1832 (Chondrichthyes: Dasyatidae) from the Mediterranean Sea. *Acta Adriatica* 44: 193-204.

Henningsen, A.D., M.J. Smale, I. Gordon, R. Garner, R. Marin-Osorino and N. Kinnunen. 2004. Captive breeding and sexual conflict in elasmobranchs. In: *Elasmobranch Husbandry Manual: Captive Care of Sharks, Rays, and Their Relatives*, M. Smith, D. Warmolts, D. Thoney and R. Hueter (Eds.), Ohio Biological Survey, Inc., Columbus, Ohio, pp. 237-248.

Hepworth, D.G., L.J. Gathercole, D.P. Knight, D. Feng and J.F.V. Vincent. 1994. Correlation of ultrastructure and tensile properties of a collagenous composite material, the egg capsule of the dogfish, *Scyliorhinus* spp., a sophisticated collagenous material. *Journal of Structural Biology* 112: 231-240.

Hernandez, S., J. Lamilla, E. Dupre and W. Stotz. 2005. Embryonic development of the redspotted catshark *Schroederichthys chelensis* (Guichenot 1848) (Chondrichthyes: Scyliorhinidae). *Gayana* 69: 191-197.

Heupel, M.R., J.M. Whittier and M.B. Bennett. 1999. Plasma steroid hormone profiles and reproductive biology of the epaulette shark, *Hemiscyllium ocellatum*. *Journal of Experimental Zoology* 284: 586-594.

Hisaw, F.L. and A. Albert. 1947. Observations on the reproduction of the spiny dogfish, *Squalus acanthias*. *Biological Bulletin* 92: 187-199.

Hitz, C.R. 1965. Observations on egg cases of the big skate (*Raja binoculata* Girard) found in Oregon coastal waters. *Journal of the Fisheries Research Board of Canada* 21: 851-854.

Hobson, A.D. 1930. A note on the formation of the egg case of the skate. *Journal of the Marine Biological Association of the United Kingdom* 16: 577-581.

Hoff, G.R. 2007. *Reproductive biology of the Alaska skate, Bathyraja parmifera, with regard to nursery sites, embryo development and predation*. Dissertation. University of Washington.

Hoffman, C.K. 1896. Beiträge zur Entwicklungsgeschichte der Selachii. *Morphologisches Jahrbuch* 24: 209-286, plates II-V.

Holden, M.J., D.W. Rout and C.N. Humphreys. 1971. The rate of egg laying by three species of ray. *Journal du Conseil International pour l'Exploration de la Mer* 33: 335-339.

Home, E. 1810. On the mode of breeding of the ovoviviparous shark, and on the aeration of the fœtal blood in different classes of animals. *Philosophical Transactions of the Royal Society of London* 100: 205-222.

Howard, M.J. 2002. Size relationships between egg capsules and female big skates, *Raja binoculata* (Girard, 1855). *Drum and Croaker* 33: 31-34.

Hussakof, L. 1912. Note on an embryo of *Pristis cuspidatus*. *Bulletin of the American Museum of Natural History*, New York 31: 327-330.

Hussakof, L. 1914. Notes on a small collection of fishes from Patagonia and Tierra del Fuego. *Bulletin of the American Museum of Natural History*, New York 33: 85-94.

Ishiyama, R. 1958. Studies on the Rajid fishes (Rajidae) found in the waters around Japan. *Journal of Shimonoseki College of Fisheries* 7: 193-394.

Jagadis, I. and B. Ignatius. 2003. Captive breeding and rearing of grey bamboo shark, *Chiloscyllium griseum* (Müller & Henle, 1839). *Indian Journal of Fisheries* 50: 539-542.

Janez, J.A. and M.C. Sueiro. 2007. Size at hatching and incubation period of *Sympterygia bonapartii* (Müller & Henle, 1841) (Chondrichthyes, Rajidae) bred in captivity at the Temaiken Aquarium. *Journal of Fish Biology* 70: 648-650.

Johanson, Z., M.M. Smith and J.J.P. Joss. 2007. Early scale development in *Heterodontus* (Heterodontiformes; Chondrichthyes): a novel chondrichthyan scale pattern. *Acta Zoologica (Stockholm)* 88: 249-256.

Johnson, M.R. and F.F. Snelson, Jr. 1996. Reproductive life history of the Atlantic stingray, *Dasyatis sabina* (Pisces, Dasyatidae), in the freshwater St. Johns River, Florida. *Bulletin of Marine Science* 59: 74-88.

Jones, T.S. and K.I. Ugland. 2001. Reproduction of female spiny dogfish, *Squalus acanthias*, in the Oslofjord. *Fishery Bulletin (Seattle)* 99: 685-690.

Joung, S.-J., C.-T. Chen, E. Clark, S. Uchida and W.Y.P. Huang. 1996. The whale shark, *Rhincodon typus*, is a livebearer: 300 embryos found in one 'megamamma' supreme. *Environmental Biology of Fishes* 46: 219-223.

Kerr, J.G. 1919. *Text-book of Embryology Vol. II Vertebrata With the Exception of Mammalia*. Macmillac and Co. Ltd., London.

Koob, T.J. 1991. Deposition and binding of calcium and magnesium in egg capsules of *Raja erinacea* Mitchill during formation and tanning *in utero*. *Copeia* 1991: 339-347.

Koob, T.J. 1999. Elasmobranch reproduction. In: *Encyclopedia of Reproduction*, E. Knobil and J.D. Neill (Eds.). Academic Press, San Diego, Vol. 1, pp. 1009-1018.

Koob, T.J. and D.L. Cox. 1993. Stabilization and sclerotization of *Raja erinacea* egg capsule proteins. *Environmental Biology of Fishes* 38: 151-157.

Koob, T.J. and W.C. Hamlett. 1998. Microscopic structure of the gravid uterus in the little skate, *Raja erinacea*. *Journal of Experimental Zoology* 282: 421-437.

Koob, T.J. and J.W. Straus. 1998. On the role of egg jelly in *Raja erinacea*. *The Bulletin, Mount Desert Island Biological Laboratory* 37: 117-119.

Koob, T.J. and A.P. Summers. 1996. On the hydrodynamic shape of the little skate (*Raja erinacea*) egg capsules. *The Bulletin, Mount Desert Island Biological Laboratory* 35: 108-111.

Koob, T.J., P. Tsang and I.P. Callard. 1986. Plasma estradiol testosterone and progesterone levels during the ovulatory cycle of the skate *Raja erinacea*. *Biology of Reproduction* 35: 267-275.

Koop, J.H. 2005. Reproduction of captive *Raja* spp. in the Dolfinarium Harderwijk. *Journal of the Marine Biological Association of the United Kingdom* 85: 1201-1202.

Kopsch, F. 1950. Bildung und Längenwachstum des Embryons, Gastrulation und Konkreszenz bei *Scyllium canicula* und *Scyllium catulus*. *Zeitschrift für Mikroskopisch-Anatomische Forschung* 56: 1-101.

Kormanik, G.A. 1993. Ionic and osmotic environment of developing elasmobranch embryos. *Environmental Biology of Fishes* 38: 233-240.

Kryvi, H. 1976. The structure of the embryonic external gill filaments of the velvet belly *Etmopterus spinax*. *Journal of Zoology* 180: 253-261.

Kudo, S. 1959. Studies on the sexual maturation of female and on the embryo of Japanese dog fish *Halealurus bürgeri* (Müller and Henle). *Report of the Nankai*

Regional Fisheries Research Laboratory 11: 41-45.
Kuenen, M. 2000. A log of captive births by an Atlantic nurse shark, "Sarah". *Drum and Croaker* 31: 22-23.
Kunze, K. and L. Simmons. 2004. Notes on reproduction of the zebra shark, *Stegostoma fasciatum*, in a captive environment. In: *Elasmobranch Husbandry Manual: Captive Care of Sharks, Rays, and Their Relatives*, M. Smith, D. Warmolts, D. Thoney and R. Hueter (Eds.). Ohio Biological Survey, Inc., Columbus, Ohio, pp. 493-497.
Kyne, P.M. and M.B. Bennett. 2002. Reproductive biology of the eastern shovelnose ray, *Aptychotrema rostrata* (Shaw and Nodder, 1794), from Moreton Bay, Queensland, Australia. *Marine and Freshwater Research* 53: 583-589.
Last, P.R. and V. Vongpanich. 2004. A new finback catshark *Proscyllium magnificum* (Elasmobranchii: Proscylliidae) from the northeastern Indian ocean. *Phuket Marine Biological Center Research Bulletin* 65: 23-29.
Lechenault, H. and J. Mellinger. 1993. Dual origin of yolk nuclei in the lesser spotted dogfish, *Scyliorhinus canicula* (Chondrichthyes). *Journal of Experimental Zoology* 265: 669-678.
Lechenault, H., F. Wrisez and J. Mellinger. 1993. Yolk utilization in *Scyliorhinus canicula*, an oviparous dogfish. *Environmental Biology of Fishes* 38: 241-252.
Leonard, J.B.K., A.P. Summers and T.J. Koob. 1999. Metabolic rate of embryonic little skate, *Raja erinacea* (Chondrichthyes: Batoidea): The cost of active pumping. *Journal of Experimental Zoology* 283: 13-18.
Lessa, R.P., V.M. Vooren and J. Lahaye. 1986. Desenvolvimento e ciclo sexual das fèmeas, migrações e fecundidade da viola *Rhinobatos horkelii* (Müller & Henle, 1841) do sul do Brasil. *Atlântica, Rio Grande* 8: 5-34.
Leuckart, F.S. 1836. *Untersuchungen über die äussern Kiemen der Embryonen von Rochen und Hayen*. L.F. Rieger & Co., Stuttgart.
Lewis, T.C. 1982. *The reproductive anatomy, seasonal cycles, and developement of the Atlantic stingray, Dasyatis sabina (Lesueur) (Pisces, Dasyatidae) from the northeastern Gulf of Mexico*. Dissertation. Florida State University.
Leydig, F. 1852. *Beiträge zur mikroskopischen Anatomie und Entwicklungsgeschichte der Rochen und Haie*. W. Englemann, Leipzig.
Libby, E.L. 1959. Miracle of the mermaid's purse. *National Geographic* 116: 412-420.
Libby, E.L. and P.W. Gilbert. 1960. Reproduction in the clear-nosed skate, *Raja eglanteria*. *Anatomical Record* 138: 365.
Lombardi, J. and T. Files. 1993. Egg capsule structure and permeability in the viviparous shark, *Mustelus canis*. *Journal of Experimental Zoology* 267: 76-85.
Lombardi, J., K.B. Jones, C.A. Garrity and T. Files. 1993. Chemical composition of uterine fluid in four species of viviparous sharks (*Squalus acanthias, Carcharhinus plumbeus, Mustelus canis* and *Rhizoprionodon terraenovae*). *Comparative Biochemistry and Physiology, Part A*, A 105: 91-102.
Long, J.H. and T.J. Koob. 1997. Ventilating the skate egg capsule: The transitory tail pump of embryonic little skates (*Raja erinacea*). *Bulletin, Mount Desert Island Biological Laboratory* 36: 117-119.
Lorenzini, S. 1678. *Osservazioni intorno alle torpedini*. l'Onofri, Florence.
Luer, C.A. and P.W. Gilbert. 1985. Mating behavior, egg deposition, incubation

period, and hatching in the clearnose skate, *Raja eglanteria*. *Environmental Biology of Fishes* 13: 161-171.

Luer, C.A., C.J. Walsh, A.B. Bodine, R.S. Rodgers and J.T. Wyffels. 1994. Preliminary biochemical analysis of histotroph secretions from the cownose ray, *Rhinoptera bonasus*, and the Atlantic stingray, *Dasyatis sabina*, with observations on the uterine villi from *R. bonasus*. In: *10th Annual Meeting of the American Elasmobranch Society*, University of Southern California, Los Angeles, Abstract 162, p. 117.

Luer, C.A., C.J. Walsh, J.T. Wyffels and A.B. Bodine. 2007. Normal embryonic development in the clearnose skate, *Raja eglanteria*, with experimental observations on artificial insemination. *Environmental Biology of Fishes* 80: 239-255.

Mahadevan, G. 1940. Preliminary observations on the structure of the uterus and the placenta of a few Indian elasmobranchs. *Proceedings of the Indian National Science Academy. B* 11: 2-47.

Marina, P., V. Salvatore, R. Maurizio, R. Loredana, L. Annamaria, L. Vincenza, L. Ermelinda and A. Piero. 2004. Ovarian follicle cells in *Torpedo marmorata* synthesize vitellogenin. *Molecular Reproduction and Development* 67: 424-429.

Marquez-Farias, J.F. 2007. Reproductive biology of shovelnose guitarfish Rhinobatos productus from the eastern Gulf of California Mexico. *Marine Biology* 151: 1445-1454.

Marshall, L.J., W.T. White and I.C. Potter. 2007. Reproductive biology and diet of the southern fiddler ray, *Trygonorrhina fasciata* (Batoidea: Rhinobatidae), an important trawl bycatch species. *Marine and Freshwater Research* 58: 104-115.

Masuda, M. 1997. Mating, spawning and hatching of the white spotted bamboo shark in an aquarium. *Japanese Journal of Ichthyology* 45: 29-35.

Masuda, M., Y. Izawa, S. Kametuta, H. Ikuta and T. Isogai. 2003. Artificial insemination of the cloudy catshark. *Journal of Japanese Association of Zoological Gardens and Aquariums* 44: 39-43.

Mathews, L.H. 1950. Reproduction in the basking shark *Cetorhinus maximus* (Gunner). *Philosophical Transactions of the Royal Society of London B* 234: 247-316.

McLaughlin, D.M. and J.F. Morrissey. 2005. Reproductive biology of *Centrophorus cf. uyato* from the Cayman Trench, Jamaica. *Journal of the Marine Biological Association of the United Kingdom* 85: 1185-1192.

McLaughlin, R.H. and A.K. O'Gower. 1971. Life history and underwater studies of a heterodontid shark. *Ecological Monographs* 41: 271-289.

Mead, R.A. 1993. Embryonic diapause in vertebrates. *Journal of Experimental Zoology* 266: 629-641.

Meehan, S.R., A.P. Summers and T.J. Koob. 1997. Active and passive ventilation of the swell shark *Cephaloscyllium ventriosum* egg capsule. *American Zoologist* 37: 112A.

Melendez, E.M. 1997. *Biología reproductiva de la raya lodera Dasyatis brevis (Garman, 1880), en Bahia Almejas, B.C.S., México*. Tesis, Universidad Autónoma de Baja California Sur.

Mellinger, J. 1983. Egg case diversity among dogfish *Scyliorhinus canicula:* A study of egg laying rate and nidamental gland secretory activity. *Journal of Fish Biology* 22: 83-90.

Mellinger, J. and F. Wrisez. 1993. Étude des éscailles primaires l'embryon de la roussette *Scyliorhinus canicula* (Chondrichthyes, Scyliorhinidae) au microscope électronique à balayage. *Annels des Sciences Naturelles, Zoologie, Paris* 14: 13-22.

Mellinger, J., F. Wrisez and M.J. Alluchon-Gerard. 1984. Biometric differences between dogfish *Scyliorhinus canicula* embryos from the Channel and Mediterranea and their yolk sacs with a new description of the hatching stage. *Cahiers de Biologie Marine* 25: 305-318.

Mellinger, J. and F. Wrisez. 1986. Developmental biology of an oviparous shark, *Scyliorhinus canicula*. In: *Proceedings, Second International Conference on Indo-Pacific Fishes*, T. Uyeno, R. Arai, T. Taniuchi and K. Matsuura (Eds.). Ichthyological Society of Japan, Tokyo, pp. 310-332.

Mellinger, J., F. Wrisez and J.C. Desselle. 1987. Transitory closures of esophagus and rectum during elasmobranch development models for human congenital anomalies? *Archives de Biologie* 98: 209-230.

Melouk, M.A. 1949. The external features in the development of the Rhinobatidae. *Publications of the Marine Biological Station Ghardaqa (Red Sea)* 7: 1-98.

Melouk, M.A. 1957. On the development of Carcharhinus melanopterus (Quoy & Gaimard). *Publications of the Marine Biological Station Ghardaqa (Red Sea)* 9: 229-251.

Metten, H. 1939. Studies on the reproduction of the dogfish. *Philosophical Transactions of the Royal Society of London* B 230: 217-238.

Michael, S.W. 2001. *Aquarium Sharks & Rays: An essential guide to their selection, keeping and natural history*. T.F.H. Publications, Inc., Neptune City, NJ.

Miki, T. 1994. Spawning, hatching, and growth of the whitespotted bamboo shark, *Chiloscyllium plagiosum*. *Journal of Japanese Association of Zoological Gardens and Aquariums* 36: 10-19.

Miyake, T., J.L. Vaglia, L.H. Taylor and B.K. Hall. 1999. Development of dermal denticles in skates (Chondrichthyes, Batoidea): patterning and cellular differentiation. *Journal of Morphology* 241: 61-81.

Mollet, H.F., G. Cliff, H.L. Pratt, Jr. and J.D. Stevens. 2000. Reproductive biology of the female shortfin mako, *Isurus oxyrinchus* Rafinesque, 1810, with comments on the embryonic development of lamnoids. *Fishery Bulletin (Washington DC)* 98: 299-318.

Mollet, H.F., J.M. Ezcurra and J.B. O'Sullivan. 2002. Captive biology of the pelagic stingray, *Dasyatis violacea* (Bonaparte, 1832). *Marine and Freshwater Research* 53: 531-541.

Monro, A. 1785. *The structure and physiology of fishes explained and compared with those of man and other animals*. Charles Elliot, Edinburgh.

Moreno, J.A. and J. Morón. 1992. Reproductive biology of the bigeye thresher shark *Alopias superciliosus* Lowe 1839. *Australian Journal of Marine and Freshwater Research* 43: 77-86.

Morris, J.A. 1999. *Aspects of the reproductive biology of the bluntnose stingray, Dasyatis say, in the Indian River lagoon system, Florida*. Ph.D. Thesis, University of Central Florida.

Mossman, H.W. 1937. Comparative morphogenesis of the fetal membranes and accessory uterine structures. In: *Comparative Morphogenesis of Fetal Membranes*, Rutgers University Press, Brunswick, New Jersey, pp. 133-246.

Musick, J.A. and J.K. Ellis. 2005. Reproductive evolution of chondrichthyes. In: *Reproductive Biology and Phylogeny of Chondrichthyes: Sharks, Batoids and Chimaeras*, W.C. Hamlett (Ed.). Science Publishers, Enfield, Vol. 3, pp. 45-79.

Nair, R.V. and K.K. Appukuttan. 1974. Observations on the developmental stages of the smooth dogfish, *Eridacnis radcliffei* Smith from Gulf of Mannar. *Indian Journal of Fisheries* 21: 141-151.

Nakaya, K. 1975. Taxonomy, comparartive anatomy and phylogeny of Japanese catshark, Scyliorhinidae. *Memoirs of the Faculty of Fisheries, Hokkaido University* 23: 1-94.

Nakaya, K. and H. Nakano. 1995. *Scymnodalatias albicauda* (Elasmobranchii, Squalidae) is a prolific shark. *Japanese Journal of Ichthyology* 42: 325-328.

Natanson, L.J. and G.M. Cailliet. 1986. Reproduction and development of the Pacific angel shark *Squatina californica* off Santa Barbara, California. *Copeia* 1986: 987-994.

Needham, J. 1931. *Chemical Embryology*. Cambridge University Press, Cambridge.

Needham, J. 1942. *Biochemistry and Morphogenesis*. Cambridge University Press, Cambridge.

Nelson, O.E. 1953. *Comparative Embryology of the Vertebrates*. McGraw-Hill Book Company, Inc., New York.

Nishikawa, T. 1898. Notes on some embryos of *Chalamydoselachus anguineus* Garman. *Annotationes Zoologicae Japonensis* 2: 95-102, pl. 3.

Oddone, M.C. and C.M. Vooren. 2002. Egg-cases and size at hatching of *Sympterygia acuta* in the south-western Atlantic. *Journal of Fish Biology* 61: 858-861.

Ohta, Y., K. Okamura, E.C. McKinney, S. Bartl, K. Hashimoto and M.F. Flajnik. 2000. Primitive synteny of vertebrate major histocompatibility complex class I and class II genes. *Proceedings of the National Academy of Sciences of the United States of America* 97: 4712-4717.

Otake, T. 1990. Classification of reproductive modes in sharks with comments on female reproductive tissues and structures. In: *NOAA Technical Report NMFS 90*, pp. 111-130.

Otake, T. and K. Mizue. 1981. Direct evidence for oophagy in thresher shark *Alopias pelagicus*. *Japanese Journal of Ichthyology* 28: 171-172.

Otake, T. 1985. The fine structure of the placenta of the blue shark *Prionace glauca*. *Japanese Journal of Ichthyology* 32: 52-59.

Ouang, T.Y. 1931. La glande de l'éclosion chez les plagiostomes. *Annales de L'Instut Océanographique. Monaco* 10: 281-370.

Parker, T.J. 1882. Notes on the anatomy and embryology of *Scymnus lichia*. *Royal Society of New England, Wellington.Transactions* 15: 222-234.

Pasteels, J. 1958. Développment Embryonnaire I. Développment des Élasmobranches ou Chondrichthyens. In: *Traité de Zoologie Anatomie, Systématique, Biologie*, P.P. Grassé (Ed.). Masson, Paris, Vol. 13, Part 2, pp. 1685-1754.

Paulesu, L.R., C. Cateni, R. Romagnoli, E. Bigliardi and W.C. Hamlett. 2000. Cytokine expression in the yolk sac placenta of the viviparous shark, *Mustelus canis*. *Placenta* 21: A.49.

Pelster, B. and W.E. Bemis. 1992. Structure and function of the external gill filaments of embryonic skates (Raja erinacea). *Respiration Physiology* 89: 1-13.

Perez, L.E. and I.P. Callard. 1992. Identification of vitellogenin in the little skate (*Raja erinacea*). *Comparative Biochemistry and Physiology*, B 103: 699-705.

Plancke, Y., F. Delplace, J.-M. Wieruszeski, E. Maes and G. Strecker. 1996. Isolation and structures of glycoprotein-derived free oligosaccharides from the unfertilized eggs of Scyliorhinus caniculus: Characterization of the sequences galactose(alpha-1-4)galactose(beta-1-3)-N-acetylglucosamine and N-acetylneuraminic acid(alpha-2-6)galactose(beta-1-3)-N-acetylglucosamine. *European Journal of Biochemistry* 235: 199-206.

Portnoy, D.S., A.N. Piercy, J.A. Musick, G.H. Burgess and J.E. Graves. 2007. Genetic polyandry and sexual conflict in the sandbar shark, *Carcharhinus plumbeus*, in the western North Atlantic and Gulf of Mexico. *Molecular Ecology* 16: 187-197.

Prasad, R.R. 1945. The structure, phylogenetic significance, and function of the nidamental glands of some elasmobranchs of the Madras coast. *Proceedings of the National Institute of Science India* 11: 282-303.

Pratt, H.L., Jr. 1988. Elasmobranch gonad structure a description and survey. *Copeia* 1988: 719-729.

Pratt, H.L., Jr. 1993. The storage of spermatozoa in the oviducal glands of western North Atlantic sharks. *Environmental Biology of Fishes* 38: 139-149.

Pratt, H.L., Jr. and J.C. Carrier. 2001. A review of elasmobranch reproductive behavior with a case study on the nurse shark, Ginglymostoma cirratum. *Environmental Biology of Fishes* 60: 157-188.

Pratt, H.L., Jr. and J.G. Casey. 1990. Shark reproductive strategies as a limiting factor in directed fisheries with a review of Holden's method of estimating growth parameters. In: *NOAA Technical Report NMFS 90*, J. Harold L. Pratt, S.H. Gruber and T. Taniuchi (Eds.), pp. 97-110.

Pratt, H.L., Jr. and S. Tanaka. 1994. Sperm storage in male elasmobranchs: A description and survey. *Journal of Morphology* 219: 297-308.

Putnam, F.W. 1870. Skates' eggs and young. *American Naturalist* 3: 617-630.

Ranzi, S. 1943. Sulle funzione dei Leucociti nell' utero del Selaci. *Archivio Zoologico Italiano* 20: 569-577.

Read, L.J. 1968. Urea and tri methylamine oxide levels in elasmobranch embryos. *Biological Bulletin* 135: 537-547.

Renfree, M.B. 1978. Embryonic diapause in mammals: a developmental strategy. In: *Dormancy and Developmental Arrest*, M.E. Clutter (Ed.). Academic Press, New York, pp. 1-46.

Renfree, M.B. and G. Shaw. 2000. Diapause. *Annual Review of Physiology* 62: 353-75.

Ribeiro, L., G. Rodrigues and G.W. Nunan. 2006. First record of a pregnant female of *Dasyatis hypostigma*, with description of the embryos. *Environmental Biology of Fishes* 75: 219-221.

Richards, S., D. Merriman and L.H. Calhoun. 1963. Studies on the marine resources of southern New England, IX: The biology of the little skate, *Raja erinacea* Mitchell. *Bulletin of the Bingham Oceanographic Collection* 18: 1-68.

Rodda, K.R. and R.S. Seymour. 2008. Functional morphology of embryonic development in the Port Jackson shark *Heterodontus portusjacksoni* (Meyer). *Journal of Fish Biology* 72: 961-984.

Romek, M. and W. Kilarski. 1993. The lipoprotein crystals in yolk platelets of a shark, *Squalus acanthias* (Selachii). *Folia Histochemica et Cytobiologica* 31: 139-145.

Rückert, J. 1899. Die erste Entwicklung des Eies der Elasmobranchier. In: *Festschrift zum siebenzigsten Geburtstag von Carl von Kupffer*, Verlag von Gustav Fisher, Jena, pp. 581-704.

Rusaouen, M. 1976. The dogfish shell gland a histochemical study. *Journal of Experimental Marine Biology and Ecology* 23: 267-283.

Sauka-Spengler, T., B. Baratte, M. Lepage and S. Mazan. 2003. Characterization of Brachyury genes in the dogfish *S. canicula* and the lamprey *L. fluviatilis*. Insights into gastrulation in a chondrichthyan. *Developmental Biology* 263: 296-307.

Saville, K.J., A.M. Lindley, E.G. Maries, J.C. Carrier and H.L. Pratt, Jr. 2002. Multiple paternity in the nurse shark, *Ginglymostoma cirratum*. *Environmental Biology of Fishes* 63: 347-351.

Scammon, R.E. 1911. Normal plates on the development of *Squalus acanthias*. In: *Normentafeln zur Entwickelungsgeschichte der Wirbeltiere*, G. Keibel (Ed.). Gustav Fisher, Jena, Vol. 12, pp. 1-140.

Schaller, P. 2006. Husbandry and reproduction of whitetip reef sharks *Triaenodon obesus* at Steinhart Aquarium, San Francisco. *International Zoo Yearbook* 40: 232-240.

Schauinsland, H. 1903. Beiträge zur Entwicklungsgeschichte und Anatomie der Wirbeltiere. I. *Sphenodon, Callorhynchus, Chamaeleo*. *Zoologica* 16: 5-32, 58-89.

Seck, A.A., Y. Diatta, M. Diop, O. Guélorget, C. Reynaud and C. Capapé. 2004. Observations on the reproductive biology of the blackchin guitarfish, *Rhinobatos cemiculus* E. Geoffroy Saint-Hilaire, 1817 (Chondrichthyes, Rhinobatidae) from the coast of Senegal (Eastern tropical Atlantic). *Scientia gerundensis* 27: 19-30.

Setna, S.B. and P.N. Sarangdhar. 1948a. Description, bionomics and development of *Scoliodon sorrakowah* (Cuvier). *Records of the Indian Museum* 46: 25-53.

Setna, S.B. 1948b. Observations on the development of *Chiloscyllium griseum* Müller & Henle, *Pristis cuspidatus* Lath. and *Rhynchobatus djiddensis* (Forsk.). *Records of the Indian Museum* 46: 1-23.

Shimada, K. 2002. Teeth of embryos in lamniform sharks (Chondrichthyes: Elasmobranchii). *Environmental Biology of Fishes* 63: 309-319.

Simpfendorfer, C.A. 1992. Reproductive strategy of the Australian sharpnose shark *Rhizoprionodon taylori* elasmobranchii carcharhinidae from Cleveland Bay Northern Queensland. *Australian Journal of Marine and Freshwater Research* 43: 67-75.

Smale, M.J. and A.J.J. Goosen. 1999. Reproduction and feeding of spotted gully shark, *Triakis megalopterus*, off the Eastern Cape, South Africa. *Fishery Bulletin (Washington DC)* 97: 987-998.

Smith, B.G. 1942. The heterodontoid sharks: their natural history and the external development of *Heterodontus japonicus* based on notes and drawings by Bashford Dean. In: *The Bashford Dean Memorial Volume: Archaic Fishes*, E.W. Gudger (Ed.). American Museum of Natural History, New York, Vol. 8, pp. 649-770.

Smith, J.W. and J.V. Merriner. 1986. Observations on the reproductive biology of the cownose ray *Rhinoptera bonasus* in Chesapeake Bay USA. *Fishery Bulletin (Washington DC)* 84: 871-878.

Smith, R.M., T.I. Walker and W.C. Hamlett. 2004. Microscopic organisation of the oviducal gland of the holocephalan elephant fish, *Callorhynchus milii*. *Marine and Freshwater Research* 55: 155-164.

Snelson, F.F., Jr., S.E. Williams-Hooper and T.H. Schmid. 1989. Biology of the bluntnose stingray *Dasyatis sayi* in Florida USA coastal lagoons. *Bulletin of Marine Science* 45: 15-25.

Southwell, T. and B. Prashad. 1910. Notes from the Bengal fisheries laboratory, No. 6. Embryological and developmental studies of Indian fishes. *Records of the Indian Museum* 16: 215-240, Plates 16-19.

Springer, S. 1948. Oviphagous embryos of the sand shark, *Carcharias taurus*. *Copeia* 1948: 153-157.

Springer, S. 1960. Natural history of the sandbar shark, *Eulamnia milberti*. *United States Fish and Wildlife Service Fishery Bulletin* 61: 1-38.

Springer, S. and G.H. Burgess. 1985. 2 new dwarf dog sharks *Etmopterus squalidae* found off the Caribbean coast of Colombia. *Copeia* 1985: 584-591.

Stehmann, M.F.W. and N.R. Merrett. 2001. First records of advanced embryos and egg capsules of *Bathyraja* skates from the deep north-eastern Atlantic. *Journal of Fish Biology* 59: 338-349.

Stevens, J.D. 1983. Observations on reproduction in the shortfin mako *Isurus oxyrinchus*. *Copeia* 1983: 126-130.

Stevens, J.D. 2007. Whale shark (*Rhincodon typus*) biology and ecology: A review of the primary literature. *Fisheries Research* 84: 4-9.

Stevens, J.D., R. Bonfil, N.K. Dulvy and P.A. Walker. 2000. The effects of fishing on sharks, rays, and chimaeras (chondrichthyans), and the implications for marine ecosystems. *ICES Journal of Marine Science* 57: 476-494.

Straus, J.W. and T.J. Koob. 1997. Protease activity in albumen during development of the little skate *Raja erinacea*. *Bulletin of the Mount Desert Island Biological Laboratory* 36: 111-113.

Sumpter, J.P. and J.M. Dodd. 1979. The annual reproductive cycle of the female lesser spotted dogfish *Scyliorhinus canicula* and its endocrine control. *Journal of Fish Biology* 15: 687-696.

Sunye, P.S. and C.M. Vooren. 1997. On cloacal gestation in angel sharks from southern Brazil. *Journal of Fish Biology* 50: 86-94.

Swaen, A. 1887. Etudes sur le développment de la Torpille. (*Torpedo ocellata*). *Archives de Biologie* 7: 537-585.

Tanaka, S., Y. Shiobara, S. Hioki, H. Abe, G. Nishi, K. Yano and K. Suzuki. 1990. The reproductive biology of the frilled shark *Chlamydoselachus anguineus* from Suruga Bay Japan. *Japanese Journal of Ichthyology* 37: 273-291.

Templeman, W. 1982. Development occurrence and characteristics of egg capsules of the thorny skate *Raja radiata* in the Northwest Atlantic. *Journal of Northwest Atlantic Fishery Science* 3: 47-56.

Teshima, K., M. Ahmad and K. Mizue. 1978. Studies on sharks-XIV. Reproduction in the Telok Anson shark collected from Perak River Malaysia. *Japanese Journal of Ichthyology* 25: 181-189.

Teshima, K. and S. Tomonaga. 1986. Reproduction of Aleutian skate *Bathyraja aleutica* with comments on embryonic development. In: *Indo-Pacific Fish*

Biology: Proceedings of the Second International Conference on Indo-Pacific Fishes, T. Uyeno, R. Arai, T. Taniuchi and K. Matsuura (Eds.), Ichthyological Society of Japan, Tokyo, pp. 303-309.

TeWinkel, L.E. 1943. Observations on the later phases of embryonic nutrition in *Squalus acanthias*. *Journal of Morphology* 73: 177-205.

TeWinkel, L.E. 1950. Notes on the ovulation, ova, and early development in the smooth dogfish, *Mustelus canis*. *Biological Bulletin* 99: 474-486.

TeWinkel, L.E. 1963. Notes on the smooth dogfish, *Mustelus canis*, during the first three months of gestation II. Structural modifications of yolk-sacs and yolk-stalks correlated with increasing absorptive function. *Journal of Experimental Zoology* 152: 122-137.

Thillayampalam, E.M. 1928. *Scoliodon, the common shark of the Indian seas*. Lucknow, India.

Thorson, T.B., J.K. Langhammer and M.I. Oetinger. 1983. Reproduction and development of the South American fresh water sting rays *Potamotrygon circularis* and *Potamotrygon motoro*. *Environmental Biology of Fishes* 9: 3-24.

Threadgold, L.T. 1957. A histochemical study of the shell gland of *Scyliorhinus canicula*. *Journal of Histochemistry and Cytochemistry* 5: 159-166.

Uchida, S., M. Toda and Y. Kamei. 1990. Reproduction of elasmobranchs in captivity. In: *NOAA Technical Report NMFS 90*, pp. 211-237.

Uchida, S., M. Toda, K. Teshima and K. Yano. 1996. Pregnant white sharks and full-term embryos from Japan. In: *Great White Sharks: The Biology of Carcharodon carcharias*, A.P. Klimley and D.G. Ainley (Eds.), Academic Press, San Diego, pp. 139-155.

Vanderbroek, G. 1936. Les mouvements morphogénétiques au cours de la gastrulation chez *Scyllium canicula*. *Wilhelm Roux' Archiv für Entwicklungsmechanik der Organismen* 47: 499-582.

Verissimo, A., L. Gordo and I. Figueiredo. 2003. Reproductive biology and embryonic development of *Centroscymnus coelolepis* in Portuguese mainland waters. *ICES Journal of Marine Science* 60: 1335-1341.

Villavicencio-Garayzar, C.J. 1993. Reproductive biology of *Rhinobatos productus* (Pisces: Rhinobatidae), in Bahia Almejas, Baja California Sur, Mexico. *Revista de Biologia Tropical* 41: 777-782.

Villavicencio-Garayzar, C.J. 1996. The ragged-tooth shark, *Odontaspis ferox* (Risso, 1810), in the Gulf of California. *California Fish and Game* 82: 195-196.

von Bonde, C. 1945. The external development of the banded dogfish or Pofadderhaai *Haploblepharus edwardsii* (Müller & Henle). *The Biological Bulletin* 88: 1-11.

Voss, J., L. Berti and C. Michel. 2001. *Chiloscyllium plagiosum* (Anon, 1830) born in captivity: hypothesis for gynogenesis. *Bulletin de l'Institut Oceanographique* 20: 351-353.

Wenbin, Z. and Q. Shuyuan. 1993. Reproductive biology of the guitarfish, *Rhinobatos hynnicephalus*. *Environmental Biology of Fishes* 38: 81-93.

West, J.G. and S. Carter. 1990. Observations on the development and growth of the Epaulette shark *Hemiscyllium ocellatum* Bonnaterre in captivity. *Journal of Aquariculture and Aquatic Sciences* 5: 111-117.

White, W.T., N.G. Hall and I.C. Potter. 2002a. Reproductive biology and growth during pre- and postnatal life of *Trygonoptera personata* and *T. mucosa*

(Batoidea: Urolophidae). *Marine Biology (Berlin)* 140: 699-712.
White, W.T. 2002b. Size and age compositions and reproductive biology of the nervous shark *Carcharhinus cautus* in a large subtropical embayment, including an analysis of growth during pre- and postnatal life. *Marine Biology (Berlin)* 141: 1153-1164.
White, W.T., M.E. Platell and I.C. Potter. 2001. Relationship between reproductive biology and age composition and growth in *Urolophus lobatus* (Batoidea: Urolophidae). *Marine Biology (Berlin)* 138: 135-147.
Whitley, G.P. 1940. Taxonomic notes on sharks and rays. *Supplement to the Australian Zoologist* 9: 227-262.
Whitney, N.M. and G.L. Crow. 2007. Reproductive biology of the tiger shark (*Galeocerdo cuvier*) in Hawaii. *Marine Biology (Berlin)* 151: 63-70.
Witschi, E. 1956. *Development of Vertebrates.* W.B. Saunders Company, Philadelphia.
Wolfson, F.H. 1983. Records of 7 juveniles of the whale shark *Rhiniodon typus*. *Journal of Fish Biology* 22: 647-656.
Wood-Mason, J. and A.W. Alcock. 1891. On the uterine villiform papillae of *Pteroplataea micrura*, and their relation to the embryo, being natural history notes from H.M. Indian Marine Survey Steamer 'Investigator,' Commander R.F. Hoskyn, R.N., Commanding. No. 22. *Proceedings of the Royal Society of London* 49: 359-367.
Wourms, J.P. 1977. Reproduction and development in chondrichthyan fishes. *American Zoologist* 17: 379-410.
Wourms, J.P. 1981. Viviparity: the maternal-fetal relationship in fishes. *American Zoologist* 21: 473-515.
Wourms, J.P. 1993. Maximization of evolutionary trends for placental viviparity in the spadenose shark, *Scoliodon laticaudus*. *Environmental Biology of Fishes* 38: 269-294.
Wourms, J.P. 1997. The rise of fish embryology in the nineteenth century. *American Zoologist* 37: 269-310.
Wourms, J.P. and A.B. Bodine. 1983. Biochemical analysis and cellular origin of uterine histotroph during early gestation of the viviparous butterfly ray *Gymnura micrura*. *American Zoologist* 23: 1017.
Wourms, J.P. and L.S. Demski. 1993. The reproduction and development of sharks, skates, rays and ratfishes: introduction, history, overview, and future prospects. *Environmental Biology of Fishes* 38: 7-21.
Wourms, J.P. and J. Lombardi. 1992. Reflections on the evolution of piscine viviparity. *American Zoologist* 32: 276-293.
Wrisez, F., H. Lechenault, C. Leray, B. Haye and J. Mellinger. 1993. Fate of yolk lipid in an oviparous elasmobranch fish, *Scyliorhinus canicula* (L.). In: *Physiological and Biochemical Aspects of Fish Development*, B.T. Walther and H.J. Fyhn (Eds.). University of Bergen, Bergen, pp. 315-322.
Wyffels, J., M. Masuda, J. Sakai and Y. Itoh. 2006. Characteristics of *Scyliorhinus torazame* egg case jelly. *Kaiyo Monthly* 45: 156-161.
Wyman, J. 1864. Observations on the development of *Raja batis*. *Memoirs of the American Academy of Arts and Sciences* 9: 31-48.
Yamaguchi, A. 2006. Reproductive biology of the longheaded eagle ray, *Aetobatus flagellum*, in Ariake Bay, Kyushu, Japan, *Joint Meeting of Ichythyologists and Herpetologists*, New Orleans, LA, USA.

Yamaguchi, A., T. Taniuchi and M. Shimizu. 1997. Reproductive biology of the starspotted dogfish *Mustelus manazo* from Tokyo Bay, Japan. *Fisheries Science (Tokyo)* 63: 918-922.

Yano, K. 1992. Comments on the reproductive mode of the false cat shark *Pseudotriakis microdon*. *Copeia* 1992: 460-468.

Yano, K. 1993. Reproductive biology of the slender smoothhound, *Gollum attenuatus*, collected from New Zealand waters. *Environmental Biology of Fishes* 38: 59-71.

Yano, K. and S. Tanaka. 1988. Size at maturity reproductive cycle fecundity and depth segregation of the deep sea squaloid sharks *Centroscymnus owstoni* and *Centroscymnus coelolepis* in Suruga Bay Japan. *Nippon Suisan Gakkaishi* 54: 167-174.

Ziegler, H.E. and F. Ziegler. 1892. Beiträge zur Entwickelungsgeschichte von Torpedo. *Archiv für Mikroskopische Anatomie* 35: 56-102.

2 Staging of the Early Development of *Polypterus* (Cladistia: Actinopterygii)

Salif Diedhiou[1] and Peter Bartsch[2]

[1]Université Cheikh Anta Diop Dakar, Senegal.
[2]Museum für Naturkunde der Humboldt-Universität zu Berlin,
Invalidenstr. 43, D-10099 Berlin, Germany.
E-mail: peter.bartsch@museum.hu-berlin.de (*corresponding author)

INTRODUCTION

Polypterus and *Erpetoichthys*, "bichirs" and the monotypic "reedfish", respectively, do belong to the family Polypteridae that only consists of these two genera and usually is given high systematic rank by reference to the order Polypteriformes or subclass Cladistia among Actinopterygii (ray-finned fishes). Morphology and development of polypterids have attracted a great deal of attention in the two centuries following their first description (*Polypterus bichir* by Geoffroy Saint-Hilaire, 1802). Recent polypterids only comprise approximately twelve valid species as a whole, with a typical 'sudano-nilotic distribution' from tropical West and Central Africa to the lower reaches of the Nile-system (Poll, 1954) and two recently described species (Britz, 2004; Schliewen and Schäfer, 2006). They exhibit a multitude of plesiomorphic characters of Actinopterygii and osteognathostome fishes and are thus considered an ancient relic group, a crucial object in the study of evolutionary biology that may inform us about phylogenetically old characters and character-combinations, which otherwise are only known from the fossil record or few other recent taxa.

The position of Polypteridae among osteichthyans is unique indeed, since no *prima facie* closely related higher taxa are known fossil or recent. Fragmentary fossil remains of polypterids date back only to the Lower Cretaceous of Africa and probably Southern America (Gayet and Meunier, 1991, 1992, 1996; Meunier and Gayet, 1996; Dutheil, 1999). But, these fossils apparently are close relatives of the existing genera, identified by autapomorphies of the group, and lack clear synapomorphies of subdivisions of either Actinopterygii or Sarcopterygii that would enable a clear hypothesis of relationship beyond the isolated classification. In recent times a sister-group relationship to Actinopterygii, first explicitly put forward by Goodrich (1928), has been clearly favoured by most authors and supported by several well investigated characters (Patterson, 1982; Lauder and Liem, 1983; Gardiner and Schaeffer, 1989; Bartsch, 1997). To lump together Acipenseriformes and Polypteriformes among Chondrostei is unfortunate, because it leaves the latter as a dumping ground of "primitive" Actinopterygii (compare Nelson, 1994, 2006). Other alternative views were mainly held by Bjerring (1986, 1988, 1991a,b for example), who argued in favour of uniqueness, a separation as 'Brachiopterygii' at equal rank with both Actinopterygii and Sarcopterygii, thus evading the statement of alternative phylogenetic hypotheses conceptually inherent in phylogenetic systematics. Admittedly, in a series of excellent morphological descriptions raising questions of unexplained character combinations and throwing doubts on seemingly established homologies, he has enforced to think over the case again. We will encounter some of his arguments later in the discussion. More recently Arnason *et al.* (2001) in their investigation of the mitochondrial genome of lower vertebrates may have revived the question by turning the phylogeny of gnathostomate vertebrates essentially upside down, leaving polypterids in a basal sister-group position to the clade of Actinistia, Actinopterygii, and Chondrichthyes. Identified as crucial objects of osteichthyan systematics, there have been several approaches to the study of the early ontogeny of Polypteriformes. The earliest, and one of the most complete and most influential studies to date, was performed by Budgett after a collecting trip to the Gambia River in 1899 and 1900, who recovered a fairly complete ontogenetic series of *Polypterus senegalus* under most difficult conditions (Budgett, 1899, 1901a,b, 1902; Kerr, 1907). More recently, breeding of polypterids in captivity has been several times successful (Arnoult, 1964; Wolff, 1992; Bartsch and Britz, 1996; Bartsch *et al.*, 1997; Schugardt, 1997), and several accounts on early ontogenetic processes and characteristics based on these, more convenient laboratory methods. These of course, have their own merits of easier access to and reproduction of certain data of developmental timing under standard conditions, and the resulting overview of the early ontogeny of two polypterid species is intended as a guideline for aimed studies of experimental embryology as well as new approaches at the ecology and reproductive biology of this interesting group of fish in the wild environment.

Materials and Methods

Breeding and Rearing in Captivity

For breeding in captivity, a group of *Polypterus senegalus*, three males and three females were available. For *Polypterus ornatipinnis*, a female and two males were used. The adult specimens of both species were maintained in glass aquaria (400 litres size) equipped with filter-devices, hiding places, and a vegetation of *Taxiphyllum barbieri* that was used as a spawning substrate. Gonad maturation was stimulated with a method used by Bartsch *et al.* (1997) and Schugardt (1997). Eggs and embryos were reared in large glass dishes. Water temperature varied between 22 and 28°C. Larvae were fed mainly with *Artemia-nauplii* in the beginning and later supplemented by *Tubifex* and chironomids.

Sampling

The determination of the developmental stages is based on the examination of 2-13 individuals of the same spawn. Those features are highlighted, which are externally recognized for the first time during the developmental sequence. Up to stage 22 few traditional and clearly defined criteria of "external development" are available. Later, a multitude of different organ systems are described, which present a complex external differentiation that can be evaluated for stage definition. To collect the developmental stages, eggs, embryos, postembryos and larvae of close external similarity were fixed together. Oocytes and ovarial eggs were taken from a fertile female and fixed in 4% buffered formalin at pH 7.3.

Light Microscopy

Some embryos and postembryos were stained by PAS to analyze the attachment glands with their mucus secretion. A portion of the specimens were embedded in paraffin via Bouin-fixation, a graded series of ethyl alcohol and chloroform with a Shandon Hypercenter Xp embedding apparatus. The other portion was embedded in a cold polymerizing plastic material (Kulzer, Technovit 7100) after passing through a graded series of ethyl alcohol at room temperature. The paraffin embedded objects were sectioned with a Leica SM2000R microtome at 8 µm and with a Leica RM216 rotary microtome at 1-4 µm for the plastic embedded material, stained with Harris' hematoxylin-eosin or methyl blue-fuchsin, respectively. All sections were documented with a Zeiss Axioskop photomicroscope equipped with a camera.

Scanning Electron Microscopy

After fixation in 4% formalin, in 2.5% or 5% glutaraldehyde, phosphate buffered at pH 7.3 (different laboratory formulations by Bardele, Zeiske, and Wourms), objects were washed in the same buffers, dehydrated in a graded series of ethyl alcohol, critical point dried with liquid CO_2 in a BA1-TEC cpd 030 critical point drying apparatus and mounted on aluminum stubs. All mounts were coated with 2 nm gold-palladium in a Polaron SC7640 sputtering device and photographed using a Leo Stereo Scan 1450 VP microscope at 10 or 20kV.

Transmission Electron Microscopy

The specimens were fixed in phosphate-buffered glutaraldehyde and sectioned at 0.7 μm (semi thin section series) and 200 ηm (ultra thin sections). They were photographed with a Siemens Elmiskop 102 transmission electron microscope at the University of Tübingen (Institut für Spezielle Zoologie).

Photos and Drawings

Photographs and graphics were made with a stereomicroscope equipped with a drawing mirror. Entire specimens were photographed with a Leica M420 photomicroscope and juvenile and adult fishes with a camera fitted with a macro-lens.

On Terminology

The terminology of early ontogeny, as applied here, contrasts with the concept of Balon (1975, 1985) which envisages the yolk-sac larva as an 'eleutheroembryo' or 'free embryonic phase' belonging to an embryonic period. Whereas there are good reasons for highlighting the developmental stages of the hatched osteichthyan offspring as embryonic in the sense of limited self-sustainability and incomplete organ function, it is avoided here lest it should imply phylogenetic implications of 'precocious hatching' as compared to directly developing chondrichthyans and amniotes. Also in favour of a uniform usage, we have adopted here the more neutral term 'postembryo' for this phase of development.

Results

Egg Envelope

The egg envelope or zona radiata of the polypterid egg, as in other actinopterygians, plays an important role in fertilization of the egg and in the interaction of the earliest developmental stages with the environment. Furthermore, it bears some characteristics that are unique to Polypteridae or are even species-specific and accordingly should be described in a short separate section here.

Two significant structures have been already described for *P. senegalus*, *P. ornatipinnis* and *Erpetoichthys calabaricus*: Micropyle and projections of the envelope surface (Bartsch and Britz, 1996, 1997; Britz and Bartsch, 1998). However, the larger body of material now available allows for a more complete description and more far-reaching interpretation. Also, ovarian eggs were studied in this context.

The ovarian egg of *Polypterus senegalus* during its growth period ('oogony', late oogenesis) is completely surrounded by a flat, monolayer of cells, the follicular epithelium separated from the remaining ovarian tissue above by a thin, acellular network, and separated from the egg-surface by a thick acellular layer, the zona radiata (Fig. 1A). There are dense circumferential layers fused and interwoven with radial (centrifugal) processes. The whole stack has a thickness of 28 µm. Accordingly, the basal membrane of the follicle cells is in direct contact with the surface of the zona radiata. It is not known whether the radial processes observed correspond to walls of the canaliculi described by Riehl in TEM-sections of the teleostean zona radiata (Riehl, 1995).

The largest ovarian eggs were supposed to be the ripest ones. Their surface exhibits a pigmentation pattern very similar to that of spawned eggs. During ripening of the oocytes prior to ovulation the zona radiata becomes modified. The cells of the follicular epithelium acquire a more roundish shape and needle-like processes at their bases become embedded in the more consolidated zona radiata. Thereby an intimate connection between follicular cells and the acellular mass is formed (Fig. 1B).

A special group of much enlarged, bottle-shaped cells of the follicular epithelium is involved in the formation of the micropyle. These cells show at their base numerous filiform processes that protrude into the zona radiata in a manner similar to other follicle cells (Fig. 1C). Beneath these cells, the zona radiata is much thinner and the large cells occupy the space of the micropylar groove, somewhat different from the histological figures given by Mark (1890) for the smaller, multilayered micropylar cells of *Lepisosteus*.

Apart from the micropyle, the egg envelope surface in spawned eggs of both species shows characteristic surface structures, i.e., bottle-shaped

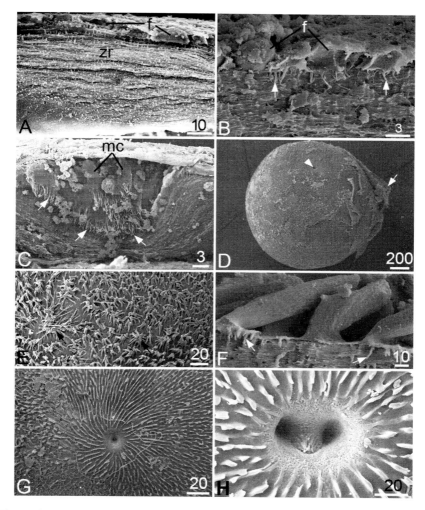

Fig. 1. SEM (Scanning electron micrograph) of ovarian and spawned eggs of *P. senegalus* (A-G) and *P. ornatipinnis* (H). A—Ovarian egg, cross section of zona radiata (zr) with follicular cells (f). B—Ripe oocyte, cross-section of zona radiata with the follicular cells (f) above it. Arrows point to the needle-like processes. C—Ripe oocyte, section through the micropyle (mc) and micropylar cells. D—Surface of spawned egg with attached debris (arrow); arrowhead points to the micropyle. E—Detail of spawned egg surface shows the bottle-shaped projections. Note the agglutination (arrow) and stretched appearance of some. F—Spawned egg, cross section of the upper part of zona radiata. Arrows point to filiform processes. G and H—Animal pole of spawned eggs show a single micropylar pit with canal in the centre (G) and magnified for *P. ornatipinnis* (H) with a double micropylar canal. Scale bar in μm.

to thread-like protrusions that are responsible for the attachment of the egg to plant material (Fig. 1D-F). Initial attachment of the spawned eggs becomes stronger by addition of more attachment points and agglutination of elongate threads to rough natural objects (Fig. 1E).

The projections of the egg envelope are most probably formed by transformation and fusion of the lateral walls of follicular cells, because the needle-like basal processes of the follicle cells in the ovarian egg (Fig. 1B, arrows) are the same as the basal processes of the projections in the spawned egg (Fig. 1F, arrows).

The surface structures around the micropyle are different from the typical villiform processes and they become more flat, forming rows and ridges that converge towards the micropyle (Fig. 1G,H). Some of these ridges reach the periphery of the micropyle groove and form a spiral pattern in *P. senegalus* (Fig. 1G). This spiral is not that pronounced in *P. ornatipinnis* (Fig. 1H). Generally, the shape and size of the projections are similar in *P. senegalus* and *P. ornatipinnis*, if the same regions of the surface of the egg envelope are compared. The projections of *P. senegalus* always seem somewhat more densely spaced, and the distal heads are thickened and club-shaped, whereas in *P. ornatipinnis* they look more like regular pillars. The shape of the micropyle is similar in *P. ornatipinnis* and in *P. senegalus*. The micropyle is a perforation of the zona radiata, and consists of a groove, the micropylar pit and a short micropylar canal. It corresponds to type II according to the classification of Riehl (1978). In *P. ornatipinnis* 2 of 5 observed specimens showed a micropyle with one pit and two canals (Fig. 1H). This 'aberration' has never been noticed in specimens of *P. senegalus*.

The micropyle and micropylar groove later become closed by contraction of the zona radiata and the viscous layer above.

Embryonic Phase

The embryonic phase in fishes begins with fertilization at the time of spawning and ends with hatching.

Egg and Cleavage

This developmental phase is characterized by the division of the zygote into numerous blastomeres or cleavage cells without any further significant external differentiation, except for blastomere size. The process of cleavage, which is holoblastic and unequal, starts shortly after insemination and finally leads to the typical hollow blastula. The eggs of *P. senegalus* and *P. ornatipinnis* divide in a rapid sequence. The cell limits appear as furrows that are well recognized externally.

Stage 1

This stage is defined as the fertilized egg cell or zygote before cleavage and accordingly extends from spawning to first cleavage. The eggs of *P. senegalus* and *P. ornatipinnis* (Fig. 2; Pl. I. 1,1*) are comparable in size, spherical, and somewhat oval in shape. They are enclosed within a primary egg envelope, the vitelline membrane or zona radiata, which is almost transparent. It delimits a narrow perivitelline space, and has glutinous properties that facilitate attachment to plant material or structures of the aquarium tank.

The maximum equatorial diameter, which includes the vitelline membrane, ranges from 2.0 to 2.2 mm in fixed eggs of *P. senegalus*. All measurements correspond well to earlier literature data of a mean of 1.4 mm in fixed eggs excluding the vitelline membrane (Bartsch and Britz, 1996), and a mean of 2.3 mm in living eggs including the vitelline membrane (Bartsch and Britz, 1996; Schugardt, 1997). The equatorial diameter, including the vitelline membrane, of fresh eggs of *P. ornatipinnis*, measured 2.2 to 2.5 mm. Bartsch and Britz (1996), however, gave a little larger size range of 2.4 to 2.7 mm, whereas Burgess (1986) reported a diameter of only 1.8 to 2.0 mm. The latter figure either excludes the vitelline membrane or refers to the fixed condition.

The eggs of *P. senegalus* and *P. ornatipinnis* differ by showing a rather sharp border between the dark brownish to blackish upper, called animal, hemisphere with the animal pole and the less pigmented lower, called vegetal hemisphere, with the vegetal pole in the former species (Fig. 2A) and a more continuous transition of clustered maternal pigmentation

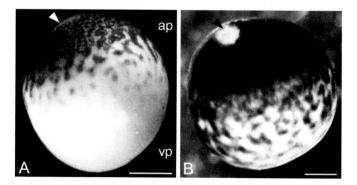

Fig. 2. LM (light micrograph) view of spawned eggs at stage 1, egg envelope removed. Observe the difference in pigmentation. A—*P. senegalus*. B—*P. ornatipinnis*. Arrowheads indicate the micropylar region. ap—animal pole, vp—vegetal pole. Scale bar: 0.5 mm.

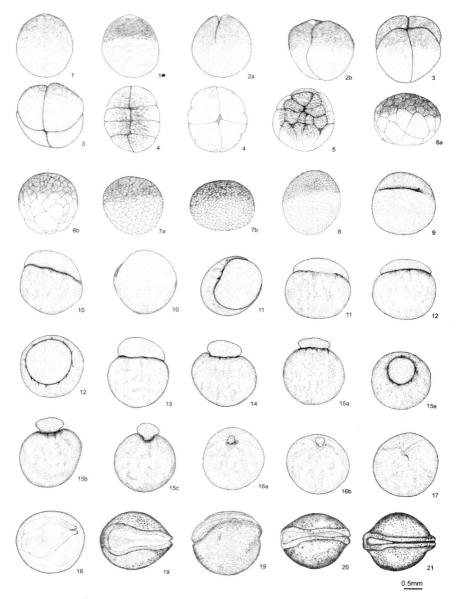

Plate I. Survey of early ontogeny of *Polypterus senegalus*, stage 1 to stage 21. 1* Fertilized egg of *Polypterus ornatipinnis*.

towards a diffuse or reticulate pattern in the equatorial region and lower vegetal hemisphere in the latter species (Fig. 2B). This gives the impression of a generally darker coloration to the unaided eye in *P. ornatipinnis*. Under the binocular microscope, however, the telolecithal polypterid egg always reveals a clear polar differentiation even though fine pigment granules are also present at the vegetal pole and between the pigment clusters in both species. Pigmentation of the egg of *P. ornatipinnis* is so dense at the animal pole that individual granules or clusters cannot be discerned (Fig. 2B; Pl. I. 1*). In both species, there is a small, circular, unpigmented area at the tip of the animal pole (Fig. 2A,B) that corresponds exactly to the micropylar region of the vitelline membrane.

Stage 2

This stage starts with the first cleavage furrow that divides the egg into two blastomeres of almost equal size. The furrow progresses from the animal to the vegetal pole (Pl. I. 2a,b).

Stage 3

The 4-cell stage is defined by the appearance of the second cleavage furrow (Fig. 3A; Pl. I. 3).

Both of the first two cleavages reach the vegetal pole, and are thus meridional, but also somewhat eccentric. Accordingly, the micropylar field (corresponding to the micropylar region of the egg) ends up in one of the four blastomeres.

Stage 4

The orientations of the cleavage planes during this stage are variable and result in 8 blastomeres. Three vertical cleavage furrows start at the animal pole, but do not reach the apex of the vegetal pole.

The fourth furrow is obliquely positioned in such a manner that a smaller and a larger area are delimited in the animal hemisphere, the former including the micropylar area (Fig. 3B,C). There are small variations in the orientation of the division plane and the formation of an animal pole micromere. *P. ornatipinnis* regularly exhibits a more strictly meridional orientation of the three vertical cleavage furrows, whereas they meet more remotely from the animal pole in *P. senegalus*. In some specimens of *P. senegalus* (Fig. 3D; Pl. I. 4), there are two synchronous divisions, oriented almost vertically on one of the first furrows at an early stage, which results in 8 almost equal blastomere areas as seen from above.

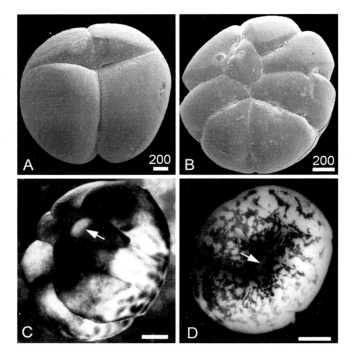

Fig. 3. Cleavage. A—SEM of *P. ornatipinnis*, early stage 3 (4 blastomeres) showing the beginning of the second cleavage furrow which cuts in vertically and at a right angle to the first. B—SEM of *P. senegalus*, stage 4 (8 blastomeres), 7 almost equal blastomeres and a smaller one. C—LM, *P. ornatipinnis*, stage 4, 7 almost equal blastomeres and a smaller one underneath the micropylar area (arrow). D—LM of stage 4. Some exceptional cases of (probably early) cleavage planes in *P. senegalus*. Note the symmetrical arrangement of cleavage furrows. Scale bar in μm except in C and D—0.5 mm.

Stage 5

At this stage the germ consists finally of 16 blastomeres. The furrows of the fourth cleavage run roughly horizontally to obliquely and are shifted towards the animal pole, well above the equator (Fig. 4A; Pl. I. 5). This results in a significant separation of the micromeres of the animal hemisphere from the macromeres of the vegetal hemisphere.

Our interpretation of this variation within and between the two species, together with the relative delay of cleavage of the vegetal hemisphere, is that of a fast initiation of cleavages, "crowded together" at an early developmental stage. Thus, the cleavages very soon become irregular and

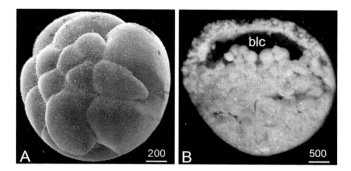

Fig. 4. Cleavage and formation of the blastula A—SEM of *P. ornatipinnis*, stage 5 (16 blastomeres). Note the significant size difference between micromeres and macromeres. B—LM of *P. senegalus*, early blastula, cut vertically through the blastocoel (blc). Note the protrusion of the larger yolk-rich vegetal cells into the blastocoel. Scale bar in μm.

asynchronous. The cell divisions that spread over the germ surface and finally result in the blastula (Fig. 4B), no longer show any significant differences between the two species.

Stage 6

The asynchronous cleavages of this stage result in the formation of a ball-shaped morula. (Pl. I. 6a,b). In contrast to the two preceding stages, these cell divisions now are complete and are initiated in the yolk-rich vegetal region. The direction of the cleavage furrows also changes resulting in polygonal cells. The typical macromeres are still significantly larger than micromeres. The latter, however, may also differ in size and in the equatorial region heavily pigmented "micromeres" may equal "macromere" size. By the end of this stage, formation of the blastocoel has started. Externally, this is recognized by a flattening of the animal pole in fixed specimens.

Stage 7

This is a blastula. The number of blastomeres has increased considerably. The blastocoel is roofed by a blastoderm, a somewhat irregular and loosely organized, but essentially monolayered cellular wall at the animal hemisphere (Fig. 4B).

Stage 8

This stage marks the end of the so-called cleavage phase of the germ. At a first glimpse, this late blastula is difficult to distinguish from the uncleaved egg of stage one, the unpigmented animal pole spot of the latter excepted (Fig. 1A,B; Pl. I. 8).

Gastrulation

In contrast to the preceding phase this developmental phase shows less accentuated cell proliferation, but mainly displays cellular movements and rearrangements. Epiboly by the animal hemisphere, i.e. overgrowth of the vegetal by the animal hemisphere, is its main characteristic. The stage determination from stage 9 to 21 refers to a pattern of gastrulation, shown externally by formation of a germring, starting as blastoporal lip ('Urmund-lippe') somewhat "north of the equator" and also by the relation between overgrown and outside remaining prospective entoderm referred to as 'yolk-plug'. Bottle cells (initiators of the involution of surface cells) are formed in the shallow blastoporal groove that eventually opens into the archenteron (Wourms and Bartsch, 1998).

Stage 9-17

Onset of gastrulation shows a small slit-like blastoporal lip, above the equator of the blastula (Fig. 5A; Pl. I. 9). During epiboly, a contraction of the cell surfaces, formation of bottle cells, and formation of folds occur (Fig. 5B). With the ingression/involution of the prospective entodermal cells involution of the prospective mesodermal cells also occurs. These first form a loose cellular congregation inside the blastocoel (Fig. 5C). At stage 10 the blastoporal lip encompasses half of the germ circumference (Pl. I. 10). At the following stage, two thirds of the final extension (Pl. I. 11) and at stage 12 a complete ring is attained (Pl. I. 12).

During the stages 13 to 17, the ring contracts until all the prospective entoderm is enclosed (Pl. I. 13-17). At stage 13 a high yolk-plug is formed. The equatorial diameter of the area of the animal hemispheric cap is a little larger than that of the yolk-plug consisting of yolky entoderm cells (Pl. I. 13). At the following stage, the equatorial diameter of the former animal hemisphere consisting mainly of prospective ectoderm is double that of the yolk-plug (Fig. 5D; Pl. I. 14). At stage 14 the vegetal part has already been far overgrown so that the blastocoel has become reduced to a small slit-like cavity. The mesentodermal cells are densely packed beneath the ectoderm (Fig. 5E). At the site of epiboly the intruding cells have lost their firm contacts, become flask-shaped, and apparently undergo active cell movements.

At stage 15, the circumference of the area of prospective ectoderm is about three quarters larger than the yolk-plug circumference (Pl. I. 15a,b,c).

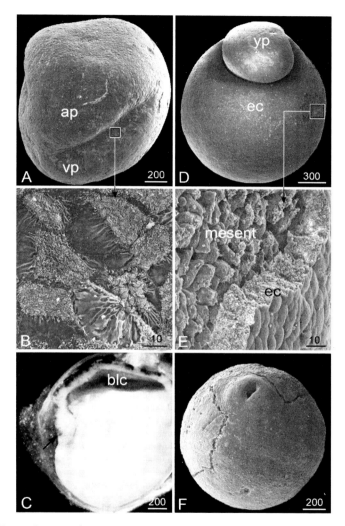

Fig. 5. Gastrulation of *P. ornatipinnis* A—SEM of Stage 9. Survey of the early gastrula at the onset of the blastopore slit formation. Animal pole (ap), vegetal pole (vp). B—SEM of details of the cell surfaces during contraction at the blastopore lip region framed in A (surface of forming bottle cells). C—LM of vertical section through early gastrula showing the involuted region (arrow) and blastocoel (blc). D—Stage 14. SEM, survey of a more advanced gastrula turned around with yolk-plug (yp) and overgrowth (epiboly) by the animal cells (future ectoderm) (ec). E—Enlargement of the fractured ectoderm (ec) portion framed in D, which illustrates the thickness of the ectoderm and structure of the underlying mesentoderm (mesent). F—SEM of Stage 17, the terminal stage of gastrulation that shows the constricted blastopore opening. Scale bar in μm.

The latter has now attained a mushroom shape. Whereas these first stages of gastrulation are a fast and continuous process within the first 24 hours, the table demonstrates considerable slowing down of the timing of yolk-plug retraction.

At stage 16, the yolk-plug has not been completely invaginated or been completely overgrown by ectoderm (Pl. I. 16a,b) and only on the future ventral side of the embryo does a small remnant of the blastocoel remain. The ectodermal covering of the late gastrula reaches a tenfold circumference of the yolk-plug until the latter is eventually retracted completely through the blastopore at stage 17 (Fig. 5F; Pl. I. 17). However, it has been observed in some specimens that a small yolk-plug remains until embryonic stages close up the future anal opening.

The phase of gastrulation is fully completed within 10 hours (at a temperature range of 24-26°C). Gastrulation has been resolved into several stages mainly because a high resolution of the process seems useful in view of its importance for the initiation of the basic morphogenetic processes in the vertebrate embryo.

Neurulation

Stage 18

Early in the neurulation phase, the ectoderm has clearly differentiated externally into an epidermal region surrounding the yolk-mass and the neuroectoderm which corresponds to the neural plate. The first sign of the developing central nervous system is a bright, translucent, "heart-shaped" medullary plate that is not sharply bordered, with its tip pointed towards the blastopore (Pl. I. 18).

The edges of the neural plate, termed neural folds, start to project and a middle longitudinal furrow sinks below the neuroectodermal surface level forming the neural groove (Fig. 6A). The cell surfaces of the neuroectoderm are smooth, arched, and the borders are slightly elevated (Fig. 6B), whereas the ectodermal cells of the future epidermal cover of the yolksac show short, curly microplicae (Fig. 6C).

Stage 19

At this stage, the flat neural plate contracts in the middle region, attains a pear shape, and is lifted above the germ surface (Pl. I. 19). A little later (Fig. 6D), the neural folds have elevated considerably and in the future head region a primary head cavity becomes more pronounced in depth. The pronephros appears as a low swelling at the dorsal area of the yolk-sac region.

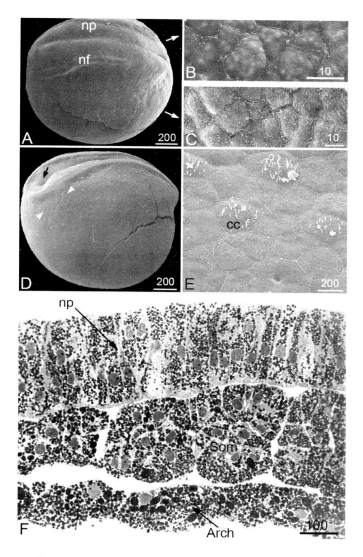

Fig. 6. Neurulation. SEM of *P. senegalus* A—stage 18, dorsolateral view of the neural plate (np) with elevated neural folds (nf), neural groove and shallow primary forebrain cavity. B—Detail of the neuroectodermal cell surface of the neural plate of A. C—Details of the lateral ectodermal surface of A showing microridges or microvillus-like protrusions. D—late stage 19, dorsolateral view. Note the deepened primary forebrain cavity (arrow). E—Enlargement of the epidermal surface region shown in D between arrowheads with ciliated cells (cc) possessing 20-30 short cilia. F—LM of sagittal section through trunk. archenteron roof (Arch), neural plate (np) and somite (Som). Scale bar in µm.

At the dorsal margin of the yolksac region large melanophores appear in groups for the first time (Pl. I. 19). They are indicative of early migrating neural crest cells. Also, ciliated epidermal cells appear for the first time in a small, paired area of the dorsal margin of the yolk-sac region (Fig. 6E). This occurs later than in some amphibians, which show a ciliated ectoderm during gastrulation (Assheton, 1896).

Stage 19 has been selected for a more detailed description of the histological differentiation of tissues and early organogenesis. Already at this stage a second epidermal layer, the subepidermis, is formed. The cells of the upper layer of the epidermis of the yolk-sac differ from those of the subepidermis in form, size, pigmentation, and differentiation of ciliated and glandular cells. The outer, epidermal cells are taller and closely packed. In surface views, the polygonal epidermal cells appear fairly uniform in size and contain many roundish yolk globules. In contrast, the inner, subepidermal layer consists of a loose chain of cells, which appear flattened in the ventral region but are more similar to the epidermal cells in the lateral body wall of the embryo.

The neuroectoderm consists of a single layer of tall prismatic cells that are tightly joined together and retain a position perpendicular to the outer surface (Fig. 6F). These cells contain many small yolk-globules and roundish to oval cell nuclei that may have different positions. The neuroectoderm is covered by a very thin monolayered epithelium. A column of mesodermal cells differentiating from the archenteron roof of the neurula corresponds to the future notochord. Even at this stage, the notochord appears in sagittal section as a cell mass at the level of the widest region of the neural plate. Its cells are antero-posteriorly flattened, and appear polygonal. They are loosely attached and contain many small yolk-globules and a large-sized roundish nucleus with an eccentric nucleolus. The voluminous embryonic gut consists of a mass of large-sized yolk cells at its bottom and a single layer of small entodermal cells at its roof. The lateroaxial mesodermal plate behind the head region is differentiated into a somitic stem and lateral plate. The former was segmented already into four somites (Fig. 6F).

Stage 20

At this stage elevation and approximation of the neural folds is more pronounced posteriorly over approximately two thirds of the neural plate length (Fig. 7A; Pl. I. 20). In the primary brain-cavity, differentiation into prosencephalon and deuterencephalon now is indicated by a transverse fold. At the level of the deuterencephalic depression the first swelling of the anlage of the external gills appears. The blastopore still remains open

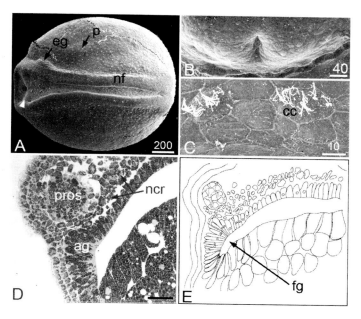

Fig. 7. Neurulation. SEM of *P. senegalus*, stage 20 A—dorsal view showing neural plate. Note a separation and differentiation of the prosencephalon (pros) by a transverse fold (arrowhead) and the first swellings of the anlagen of the external gills (eg) and the pronephros (p). B—Frontal aspect of the blastopore region of A. C—Close up of the epidermal surface of the yolksac of A showing multiciliated cells (cc) with elongated cilia. D and E—LM and corresponding graphic survey of the internal development of the foregut (fg) and attachment gland (ag). ncr—neural crest. Scale bar in μm except in D—0.1 mm.

(Fig. 7B) or it may be closed by a small yolk-plug remnant. The ciliated epidermal cells are distributed over the entire ectodermal surface, except for the neural plate. The number of cilia appears to vary to some extent, for in the neurula shown in Fig. 7C only 20 are observed as a mean maximum, fewer than in the specimen depicted in the preceding stage and in contrast to the general developmental tendency observed in later stages.

On each side, internally and rostrolaterally, a horizontal archenteron bulge forms, from which the attachment glands derive (Fig. 7D,E). The entodermal cells which give rise to their anlage are now distinguishable. They differ from those of the archenteron roof by their small size and dark staining with methylblue. Large-sized, roundish, and lightly coloured nuclei contain an eccentric nucleolus.

Stage 21

At this stage the neural folds approach each other along the posterior two thirds of their length (Fig. 8A; Pl. I.,II. 21). The neural folds are further elevated in the head region arching over the deep prosencephalic and deuterencephalic cavities. The most caudal part of the neural anlage is somewhat elevated from the epidermal yolk-sac cover forming a rudimentary tail-bud above the blastopore. In the transitional area from neural folds to the anterior yolksac, there are paired lateral swellings recognized externally as the anlagen of the attachment glands.

The progressive formation of the neural folds, the neural crest, and further differentiation of the attachment glands as well as of the notochord and of the intestine were studied in more detail. During the upward movement of the neural folds leading to closure of the neural vesicle and neural tube, the neural crest cells appear on each side in the contact region between neural folds and epidermis of the head region (Fig. 8B). The cells of the neural crest first migrate ventrally and later caudally. At first, they lie mainly between the lateral walls of the neural groove and the epidermis.

The differentiation of the notochord progresses caudally. Its anterior tip is located at the level of the deuterencephalon and is surrounded by the mesodermal head plate, the roof of the gut, and dorsally by the neural furrow (Fig. 8C). From this level to the level of the last somite the notochord displays its typical shape of a cylindrical cell column. More posteriorly it looses its characteristic shape and merges into the mesodermal material. The mesoderm plate is differentiated along the rostral to caudal direction into six somites which contact the lateral plate by a nephrotome. The mesodermal mass of the lateral plate grows ventrolaterally. The cellular mass of the mesodermal head plate extends laterally and fills the space between the ectoderm and the entoderm in the head region. This mesodermal matter accompanies the entire head part of the notochord and surrounds the brain ventrally and laterally. Later it is assumed to form the pericardium, heart, the parachordal neurocranial base, vessels, and visceral muscles of the head. Along with the differentiation of the neural plate in the head and trunk regions, the anlage of the gut also differentiates into anterior, middle, and posterior parts. The central mass of the ventral entoderm expands and almost fills the entire gut lumen. Six somites have now differentiated.

Internally, the previously mentioned bulge of the attachment glands becomes more pronounced at the front of the wall of the gut (Fig. 8D,E). Its cells are devoid of yolk globules, are taller and become club-shaped. In sagittal section, these cells display large-sized central nuclei; this is in contrast to the remaining entoderm the nuclei of which occupy the entire width of the cells.

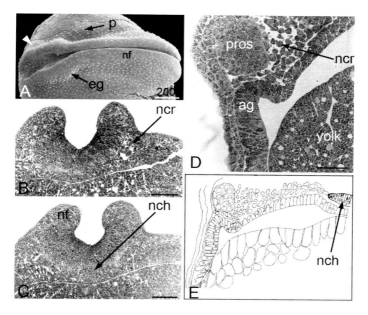

Fig. 8. Neurulation of *P. senegalus*, stage 21 A—SEM of dorsal view of the neural folds (nf) almost in contact (arrowhead). B and C—LM of cross sections. B—Prosencephalic region. Neural crest (ncr). C: Deuterencephalic region. Neural fold (nf), notochord (nch). D-E—LM and corresponding graphic survey of the internal development of the foregut and attachment gland (ag). nch—notochord, ncr—neural crest, pros—prosencephalon. Scale bar in μm except in B, C and D—0.1 mm.

Stage 22

This stage is the last of the neurulation phase. It is characterized by complete closure of the neural folds that results in a neural tube with the prosence-phalic and deuterencephalic cavities in front. (Fig. 9A; Pl. II. 22). The posterior end of the embryonic anlage is lifted up considerably from the underlying epidermal cover of the yolksac region. Therefore, this stage deserves the name "early tailbud stage". In contrast, the future head region is still more flat and shows no lifting upwards. Large melanophores are uniformly scattered all over the embryo. In the specimen figured above, for example, the head and the tail region and dorsal parts of the yolksac are provided with a higher density of melanophores than the other parts of the embryo (Pl. II. 22). Ciliated cells are distributed all over the late neurula including the epidermal covering of the neural folds. As a concluding remark it may be said that uplift and closure of the neural folds are a relatively slow process, whereas migration of neural crest cells, differen-

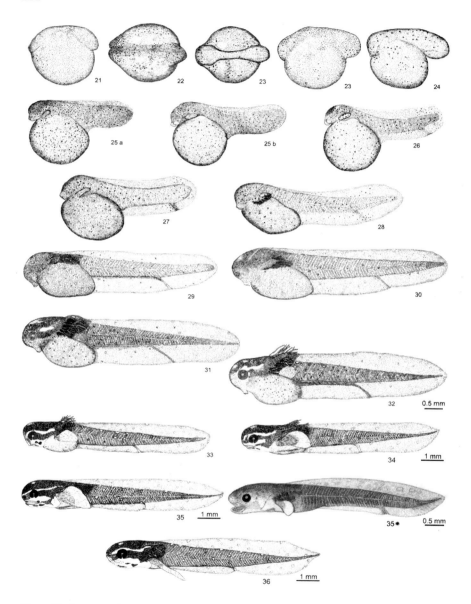

Plate II. Survey of early ontogeny of *Polypterus senegalus*, stage 21 to stage 36. 35* Last yolksac larval stage of *Polypterus ornatipinnis*.

tiation of ciliated epidermal cells, formation of the anlagen of attachment glands, external gills, and pronephros body are initiated early at the onset of neurulation and proceed rapidly well before the completion of neurulation.

Organogenesis

Stage 23

The embryonic body anlage has changed into a club-like shape with the prominent growth of the attachment glands and external gill anlagen on each side of the brain vesicle (Fig. 9B; Pl. II. 23). The embryo bulges considerably above the yolksac and its contour is clearly defined, but the borders of the aforementioned organ anlagen are not well defined externally. The embryo is more elongate and equally projects well beyond the yolksac both cranially and caudally, even before the tail starts a pronounced growth. Posteriorly, the head region is delimited by a narrowing at the level of the external gill anlage that corresponds to the posterior border of the deuterencephalon and accordingly turns out to be an important external marker during the subsequent stages: the head and the body of the embryo can be delimited from each other from this stage onwards. The outlines of the brain vesicle and neural tube are visible under the translucent epidermis in some embryos. The cilia-bearing cells are more dense on the head and body regions of the embryo. (Fig. 9C). New cilia also emerge in between the bundles of older cilia (Fig. 9D).

In the specimens of this stage of *P. senegalus* (Pl. II. 23) only a few, large-sized melanophores are present. These cells are usually more abundant and more evenly distributed in embryos of the same stage of *P. ornatipinnis*. The quantity and distribution of these melanophores vary much among individuals.

Histological sections demonstrate that it is the epidermal regions of the neural folds, already touching each other in stage 22, that fuse and form a continuous layer of skin above the neural tube in stage 23 (Fig. 9E-H). The external ectodermal layer hardly changes during the subsequent development, but the subepidermal layer is involved in formation of a variety of thickened placodes of sensory organs in the head region, e.g., anlagen of lens, olfactory organ, otic vesicle and lateral line organs.

The anlage of the central nervous system consists of a wide neural vesicle (Fig. 9E,F) and a narrow neural tube with monolayered cellular walls. The prismatic cells are arranged with their long axis perpendicular to the lumen. The margins of the neural folds at first are not fused and, accordingly, the prosencephalon still has the shape of an open trough in transverse section. During closure of the neural folds, some neural crest

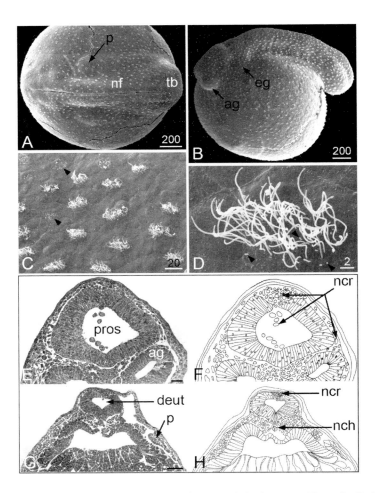

Fig. 9. Neurula and embryo. A—SEM of *P. ornatipinnis*, stage 22, early "tailbud" stage. Dorsal view of the completely closed brain vesicle and neural tube. Note the trace of contact zone between the neural folds. B-H—SEM of embryo of *P. senegalus* stage 23. B—Close anterolateral view. C—Magnified epidermal surface of the trunk of B illustrating the brush-shaped bundle of cilia and the growth in density of ciliated cells by addition of new cells (arrowheads). D—Detail of a cilia bundle of the ciliated cells in C. Note the high number of cilia per cell and the emergence of new cilia in between (arrowheads). E and F—Cross section through the prosencephalon (E, LM) and the corresponding graph (F). Note the neural crest cells included between the covering epidermis and the thick walls of the brain. G and H—Cross section through the deuterencephalic region. ag—attachment gland, deut—deuterencephalon, eg—external gill, nch—notochord, ncr—neural crest, nf—neural fold, p—pronephros body, pros—prosencephalon, tb—tailbud. Scale bar in μm except in E and G—0.1 mm.

cells have shifted into the brain cavity. Some of these cells also move caudally inside the neural tube. Probably, these are prospective glia cells and pigment cells of the central nervous system. Early after the complete closure of the neural vesicle and neural tube the anlage of the eye develops as a prosencephalic evagination. This extends at its distal part and forms the eye vesicle, which presents a histological and cellular structure similar to the brain. The notochord is cylindrical throughout its entire length, its tapering anterior end reaches up to the infundibulum. The hypochorda is visible beyond and extends from the level of the deuterencephalon backwards and like the notochord it merges into the cellular mass of the tail bud. The roof of the oral cavity starts to thin out.

Stage 24

The tail projection has attained a length equivalent to about half the diameter of the yolksac (Fig. 10A,B; Pl. II. 24). The trunk and the tail become laterally flattened and the thinning of the dorsal border shows the first trace of the embryonic fin-fold. The external gills become elliptical in shape and are positioned anteroventrally-posterodorsally. The epidermal ciliated cells increase in density. The paired knobs of the attachment glands have slightly separated from the yolksac epidermis. A conspicuous characteristic of the neural anlage at this late stage is the ventral bend of the brain to form a right angle (Fig. 10C). At this basikyphosis the lateral constriction of the prospective diencephalic region of the brain vesicle begins, which is accentuated by the outgrowth of the fore- and hindbrain. The anlage of the epiphysis is visible as a slight swelling at the roof of the prosencephalon. Ventrally, across the anlage of the epiphysis, the first sign of the chiasma opticum and the anlage of the neurohypophysis, as a dome-shaped thickening at the bottom of the prosencephalon, becomes visible.

Over its entire length the ventrolateral walls of the neural tube are thicker than the dorsal walls. Accordingly, the canal of the neural tube is triangular in cross section with a dorsal base. In the tailbud region it becomes vertically cleft-shaped. Towards the deuterencephalic region, the dorsal wall enlarges and forms a thin triangular roof, the future tela choroidea of the rhombencephalon. The walls of the prosencephalon have thickened dorsolaterally and ventrolaterally to such a degree that the lumen becomes cross-shaped. The telencephalon of Polypteridae develops in a similar manner as in the other Actinopterygii, and in contrast to Sarcopterygii (Nieuwenhuys *et al.*, 1969; Senn, 1976; Nieuwenhuys, 1982) by thickening of the lateral walls and eversion, maintaining a thin tela telencephali. This, however, will first become obvious in the following stages.

At this stage, 10 to 12 somites are counted in sagittal section. The

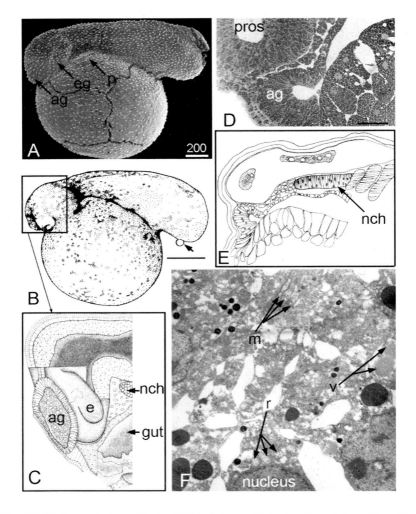

Fig. 10. Embryo at stage 24. A—SEM of *P. ornatipinnis*, viewed from the side, showing the elliptical anlage of the external gill. (Yolk-sac fractured). B-F—*P. senegalus*. B—Drawing of embryo. Arrow points to a remnant of the yolk-plug. C—Graphic reconstruction of head region of B from cross sections (0.7 µm), slightly diagrammatically represented. Note the ventral bend of the brain at a right angle, the eye, and the conspicuous attachment gland. D-E—LM and corresponding graphic survey of the internal development of the foregut and attachment gland. F—TEM (Transmission electronmicrograph) (200 ηm) of the attachment gland of specimen in B, magnified 3000x. ag—attachment gland, e—eye, eg—external gill, m—mitochondria, nch—notochord, p—pronephros body, pros—prosencephalon, r—ribosome, v—vesicle. Scale bar: A—in µm, B—0.5 mm, D—0.1 mm.

anlage of the attachment glands becomes a rounded vesicle (Fig. 10D,E), which, in some specimens of this stage already has separated from the entodermal head gut. It is by now integrated into the ectoderm and subepidermal layer of the forehead, but still is covered by a thin epidermis.

Apparently, the production of the mucous secretion has started already. The thicker part of the club-shaped cells of the anlage of the attachment glands is loaded with yolk in contrast to the remaining hyaline neck part. The latter is filled with densely packed vesicles that appear dark red colored by PAS-staining. The electron microscopic study of a specimen of this stage has permitted a more exact determination of the cellular content: The cytoplasm of the neck part contains a higher concentration of organelles, i.e., many packed mitochondria with widely spaced cristae, endoplasmatic reticulum and several free ribosomes as well as the afore mentioned vesicles (Fig. 10F). Their content appears as a greyish, unstructured matter, much less electron-dense than yolk-globules. The olfactory organ develops from a subepidermal thickening of the embryonic forehead and constitutes a distinctive placode at stage 24.

Stage 25

The embryonic body becomes more flattened laterally. Due to outgrowth the embryonic body has increased in total length and has reached about twice the diameter of the yolksac (Fig. 11A; Pl. II. 25a,b). Large-sized melanophores have spread over the entire embryo, but are most dense in the body region. The anlagen of the external gills have extended and are bud-shaped. The translucent membraneous embryonic finfold appears at the dorsal border of the embryonic body and extends over both the body and tail region. In an advanced representative specimen of *P. senegalus* (Pl. II. 25b), the finfold develops caudally, arches around the entire tip of the tail and ventrally ends at the cloacal aperture. About 12 somites have differentiated.

The attachment glands lie in an oblique anteroventral position and are still closed at their rounded distal end. During this stage they differentiate more clearly from the head, become more cylindrical in shape, and diverge laterally. In *P. ornatipinnis*, by contrast, the opening of the secretory canal of the attachment glands was already observed in most individuals at this stage (Fig. 11B). This process usually occurs in *P. senegalus* at stage 26. Bundles of cilia of epidermal cells now usually consist of 100 to 120 cilia.

The anterior wall of the gut extends cranially underneath the brain, and forms a diverticulum that approaches the posterior part of the mouthcavity. However, a space still remains between the stomodaeal trough and the entodermal front wall of the gut, which is filled with mesenchyme. Shortly after hatching the preoral mouth trough breaks through into the entodermal lumen.

Stage 26

The unpaired epidermal embryonic finfold is fully formed (Fig. 11C; Pl. II. 26). Its surface has probably become more important for gas exchange and first embryonic movements. The differentiation of somites reflects the growth in length of the embryonic body. In the specimen of *P. senegalus* (Pl. II. 26, 3.1 mm TL), 12 somites are externally visible. However, 24 differentiated somites are counted in sagittal section. Large-sized melanophores appear for the first time in the finfold, particularly in the posterior part. From this stage onwards, in the posterior brain region, the tectum of the rhombencephalon becomes visible dorsally as a thin, elevated, translucent sheet, below which the characteristic shape of the tegmentum is recognized. The cloacal tube starts to form at the cloacal aperture and traverses the ventral finfold. Early in stage 26 of *P. senegalus*, the epidermis still remains intact at the distal end of the attachment glands and only a slight depression is observed. In somewhat older embryos, either a piece or cells of the epidermis may disappear at the central part of the organ (Fig. 11B) as is the case in stage 25 in *P. ornatipinnis*. Thereby, the subepidermal layer, formed in preceding stages, becomes exposed. This probably involves a process of thinning and degeneration of the apical epidermis. The subepidermal layer, the anterior wall of the vesicle of the attachment gland, becomes deepened. The anterior wall then starts to invaginate into the vesicular cavity, thereby forming a double-walled cup and in further course

---

Fig. 11 SEM of embryo, A—*P. ornatipinnis*, stage 25, general side view. External gill (eg) bud-shaped (black arrow), attachment gland anlagen cylindrical in shape (frame). Emerging finfold (white arrow). B—Frontal aspect of left attachment gland (framed in A). Initial opening of the secretory canal by obliteration (degene-ration) of cells of the epidermis (arrowhead). C—*P. ornatipinnis*, stage 26, general side view. Embryonic fin fold (arrow) is fully formed. D—Frontal view of the distal end of an attachment gland of *P. ornatipinnis* in late stage 26. It is wide open externally and the subepidermal tissue starts to give way at the bottom of the secretory canal to form a glandular lumen. E-H—LM of embryo of *P. senegalus*, stage 26. E—Slightly oblique section through head, anterior parts of the trunk, and yolksac. F—Median section through middle part of the trunk. G—Survey of the internal development of the foregut and attachment gland and corresponding graph. H. ag—attachment gland, eg—external gill, epi—epiphysis, fg—foregut, h—heart, inf—infundibulum, nch—notochord, ncr—neural crest cells, oc—ocellum, pros—prosencephalon, Rp—Rathke's pouch, Sc—secretory canal, som—somite. Scale bar in μm except in E, F and G—0.1 mm.

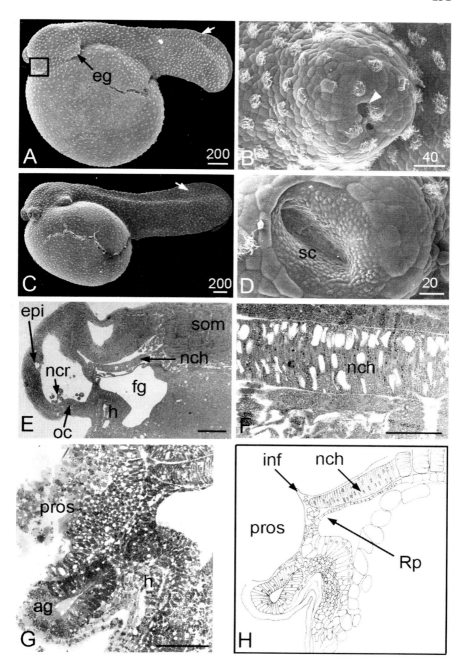

Fig. 11. See previous page for caption

of development the gland acquires a canal- or sac-like shape, the lumen of which, however, no longer communicates with the foregut tube and prechordal mesoderm region. Completion of this incorporation of the attachment organ into the epidermis and external opening is seen at stage 26 of *P. ornatipinnis* (Fig. 11D) or of *P. senegalus* (Fig. 11G,H). Owing to individual variation, the exact time of the extrusion of the secretion cannot be precisely determined.

The arrangement of ciliated cells and bundles of cilia has changed insofar as they now more precisely form an arc-shaped series and mainly run in an anteroventral to posterodorsal direction.

Histologically, the formation of the epiphysis and the optic chiasma are more conspicuous (Fig. 11E). The thinning of the tectum is more pronounced, and for this reason it sinks into the rhombencephalon in fixed specimens. The tegmentum is thicker at its median level than in the remaining parts of the brain. Some vacuoles, filled with a clear liquid, appear inside the inner notochordal cells (Fig. 11F) while at the same time, the number of yolk globules in these cells has diminished.

Stage 27

The tail starts to increase in length. The major variation in length probably is caused by the rather fast growth from this stage onwards (Fig. 12A,B; Pl. II. 27). The yolksac is somewhat reduced in size and has become rather egg-shaped with the narrower pole facing caudally. The externally visible somites number 17 to 22 and they are still confined to the preanal region. They have acquired a slightly concave up to angular shape that indicates the onset of myomere differentiation. The first slow movements of the trunk and tail are now noticed. Later, stronger beats of the tail will result in rotational movements of the embryo inside the egg envelope. Sometimes, they are either spontaneous or elicited by disturbances or positional changes. The attachment glands are wide open externally and the deeply invaginated subepidermal tissue starts to open up at the bottom to form a glandular lumen in *P. senegalus* (comp. Fig. 11D). In some *P. ornatipinnis* specimens formation of the glandular lumen and canal has been already completed. The attachment glands first become functional at stage 27 to 28; and first extrusion of a secretion occurs sometimes before hatching. The external gill rudiment displays up to three constrictions, which later will form primary branches of the external gills (Fig. 12A,B). These are filled interiorly with mesenchyme that contains many yolk globules. The thickening of the subepidermal layer is also conspicuous here. The posterior intestine and cloacal tube are now quite distinct within the ventral finfold. Large melanophores are more densely spaced and form groups. Because of the denser pigmentation the epidermis has lost its transparency in the head region. In contrast, in *P. ornatipinnis* the ground colour is still more light brownish and the translucent skin allows for better observation of vessels. The first pulsations of the heart are observed. At this stage, the peripheral vascular system of the embryos starts to differentiate. A functional yolksac circulation is already in existence and the earliest circulation through this venous network is observed. First, unstained embryonic blood cells are seen floating around the yolksac with the rhythm of the heartbeat and also inside the branches of the external gill anlage (Fig. 12B).

Postembryonic Phase

This phase begins with hatching during stage 28. It involves 8 stages and ends with the earliest onset of exogenous feeding.

Stage 28

This stage demarcates the time of hatching. Differentiation of somites has reached the postanal body region. In most living specimens, 33 to 35 myomeres are visible externally in both species (Pl. II. 28). The somites have enlarged dorsally above the neural tube.

In fixed specimens, however, because of the opaqueness of the body, only 21 to 22 myomeres can be discerned. The hatched postembryos vary from 4 to 4.9 mm in total length. The embryo, narrowly bent inside the egg envelope (Fig. 12A) straightens out immediately upon hatching (Fig. 12D). Timing of hatching usually varies by some hours, even within one cluster of eggs. In *P. ornatipinnis*, hatching occurs 65 to 78 hours after spawning. Approximately the same hatching interval was observed in *P. senegalus*, but at lower mean temperatures. The postanal body region is much elongated and comprises one third of the total length. The head bulges out more rostrally and the attachment glands have become more vertically oriented. Ciliated epidermal cells have become rare at the distal end of the attachment organ (Fig. 12E). The attachment glands are wide open, have a large glandular lumen, and show secretory activity from stage 27 onwards (Fig. 12F). The coloration of the postembryos is dark brown. The body appears to be sunk into the yolksac. Primary branches of external gills are further differentiated and elongated.

The vena caudalis is visible underneath the notochord and dorsal aorta at the inner border of the postanal fin fold. It runs anteriorly from the posterior somites, bends left of the rectal tube, and rostrally becomes the vena subintestinalis, which takes its course along the dorsal margin of the preanal finfold. In the yolksac region the vena subintestinalis gives off many branches that extend over the whole surface to form the yolksac circulation (venae vitellinae). These venae vitellinae unite again into a

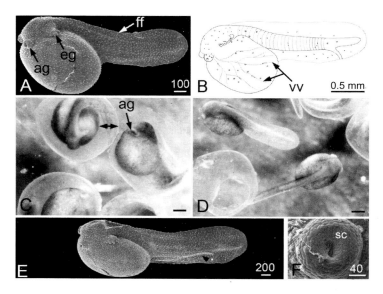

Fig. 12. Embryos of *P. ornatipinnis* A-B—Stage 27 A—SEM, general lateral view which shows two commencing constrictions of the ciliated external gills. B—Sketch giving a survey of the vascular system of the yolksac of C-F—Stage 28. C—LM of living embryos in different states immediately before hatching. Note the enlargement of the perivitelline space (double–headed arrow) and strongly bent embryos inside the egg envelope. D—LM of stage 28, just hatched living postembryos. E—Postembryo just after hatching, general side view. Note the development of the gut-tube (arrowhead). External gill shows a somewhat palmate arrangement of short lobes growing out. F—Detail in frontal view of the left attachment gland in E. Note the wide-open secretory canal (sc) and lumen. ag—attachment gland, eg—external gill, ff—finfold, sc—secretory canal, vv—venae vitellinae. Scale bar in µm except in B, C and D—0.5 mm.

superficial sinus at the anterior border of the yolksac leading towards the heart (sinus venosus). The number of blood cells has increased, which is particularly noticeable in the vessels of the external gills.

Stage 29

The most significant feature of this stage is the external appearance of the position of the pineal organ. It is located underneath a small, unpigmented, diffusely bordered spot on top of the head (Fig. 14G; Pl. II. 29). For the first time the anlage of the eye is externally visible as a darker concentration of pigment above the attachment glands. The total length, which still varies much among specimens, is between 4.9 and 5.8 mm. The number of somites has increased to 43 and 44. The ground pigmentation has become more intense in the head region. Lipophores and guanophores now appear in the head region and external gills and produce a characteristic pigment pattern. Inside the ventral finfold, the posterior digestive tract is well separated from the musculature of the trunk. The attachment glands have lost their divergent orientation and lie closer to the midline and are directed posteroventrally (Fig. 13A).

The depression behind the origin of the attachment glands demarcates the primary nasal pit and beyond the position of the olfactory organ. The ciliated epidermal cells have reached their maximum density particularly in the head region. A deep groove or invagination between head, yolksac, and the paired attachment glands indicates the stomodaeum (Fig. 13A).

The dorsal aorta is now quite visible and directs the blood towards a point close to the last somite of the tail. From here, blood turns around and flows inside the spacious caudal vein and moves more anteriorly through subintestinal and cardinal veins that diverge at the level of the cloacal tube towards the yolksac. The large afferent artery of the external gill (arteria hyoidea), originating at the front of the postembryonic heart, can be seen to curve dorsolaterally towards the stem of the gill branches.

The circulating blood is still colourless. The superficial veins of the yolksac form broad lacunae and direct the blood towards the heart. The entire blood circulation still does not display a continuous flow, but rather follows the rhythm of the heart beat.

Stage 30

The pectoral fin emerges as a small, flattened, semicircular protrusion that stands vertically on the dorsal face of the yolksac in a position parallel to the sagittal axis of the embryo (Fig. 13B). A thin pigment layer has formed in the eye and the lens is faintly visible. The bay of the stomodaeum further invaginates and the epidermis at its bottom starts to give way, thus forming the open communication with the pharynx (Fig. 13C). Primary branches of the external gills have elongated considerably, and beneath the first series more primary gill branches have emerged, (Fig. 13D, arrows).

Again, the body of the postembryo has become more elongated and its total length is between 6.0 and 6.6 mm. The representative specimens depicted in plate II (Pl. II. 30) are three days 12 hrs old (6.1 mm). The myomeres have already reached their final number (*P. senegalus* ca. 54), though this is not always apparent externally. Only 36 to 40 myomeres are visible in living specimens of *P. senegalus* and *P. ornatipinnis*. From this stage onwards further growth in length is accordingly due to myomere growth and more intense folding in a craniocaudal direction.

The afferent and efferent vessels of the external gills have formed and

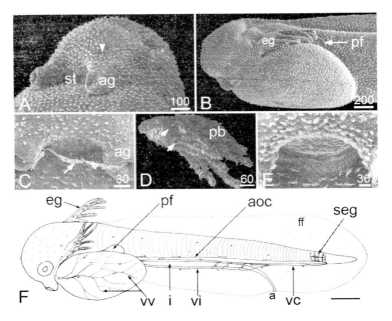

Fig. 13. SEM of postembryos of *P. senegalus* A—Stage 29, close-up, ventrolateral view of the densely ciliated head region showing the recess of the stomodaeum (st). Attachment glands are entirely devoid of ciliated cells. Arrowhead points to the primary nasal pit. B—Stage 30. Lateral view of head and anterior abdominal region. Note the skinny rudiment of the pectoral fin fold behind the external gill. C—Stage 30. Ventral view of the attachment glands and mouth region. Epidermis at the bottom of the stomodaeum starts to give way communi-cating with the pharynx. Attachment glands are devoid of ciliated cells and show secretion (arrow). D—Stage 30. External gill showing primary branches (pb) and first outgrowths of secondary branchlets (arrows). E—Stage 31. Ventral view of the mouth region. Opening of the mouth by lateral extension of the gap at the bottom of the recess. F—Stage 30. Sketch of a survey of the vascular system in *P. ornatipinnis*. a—anus, ag—attachment gland, aoc—aorta caudalis, eg—external gill, ff—fin fold, i—intestine, pb—primary gill branches, pf—pectoral fold, seg—intersegmental vessel, st—stomodaeum, vc—vena caudalis, vi—vena intestinalis, vv—venae vitellinae. Scale bar in µm except in F—0.5 mm.

the blood circulation is observed inside them. In cross section these vessels occupy almost the whole diameter of these branches with little connective tissue investment. The primary branches of the external gills are covered by a thin, monolayered epidermis of flattened cells.

Formation of erythrocytes is now significant and their number has increased to such an extent that individual blood cells are no longer observed. Also, circulation has now become more continuous in the major vessels (vena caudalis, venae cardinales and vena subintestinalis). Only close to the heart, in the aorta dorsalis, inside the network of the yolksac veins and inside the external gills does the heart beat rhythm of the blood flow persist. Intersegmental vessels have formed and are visible in the posterior part of the caudal region between the last somites. Subintestinal vein and cardinal veins are interconnected by anastomoses across the posterior intestine.

Stage 31

The depicted postembryo has reached a total length of 6.5 mm at an age of 5 days 6 hours (Pl. II. 31). The postanal region has reached almost the same length as the abdominal region and the yolksac is further reduced in size. The attachment glands are now in a more oblique position, point backwards, and are in contact with the yolksac. This seems to have been caused by further reduction of the yolksac and a rostral outbulging of the head before jaw formation occurs. There are secondary ramifications of the external gills found at all of the primary branches. The secondary branchlets have little pigment. Some have reached the same length and generally have the same diameter as the primary branches. The pectoral fin fold becomes ellipsoid in shape by dorsocaudal prolongation. The process of the opening of the mouth advances by lateral extension of the gap at the bottom of the mouth cavity (Fig. 13E). More and more of the entodermal cells of the pharynx become visible. Scattered ciliated (and probably ectodermal) cells are also visible inside the oral cavity. Pigmentation and structure of the eye display more contrast and the bright lens is well recognized. There are still many large melanophores irregularly scattered over the head, yolksac and abdominal region. The first elements of the very distinct colour pattern of the head region, characteristic of *P. senegalus*, start to form. Front and top of the head are dark-brownish, except for the bright areas of the pineal spot and an elongate stripe in front of the base of the external gill, above and behind the eye (Pl. II. 31). Pigmentation here mainly consists of guanophores and iridiophores. Below the eye, including the attachment glands, lower cheek, and yolksac area, the embryo is generally brighter.

In *P. ornatipinnis*, the number of differentiated myomeres externally visible is much better and 43 to 44 may be counted at this stage. The total number of somites present, however, appears to be much higher and, as in *P. senegalus*, it is roughly equivalent to the adult myomere count of 64 to 65 in *P. ornatipinnis*. In this species, a reflecting, whitish pigment, probably consisting of evenly scattered lipophores, covers the whole unpaired fin fold. Such pigmentation, in the form of irregularly distributed larger spots, occurs also in *P. senegalus*, but not before stage 32.

In the brain region the neural crest cells multiply quickly. They form distinct condensations of mesenchymatous cell masses that are now surrounded by several concentric layers of cells, the primordia of ganglions and skeletal tissue.

The olfactory sac starts to compartmentalize successively and the diverticles and folds of its epithelium form (see Bjerring, 1988 in a later larva; Pfeiffer, 1968 for the adult situation). At hatching (stage 28), the olfactory organ is still very poorly developed. However, in the larval phase it occupies a large space in the well developed ethmoidal region of the skull. At the juvenile and adult stage the Polypteridae are exceptionally macrosmatic (Pfeiffer, 1969).

Analogous to the formation of the lens and different from the invagination of the superficial epidermal and subepidermal olfactory placode in Sarcopterygii and some other bony fishes (comp. Salensky, 1881; Semon, 1901; Kerr, 1909; Zeiske *et al.*, 1997, 2003; Hansen *et al.*, 1999), the olfactory organ in *Polypterus* originates first by a massive growth of the subepidermis. As a result of cell death and departure of cells a central cavity originates, which becomes connected to the surface at stage 32. The nervus olfactorius, the axon fiber bundles of the olfactory sensory neurons, connect to the brain before and during the early development when the olfactory sac lies very close to the telencephalon.

Along with the increasing number of vacuoles inside the inner notochordal cells, the plasma with the nuclei has shifted to the periphery. These cells become larger, more closely packed and pressurized, and consequently attain a polygonal shape.

The quantity of the yolk globules in the cells of the attachment gland is much diminished and two types of cells can now be distinguished: viz. the basal and the wall cells. Their cell-junctions ('Schlussleisten') close up the glandular epithelium that faces the glandular cavity. The basal cells are short and form the proximal part of the organ. Their oval to roundish nuclei, with a centric nucleolus, lie near the basal lamina and occupy almost the entire width of the cells. These basal cells take up a strong distal stain with the PAS-reaction. Without a doubt, they are derived from the previously described embryonic entodermal cells. In contrast to the basal cells, the wall cells of the attachment glands are long and slender. They appear to arch distally into the lumen of the organ. Most nuclei of these cells also lie basally. Some, however, may be positioned up to the middle of the cell. These cells are uniformly stained by the PAS-reaction, but remain always less intensely stained than the secretory mucous basal cells.

The ciliated epidermal cells have reached their maximum density. The cells bear 200 to 300 cilia, and the cilia are more elongated than in the preceding stages.

Stage 32

The postembryos have attained a total length of 7 to 9.25 mm at this developmental stage.

In representative specimens of *P. senegalus* pigmentation of the head has a richer contrast and the temporal, whitish streak extends above the eye and onto the external gill stem [PL.II. 32; 6 days 6 hrs AF (after fertilization); 7.1 mm TL (total length)]. At this stage, melanophores produce dark blotches inside the pigmented layer of the eye, which in addition appears in general much darker. In stage 32 embryos of *P. ornatipinnis* a reticulate pattern of dark pigment appears on the head, and the pigmented layer of the eye is jet-black. Some large melanophores also appear on the body, somewhat later than in *P. senegalus*. The number of myomeres that can be counted externally reaches 54 to 56. The ciliated cells bear 200 to 300 cilia, and the cilia are more elongated than previously. Their density is reduced in the head region, except for the areas of the eye, and around the primary nasal pit and mouth opening (Fig. 14A).

The attachment glands are now clearly set apart from the front of the yolksac, and are positioned at the level of the eye. They are still functional and are situated close together at the site of the future upper lip. The development of the mouth and labial folds commences lateral to the origin of the attachment glands (Fig. 14A,B). Invagination of the stomodaeum is also progressing and the preoral as well as the oral epidermis of upper and lower jaws start to separate from each other. With growth of the mouth cleft, a small, arch-shaped mouth angle is formed laterally to the attachment glands. Here and in the following stages, the lip folds originate that will extend further cranially on the upper and lower jaws. From the first invagination at stage 28, the single primary nasal pit is formed. This has now deepened and communicates with the cavity of the olfactory placode, the lumen of the primary nasal sac (Fig. 14A,B, arrows).

The opercular flap and opening start to form at the base of the external gill stem. The stem in the external gill has elongated, whereas the primary branches and secondary twigs that partly cover and reach beyond the pectoral fin anlage have not done so. The circularly shaped pectoral finfold develops into a thin (peripheral) margin and an extended (proximal) plate. Its base is still elongate and broadly inserted at the yolksac, whereas its cephalic edges approach the body. The posterior intestine has further differentiated and has undergone spiralization. The small veins of the yolksac have multiplied. The brightly colored red blood flows towards the heart in a slow, rhythmic current. Inside the external gill filaments a rhythmic current is also observed, but with short pauses. In the posterior caudal region, intersegmental vessels are more abundant than when first observed at stage 31. In addition, the first capillary networks have formed inside the posterior dorsal median finfold. The terminal loop of the dorsal aorta and caudal vein is now close to the tip of the notochord.

Stage 33

The paddle-shaped pectoral fin is much extended. Its muscular base has differentiated and is now repositioned close to the external gill base as a result of posterior expansion of the opercular gill cover and regression of the yolksac. It is clearly oriented anterodorsally-posteroventrally (Fig. 14D; Pl. II. 33). The first active respiratory movements are now observed.

The attachment glands start to atrophy and their openings contract, concomitantly with the formation of the labial folds. At this stage, there is only a remnant of the attachment glands present on the upper lip at the level of the anterior margin of the eyes. Their cells become thinner, but still contain some yolk globules. A black pigment accumulates between the cells of the reduced attachment organ,. The labial folds at the margin of the mouth are distinctly formed at this stage and have a simple angular shape.

The primary nasal openings start to expand anteriorly and posteriorly and the dorsal and ventral margins approach each other around the middle of the elongated nasal slit to form anterior and posterior nasal openings (Figs. 14D, 15B).

Fig. 14. Postembryos. A-C—SEM of *P. ornatipinnis*, stage 32 A—Lateral view of head and abdominal region, conspicuous external gill which shows the formation of the secondary branches in the upper primary branches, primary nasal pit (arrow), mouth angle (arrowhead), and regressive attachment gland (ag). B—Ventral view of head and anterior abdominal region. Note the mouth cleft as a small arch-shaped mouth angle (arrowhead). Arrow points to the primary nostril. Regression of the attachment glands with secretion (double-arrows). C—magnified primary nostril pointed at by the arrow in B. D-G—*P. senegalus* of stage 33 D—SEM showing lateral surface view of head and anterior abdominal region with lateral line organs (black and white dots) E—SEM of lateral surface of posterior part of the body and tail of the specimen in D. F—LM. Whole lateral view of living specimen short time before onset of extrinsic feeding. G—Same specimen as in F. Dorsal view of head and anterior parts of the body. H—Sketch giving a survey of the vascular system in *P. senegalus.* ag—attachment gland, e—eye, eg—external gill, eth—ethmoidal commissure, ff—fin fold, io—infraorbital lateral line, m—mandibular lateral line, mpl—mandibular pit line, opf—opercular fold, pf—pectoral fin, po—preopercular lateral line, ps—pineal spot, so—supraorbital lateral line, st—stomodaeum, ur—urostyle, vf—vessels of the fin-fold, vpl—vertical pit line. Close above the black and white point signatures lie the lateral line neuromast organs and pit line organs, respectively. Scale bar in µm except in F, G and H—1 mm.

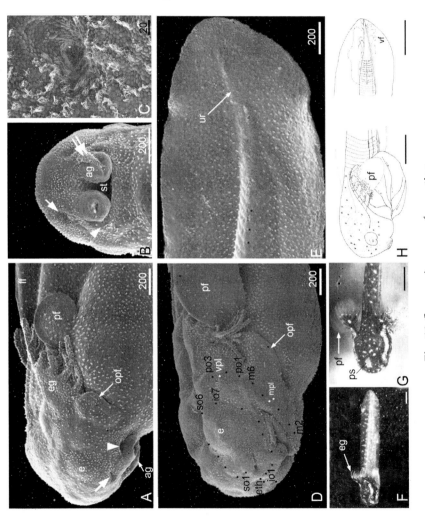

Fig. 14. See previous page for caption

The specimen of *P. senegalus* illustrated in Plate II. 33 is approximately 7.8 mm in length and is aged 7 d 6 hrs. The posterior end of the tail has begun to arch upwards, as is characteristic of this stage, i.e. "flexion stage" (Fig 14E). The opercular gill cover is translucent with a distinct pigment pattern of melanophores. There is now a remarkable, sharply defined, bright colour pattern of the head. A whitish stripe that extends from the upper jaw, runs closely above the eye and reaches back into the main stem of the external gills (Fig. 14F,G). The translucent pineal spot on top of the front of the head is distinct, but other parts of the upper head are rendered dark by densely spaced melanophores. In this specimen, many guanophores that form reflective spots on the fin folds and body are apparent (Fig. 14E). As compared to the previous stage, the ventral fin fold is diminished in height. The general epidermal ciliation has changed from its formerly more even distribution. In the head region, in particular, cilia are absent in the area closely surrounding the neuromast organs of the lateral line system; however, they remain in between. The density of ciliation is exceptionally high around the nasal opening, on the eyes, on the labial folds, the external gills and on the isthmus region of the gill cover (Fig. 14D). At this stage, all neuromast organs of the head are present (compare the chapter on the lateral line system of *P. senegalus*).

The number of myomeres in *P. ornatipinnis* is either 64 or 65. The colour pattern of *P. ornatipinnis* is distinctly different from the emerging pattern characteristic of *P. senegalus* as described above. Postembryos of *P. ornatipinnis* display only a mottled or reticulate pattern of greyish to brownish coloration with scattered asterisks of dark melanophores but no differentiation into larger spots or stripes.

The spiral-shaped convolution of the intestine and differentiation of the region of the stomach continues through stage 33. The veins of the yolksac are less abundant, but they have formed larger vessels that seem to penetrate more deeply into the yolk. For the first time capillary networks emerge inside the ventral caudal fin fold beneath the urostylar end of the tail (Fig. 14H). The blood stream has become more continuous here.

Stage 34

The head has elongated considerably. The events associated with the opening of the mouth, the formation of the nasal openings and regression of the attachment glands are almost finished. Inside the oral opening, splitting of the stomodaeal membrane has progressed further. The pattern of ciliation is modified in comparison to the preceding stage. A large area around the neuromast organs is devoid of cilia and no cilia remain between the neuromast organs, which is the prospective area of the formation of the lateral line canal. In addition, after the emergence of pit lines, the ciliated fields are interrupted in the respective preopercular and gular regions. It is generally observed that interruption of the ciliation appears synchronously with the appearance of additional neuromast organs of the main lateral line and pit lines. Ciliation remains unmodified up to stage 35: it is dense in the orbital, outer border of the primary nasal opening as well as inside the duct and gular regions (Fig. 15A,B,C).

Development of the opercular membrane and slit has also proceeded and the opercular folds are attached only far cranially to the gular region, which results in the openings also reaching far cranially to the isthmic region (Fig. 15A). In this stage active breathing movements by rhythmic buccal expansions and movements of the gill cover are first observed. Pronounced constriction of the nasal slit has resulted in a ∞-shape of the nasal opening (Fig. 15B). The labial folds of the mouth have differentiated much further, extended cranially, and are deeply embedded into labial furrows (Fig. 15A,B). The mouth is fully opened. Vomeral and coronoid teeth and one line of teeth of the outer dental arcades have formed (Fig. 15C, cf. Wacker *et al.*, 2001). The posterior end of the tail now becomes pointed (compare Fig. 15F).

Externally, the pectoral fins have finally differentiated into a muscular peduncle and a flat distal fin fold (Pl. II. 34). Internally, this coincides with formation of the skeletal support, viz. cartilaginous basals in the shape of a proximal plate and the actinotrichs distally inside the portion of the finfold.

In the specimen of *P. senegalus* illustrated in Plate II. 34 (8 days and 6 hrs AF, 8.6 mm TL), the yolksac has regressed to a small remnant that bulges out only a little and is visible as a greenish mass of yolk through the whitish, translucent skin of the belly.

Because of the generally dark ground colour of the embryos, most vessels are not very well visible in the head and trunk regions. However, the heart and its compartments (atrium, ventricle, and bulbus cordis) are quite well visible behind the transparent operculum. Also, the prominent hyoid artery that first was apparent at stage 31 and forms the afferent and efferent vessels of the external gills, is still seen inside the opercular fold. Inside the finfold of the caudal region additional vessels form an extended capillary network.

Stage 35

This developmental stage concludes the embryonic period. One of the main characteristics of this stage is the definitive separation of the primary nostril into anterior incurrent and posterior excurrent nasal openings (Fig.

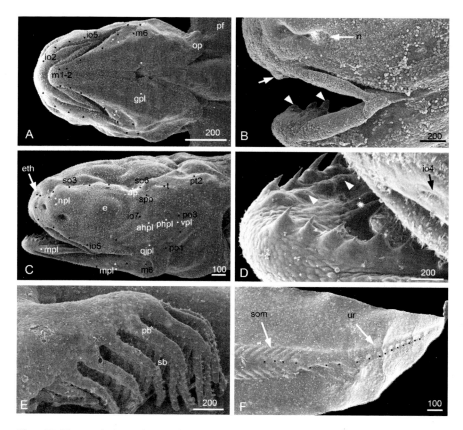

Fig. 15. Postembryos. SEM of *P. senegalus*. A-B—stage 34. A—ventral view of the head. B—Lateral view of the nasal and mouth regions. Note the emergence of the primary teeth of the lower jaw indicated by arrowheads and the small remnant of the attachment glands (arrow). C-F—Stage 35. C—Laterodorsal view of the head. D—Anteriormost part of the lower jaw with developed teeth. The arrowheads indicate the onset of the second group of primary teeth. Asterisk: coronoid fangs. E—Dorsolateral view of the external gill, which has reached its maximal development. F—Lateral surface of the posterior part of body and tail. ahpl—anterior horizontal pit line, ap—anterior pit line, e—eye, eth—ethmoidal commissure, gpl—gular pit line, io—infraorbital lateral line, m—mandibular lateral line, mpl—mandibular pit lines, n—nostril, op—opercular fold, pb—primary branches, pf—pectoral fin, phpl—posterior horizontal pit line, po—preopercular lateral line, qjpl—quadratojugal pit line, sb—secondary branches, so—supraorbital lateral line, som—somite, t—temporal lateral line, pt—posttemporal lateral line, ur—urostyle, vpl—vertical pit line. Close above black and white point signatures in C and F, lie the lateral line neuromast organs and pit line organs, respectively. Scale bar in µm.

15C). The pectoral fins have acquired a large surface area and the extended opercular folds cover their peduncular bases. The endoskeleton of the pectoral fin and pectoral girdle starts to become cartilaginous. First it consists of a continuous roundish plate with a narrow, proximal articulatory region and thickened margins; pro- and metapterygial elements derive from the latter (cf. Budgett, 1902; Kerr, 1907; Braus, 1909; Bartsch and Gemballa, 1992).

The postembryo of *P. senegalus* illustrated in Pl. II. 35 (9days 6hrs AF; 9.2 mm TL) demonstrates another characteristic of the species, viz. the dark melanophore streak on the cheek is continued forward of the eye in the ethmoidal region to the tip of the snout. This coincides with a further preorbital elongation of the head.

More than 50 myomeres are formed in stage 35 of *P. senegalus* and the number characteristic of the adult has almost been reached. The lateral plates have met ventrally and form a slit-like coelom between the splanchnopleura of the inner organs and the somatopleura. The hypaxonic expansion of the body musculature, started at stage 32, has been completed and the intestine is finally covered by the myomeres both laterally and ventrally, instead of running along the base of the preanal fin fold. Development of the stomach and associated glands of the intestine results in a shift of the remaining yolk toward the right side. At the end of the anal or cloacal tube, a small indentation of the finfold occurs. The cloacal tube is short and is now more vertically oriented. The formation of the sheath of the notochord is visible as a faint external envelope. In this stage, also the cartilaginous neural arches start to form. Towards the end of the postembryonic phase, the ventrolateral regions of the neural chord start to differentiate into the substantia alba, the compacted axons, as contrasted to the substantia grisea, the region of the perikaria of the motorneurons.

In the eye, the development of the retina has been completed. The cornea and the iris are now completely formed. With respect to the observed behaviour (reaction towards mobile objects), one can conclude that the sense of vision is fully functional now. There is a long time period between the onset of formation of the eye cup in stage 24 and the first sign of retinal differentiation in *P. senegalus*. This results in hatched embryos (stage 28) that only have a faint retinal layer formation. The rods and the cones of the retina will be formed first, after the step of yolksac larvae attachment changes to the free swimming step. The relatively slow development of the eye in hatched *P. senegalus* (and *P. ornatipinnis*) should be assessed in relation to the prolonged passive developmental phase outside the egg envelope. Only during the passive attachment step does the eye advance to a functional stage immediately before the onset of the initial extrinsic feeding behaviour.

On the Development of the Lateral Line
System during Stages 33 to 35

Arrangement of the neuromast organs in the head region becomes almost complete during stage 33. Therefore, it deserves some more detailed description. The main lateral lines are dealt with first, followed by that of pit-line arrangements.

The supraorbital lateral line, which consists of six neuromast organs in stage 33, corresponds already to the number observed in later larvae and juveniles (Fig. 14D). Moreover, the seven neuromast organs of the juvenile infraorbital line have formed. The most anterior infraorbital neuromast organ lies at the level of the first supraorbital neuromast organ, immediately beneath the olfactory opening. With prolongation of the snout, the first forming neuromast organ of the supraorbital lateral line is shifted rostrally in stage 34, so that it forms the anteriormost organ of the line (Fig. 15C). Likewise, the most rostral neuromast organ of the infraorbital lateral line is displaced anteriorly to the ∞-shaped olfactory opening.

The ethmoidal line consists of one neuromast organ on each side that appear ventrally and medially of the first organ of the supraorbital line in stage 33 (Fig. 14D). It remains in the same relative position, located anteriorly and medially above the anterior nasal tube. This organ has been termed by Pehrson (1947, 1958) as an anteriormost infraorbital neuromast organ, which is correct as judged by the innervation trough of the lower branch of the upper lateral line nerve or ramus buccalis. Together with its pendant on the other side it forms the so-called ethmoidal commissure around the rostrum or tip of the snout, which is a very constant component of the head lateral line system in osteognathostomes. Piotrowski and Northcutt (1996) have termed another organ, the infraorbital organ, immediately posteroventral of the nasal tentacle in the juvenile, 'ethmoidal neuromast organ' (Piotrowski and Northcutt, 1996, p. 65, fig. 5). At stage 33, the mandibular neuromast line is complete with six organs, an arrangement that remains the same up to stage 34 (Fig. 14D; compare with Fig. 15A). A dorsal prolongation of the mandibular lateral line is also formed by the preopercular line of three neuromast organs that are quite constant in position up to stage 35 (Fig. 14D; compare with Fig. 15C).

At stage 33, the main lateral line of the trunk comprises 18 neuromast organs, a little above the horizontal septum. The posteriormost one is positioned at the level of the fourth myomere as counted from the urostyle (Fig. 14E). Between the older neuromast organs present in the preceding stage, seven additional organs form at stage 34. Moreover, the main lateral line is prolonged posteriorly in the caudal region by the addition of thirteen more neuromast organs. Among the latter, the three most anterior ones are obliquely arranged in an anterodorsal to posteroventral direction and the remaining ones follow close to each other, immediately beneath and upwards along the urostyle, thus reflecting the upward curve and asymmetry of the urostylar region of the caudal fin. At stage 35, the main neuromast line of the trunk has increased in number to 42 organs. 29 of these are located in the trunk region and 13 along the urostyle (Fig. 15F). The location of the latter at the ventral side of the posterior caudal peduncle is characteristic of many actinopterygian groups that show an epicercal caudal fin (Chondrostei, for example: See Bartsch, 1988; Grande and Bemis, 1991).

In juveniles (three months old), the mandibular canals on both sides have become interconnected in the symphysial region. Externally, only one median pore is visible here. This seems to be the usual feature in adult polypterids (see Allis, 1922), but is neither shown in Webb and Northcutt (1991) nor in Piotrowski and Northcutt (1996).

In some specimens of stage 35 the additional, dorsolateral line now extends over two thirds of the abdominal region and consists of 12 neuromast organs. A third, dorsalmost line, just below the dorsal fin insertions, is formed only during later larval stages. All neuromast organs of the trunk in polypterids are located in grooves and are not enclosed in canals as are the headlines (Müller, 1844; Allis, 1922).

The vertical pit lines and the so-called mandibular pit lines are the only pit-organs to appear at stage 33 (Fig. 14D; compare with Fig. 15C). The latter consists of three pit-organs, one of which is situated far anteriorly at the border of the lower lip and appears later in stage 35. The other two become a continuous angular row of pits in the juveniles.

The sphenotic and temporal lateral line organs appear first at stage 34. The number of the neuromast organs in most lines now corresponds to the number of canal pores of the juveniles. Two neuromast organs form the transversal or occipital lateral line as a medial commissure in the post-temporal region. In addition, three post-temporal organs develop that continue the temporal and otic head line towards the lateral line of the trunk.

The two pit-organs of the nasal pit line lie almost above one another and beneath the second neuromast organ of the supraorbital lateral line. This position which is quite constant is maintained in juveniles. The anterior pit-line consists of one pit organ situated in front of the sixth supraorbital neuromast organ. The middle and posterior pit-lines are externally visible at stage 34. The quadratojugal pit line consists of one pit-organ situated above and close to the sixth mandibular neuromast organ. The two horizontal pit-lines, the anterohorizontal pit-line and posterohorizontal pit-line, each consists of one pit-organ at first. In juveniles, the middle and posterior pit-lines are fused and form a V-shape structure rostrally to the second preopercular neuromast organ. The gular pit-line is the last to appear at this stage.

Larval Phase

The larval period includes the onset of exogenous feeding, differentiation of the median and abdominal fins, scale formation, and ends with the complete regression of the fin-fold. In *Polypterus senegalus* a change from the larval to the juvenile and adult coloration marks the end of this period. There is a slower, continuous transition of larval to juvenile and adult colour pattern in *P. ornatipinnis*.

Stage 36

This stage corresponds to the onset of the larval period (Pl. II. 36; 11days 20 hrs AF; 9.8 mm TL). It is characterized by the transition from endogenous nutrition by the supply of yolk to extrinsic feeding. Functional teeth have already emerged before and yolk is resorbed completely or reduced only to a tiny remnant. The attachment glands have completely disappeared. All somites of the preanal region are arranged in the typical inverted Σ-shape.

Survey of Post-staging Larval Development

Processes that are characteristic of this period involve the differentiation of the unpaired and abdominal fins, formation of squamation, inclusion of head neuromast organs into canals, reduction of external gills, differentiation of lungs and transition to atmospheric air breathing, and change to the juvenile and adult coloration.

The borderline between the dorsal and caudal fin is apparent in 18 to 20 days old specimens, when the formation of the skeleton of the dorsal fin starts (Fig. 16A). This proceeds in a posteroanterior direction. The first caudal lepidotrichia are already formed when the preanal finfold starts to regress, but the major part of the dorsal larval finfold remains intact. In addition, the extended urostylar notochord produces a pointed and somewhat upwardly directed outgrowth that demarcates the border between the dorsal and the caudal divisions of the fin-fold now, and of the composite caudal fin of later stages (cf. Budgett, 1902; Schmalhausen, 1913; Sewertzoff, 1924; Bartsch *et al.*, 1997).

The larvae, at first rather passive ambush hunters of small planktonic organisms, subsequently become more active in searching for prey and become strongly aggressive toward each other. Bite wounds inflicted on

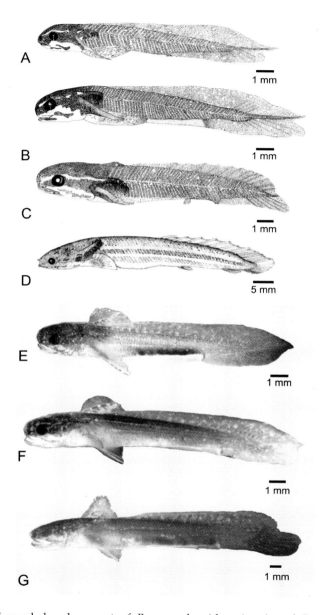

Fig. 16. Larval development of *P. senegalus* (drawings) and *P. ornatipinnis* (photographs). A, E—Early finfold larva (14 and 15 days, respectively). B, F—Later finfold larva (20 and 18 days, respectively). C, G—Pterolarva (25 days). Note: the elongate caudal finrays; the emerging pelvic fin. D—Transition from larval to juvenile phase in *P. senegalus* (60 days).

the fins and external gills are frequent in larvae 25 days and older (Fig. 16B). The late fin-fold larvae are often observed swimming slowly and are intermittently propelled by a whirling undulation of their pectoral fins. The pectoral fins are also used as supports for the elevated anterior body while the larvae are resting on the ground.

In the 25 day old larva the hypochordal lepidotrichia have grown much longer than the posteriormost dorsal lepidotrichia (Fig. 16C,G). Consequently, the original caudal compartment is still externally visible as an elongated tail. The translucent pineal spot on top of the head has disappeared. A white, longitudinal streak is conspicuous along the lateral line of the body. The preanal finfold has completely disappeared. Small paired lobes, the pelvic fins, emerge rostrally to the aperture of the anus. The formation of 13 dorsal spines is attained in *P. senegalus*, the number of which will remain constant in the adult specimens. Nasal tentacles have formed as a tubular projection of the anterior incurrent nasal opening (Fig. 16D). Elongation of the head has become conspicuous and the preorbital region forms a pronounced snout. Remnants of the ventral fin-fold regress as the anal fin forms. The prolonged urostyle is no longer externally visible in late larvae. In *Polypterus senegalus* an additional narrow longitudinal band becomes visible dorsolaterally on the body. It extends parallel to the early-formed streak of the lateral line, before the marked change into the grey juvenile coloration occurs. These colour patterns correspond in position to the dorsolateral and lateral rows of neuromast organs of the body respectively. In *Polypterus ornatipinnis* a rather continuous transition from the dusky larval coloration with scattered bright spots towards the juvenile and adult reticulate black and white pattern is observed. In both species, external gills in variable form may last well into the juvenile period.

DISCUSSION

The Egg and Egg Envelope

Polypterid eggs do not differ from the majority of other actinopterygian eggs in their general construction,. They are covered and protected by a noncellular envelope, the zona radiata. In *Acipenser ruthenus*, however, there are several micropyles present, usually 7 (Salensky, 1881). Mature eggs of paddlefish *Polyodon spathula* possess four to 12 micropyles at the animal pole (Linhart and Kudo, 1997). Several micropyles also have been described in other sturgeon species (Cherr and Clark, 1982, 1985). Spawned eggs of acipenserids possess an additional jelly coat similar to that of amphibian eggs (Cherr and Clark, 1982). The egg of *Lepisosteus* has only one micropyle and possesses attachment structures on the surface (Mark, 1890). Also the eggs of teleosteans invariably possess only one micropyle (Riehl, 1991).

Just after the penetration oft the first sperm in *Polypterus* the *"plasmapfropf"*, the spherical plasmatic plug, mechanically closes the micropylar canal. Closure of the micropyle inhibits sperm and external pathogens from penetrating into the perivitelline space and seems to be involved both in the permanent prevention of polyspermy and protection of the developing embryo from bacterial infection. The plasmatic plug is a clear, external indication of successful insemination. The closure of the micropylar canal is widespread among teleostean fish (Kudo, 1980; Iwamatsu *et al.*, 1991; Yamamoto and Kobayashi, 1992) in contrast to e.g. trout, salmon and herring eggs, in which polyspermy is known (Yanagimachi *et al.*, 1992). Of interest with respect to phylogeny and function in *Polypterus* and *Erpetoichthys* are also the attachment structures described for the eggs. These structures are common and well-known and often used for the identification of species in teleostean eggs (see e.g. Wickler, 1956; Arndt, 1960; Götting, 1964; Riehl, 1978, 1991, 1996; Laale, 1980; Groot and Alderdice, 1984; Johnson and Werner, 1986; Mooi, 1990; Patzner and Riehl, 1992; Britz *et al.*, 1995; Britz, 1997; Riehl and Patzner, 1998; Chen *et al.*, 1999 for overview of different structures and groups). Non-structured gelatinous envelopes are found not only in Sarcopterygii, but also in *Acipenser* (Cherr and Clark, 1982, 1985) among Chondrostei and e.g. *Silurus glanis* among Teleostei (Ostariophysi, Siluriformes). It was demonstrated in both of the *Polypterus*-species investigated that the degenerated follicle cells remain as a component of the zona radiata and form the mucopolysaccharid matter in contact with water. Furthermore, we were able to prove that there are specialized follicle cells responsible for the formation of the sperm canal of the zona radiata. This is known in some Teleostei (Riehl, 1991), for the egg and follicle of *Lepisosteus* (Mark, 1890) and, spectacularly enough, in Myxinoidea (Morisawa, 1999). Micropyles are not known to occur in sarcopterygians or other Gnathostomata and thus offer an additional, potential synapomorphy of Actinopterygii (Bartsch and Britz, 1997).

Cleavage Patterns and Gastrulation

The early pattern of development is characterized by a total unequal cleavage of the egg that resembles cleavage of amphibian and dipnoan eggs among Sarcopterygii and those of Chondrostei among Actinopterygii (compare, e.g., with Semon, 1901; Budgett in Kerr, 1907; Kerr, 1909; Bartsch *et al.*, 1997). Formation of the germ is not restricted to a blastodisc as is the case in teleostean eggs (e.g., Kupffer, 1868; McElmann and Balon, 1979; Shardo, 1995; Meijide and Gerrero, 2000; Horie *et al.*, 2002; Arezo *et al.*, 2005). Among other "primitive" actinopterygians, e.g., *Lepisosteus*, meroblastic, partial cleavage, germ-disk formation and overgrowth of the yolk mass by the germ-disk edge (epiboly) occur. Moreover, at the posterior pole of the bilaterally symmetrical embryonic shield a distinct gastroporal lip is found, around which the prospective mesodermal cells become involuted (compare Mark, 1890; Dean, 1895a; Long and Wourms, 1991; Long and Ballard, 2001). In *Amia calva*, by contrast, an extremely unequal cleavage produces several giant yolk cells at the vegetative pole (compare Dean, 1896; Eyclesheymer and Wilson, 1906; Ballard, 1986a,b). Gastrulation in *Polypterus* follows the same principles as the amphibian gastrula and the pattern found in other sarcopterygians, i.e., there is an involution of the prospective mesoderm around an upper blastopore lip (see, e.g., Goette, 1875; Semon, 1901; Kerr, 1909; Tahara, 1959; Harrison, 1969; Keller, 1980; Kemp, 1982). Initial gastrulation in *Polypterus* is typically associated with apical constriction and formation of bottle cells in the region of the blastopore slit. The mesodermal cells lose their epithelial arrangement and initially form a rather unstructured mesodermal cell mass inside the blastocoel (Wourms and Bartsch, 1998). Eggs of sturgeon and paddlefish possess also a holoblastic cleavage pattern and a very similar type of gastrulation (Salensky, 1881; Dean, 1895a; Sawadski, 1925a,b; Dettlaff and Ginsburg, 1954; Ballard and Needham, 1964; Bemis and Grande, 1992).

It is evident that the illustration and concept of an equatorial, third cleavage in polypterid eggs depicted by Budgett in Kerr (1907) is not correct in the two *Polypterus* species investigated here. Moreover, it has been observed that the smallest blastomeres are situated centrally at the animal pole only after a fourth cleavage in both species, whereas prior to this there are two synchronous meridional or vertical divisions. Subsequently asynchronous and irregular cleavages take place, but with a significant delay in the yolk-rich vegetal region. As demonstrated, early cleavages may show some intra- and interspecific variation.

Dermal and Epidermal Structures: Neurulation, Subepidermal Placodes, Ciliation, Pigmentation

The neural tube arises relatively slowly during neurulation as an infolding of the neuroectoderm and fusion of the approaching neural folds, whereas the migration of the neural crest cells in the head region occurs very rapidly and early, long before the end of neurulation. Shortly after the migration of neural crest cells large melanocytes are visible on the embryonic anlage and yolk-sac surface. Olfactory and otic placodes develop very early. First, they originate as a solid cell mass only from a subepidermal layer and form the cavity of the corresponding organ with its sensory epithelium (nasal sac, ear and labyrinth vesicles) beneath the upper epidermal layer, before opening to the external environment *via* primary nasal pits or ductus endolymphatici, respectively. A similar mode of formation occurs in the

Neopterygii and Teleostei (cf. Dogiel, 1887; Gawrilenko, 1910; Verraes, 1976; Hansen and Zeiske, 1993, 1995; Zeiske *et al.*, 1996, 1997; Arvedlund and Takemara, 2006).

At stage 20 ciliated cells are distributed over the entire germ, except for the dorsal area of the neural plate. The movements of the bundles of small cilia produce a water-current inside the egg membrane. The density of the cilia is particularly high around the nasal pits, edges of the mouth, eyes, and external gills. In juveniles and adults, ciliated areas are retained and have a ventilatory function, e.g., around the openings of the lateral line canal (Webb and Northcutt, 1991) and in the respiratory epithelium of the nasal sac (Pfeiffer, 1968; Theisen, 1970; Verraes, 1976; Bjerring, 1988).

Species-specific differences in pigmentation are found even in the egg cell. The assumption, however, that the structure of the surface of the zona radiata or vitelline membrane can always be distinguished (Britz and Bartsch, 1998), has not been clearly confirmed in the two species, because the attachment filaments change quickly after spawning and contact with water and environmental structures.

Attachment Organs

The postembryos possess paired complex attachment glands at the upper lip region in the anterior part of the head ("*Klebdrüsen*", "*Haftorgane*" "attachment organs", "cement glands"). The structure and function of these postembryonic or early larval organs have attracted interest of many researchers in the past (Kerr, 1907 for polypterids; Lieberkind, 1932; Schindler, 1935 and Jones, 1937 for teleostean fishes).

The hatched postembryos are capable of axial locomotion as soon as the myomeres have differentiated and show escape and avoidance reactions upon massive disturbances. Prior to the colour change, eye and pectoral fin development in late postembryonic stages, the yolksac larvae, are rather inactive and remain attached to plants and objects above the ground. Camouflage, avoidance of predation and anoxic conditions close to the bottom and also protection against drift in currents of rain affecting flat-water zones, can be identified as biological roles of these organs. The attachment glands are open externally before hatching and their cells become active in secretion. A peculiarity of polypterids is that the anlagen of these organs develop as diverticulae of the foregut, secondarily integrated into the epidermal region of the upper lip. Kerr (1907) proposed this process of formation, but on the basis of a very widely separated staging series. The PAS-staining and the TEM-investigations have proven that from stage 28 onward these cells contain secretory matter, mostly mucopolysaccharides. Close similarity exists to the unpaired, large, frontal attachment organ fields in *Lepisosteus* and *Amia calva* (Agassiz, 1879; Dean, 1895a, 1896;

Eyclesheymer and Wilson, 1908; Reighard and Phelps, 1908; Lindahl, 1944). In these species a participation of "mesoblast material" was also demonstrated. The attachment glands in dipnoan and anuran larvae are comparable (Kerr, 1909; Lieberkind, 1937; Brien and Boullion, 1959; Picard, 1975). However, they are always situated ventrally, beyond the mouth opening. The paired rostrolateral attachment glands in *Esox lucius* (Nüesch, 1958; Braum *et al.*, 1996) and *Hepsetus odoe* (Budgett in Kerr, 1907), among teleosteans, are organs that display a comparable complex formation. As far as known, attachment glands in Urodela and Teleostei are small, purely epidermal formations at different positions of the head or trunk (Lieberkind, 1932, 1937; Ilg, 1952; Peters, 1965; Britz *et al.*, 2000; Martinez-Palacios and Ross, 2004).

In Chondrostei, presence of attachment glands has not been unequivocally proven. Interpretations (Salensky, 1881) are due to confusion of the barbel anlage with gland fields of the hatching gland cells (for *Polyodon spathula*, see Bemis and Grande, 1992). The latter are found also in the epidermis of the embryos of polypterids, where they are diffusely distributed over the entire epidermis, and are particularly dense in the anterior region.

Glandular cells of this type and enzymatic destruction of the vitelline membrane are widely distributed among lower vertebrates (Wunder, 1935; Willmess and Denucé, 1973; Hagenmaier, 1974; Yoshizaki, 1979; Yoshizaki *et al.*, 1980; Schoots *et al.*, 1982; Iuchi *et al.*, 1982). In polypterids a gradual disintegration is clearly observed before the mechanical breakthrough of the egg membrane by the hatching embryo occurs.

The attachment glands of *P. senegalus* show a similarity to those of *Lepisosteus*, with respect to formation and development with a conspicuous entodermal component (Lindahl, 1944) and *Amia calva* (Eycleshymer and Wilson, 1906; Reighard and Phelps, 1908). By contrast, the attachment organs in Dipnoi and Amphibia, as well as in Teleostei, seem to have an exclusively ectodermal origin (Kerr, 1909; Lieberkind, 1937; Britz, 1997; Britz *et al.*, 2000). In lepidosirenid dipnoans and anurans, the attachment organ is a ventrally placed disc close to the yolksac and heart anlage (Brien and Boullion, 1959). The attachment glands are often consolidated into a single unpaired organ, as is the case in *Lepisosteus* and *Amia*. Paired attachment glands, such as those of teleosts, e.g., *Esox lucius, Betta splendens, Pterophyllum scalare* and *Herotilapia multispinosa* (Kunz, 2004), and Polypteridae, appear to be an exception. However, the possible connection or median position in some specimens of *P. ornatipinnis* (compare Fig. 12C) indicates that this difference is not large. We conclude that labial attachment organs of the polypterid, lepisosteid, and amiid type do belong to the basic pattern found in Actinopterygii.

Locomotory System and Fins

The differentiation of the pectoral into a muscular peduncle and fan is externally conspicuous. The differentiated pectoralis of Polypteridae is quite unusually built and many features have been compared to the paired extremities of Sarcopterygii. Some characteristics are difficult to compare in general to those of other bony fishes, e.g., course of the innervation via the plexus brachialis (Jessen, 1973). In the early ontogenesis, however, no dramatic differences appear to exist compared to the development of the pectoral limb bud of other primitive Actinopterygii or of Teleostei (Kryzanovsky, 1927; Grandel and Schulte-Merker, 1998). In the late larval stages, differentiation of the cartilaginous plate into a proximal, medial, and distal series of radials occurs. The results of Budgett (1902), Severtzoff (1924) and Bartsch and Gemballa (1992) concerning the development of the unpaired fins are mainly confirmed. A shortened epicercal caudal fin forms that is well separated from a dorsal fin and its endoskeletal supports; the latter in turn differentiates posteroanteriorly into the characteristic finlets during the larval period. Formation of the pelvic fins is much delayed until the early larval stage.

On Some Interpretations of Ecology and Reproductive Biology

The breeding of *Polypterus senegalus* and *Polypterus ornatipinnis* has provided considerable material as well as morphological and ethological data on early ontogenetic stages. Contrary to former studies, based on sporadic breeding in large aquarium tanks, the eggs, embryos, postembryos and larvae dealt with here, were reared and recovered with good predictability. Reproductive behaviour and the general course of ontogenesis are very similar in both species, *P. ornatipinnis* and *P. senegalus*, even though the adults differ very clearly with respect to their morphological and ethological characteristics. The characteristic adult colour and activity pattern are first established relatively late in ontogeny, i.e., by the late (ptero-) larval phase. Nevertheless, the postembryos and apterolarvae of both species do show sufficient external morphological differences in detail for exact species determination (cf. Britz and Bartsch, 1998). Aside from the described differences in colour pattern of the postembryos and larvae, there are the higher myomere count and the wider, more robust shape of head of *P. ornatipinnis* that distinguish the late embryonic and larval phase of the two species. Difference of pigmentation is even found in the egg-cell, but it has not been clearly confirmed that the surface structure of the zona radiata is always a distinguishing character (Britz and Bartsch, 1998). The adhesion structures rapidly change after spawning, when in contact with water and environmental objects.

A high number of interdependent and transitory embryonic and larval features are interpreted as "early ontogenetic adaptations" to the habitat and life style of the early developmental stages (Bartsch et al., 1997). Attachment structures of eggs and adhesion organs of postembryos have a significance for certain ecological functions, viz. avoidance of drift away from the spawning habitat and nursing ground of postembryos and keeping a position in the water column during the more inactive ontogenetic stages. The gas exchange occurs across the body surface because the gills develop later. Favourable water circulation normally exists in water column and improves the respiration (Braum et al., 1996). Special respiratory organs, in the broadest sense, are important as long as embryos do not perform active respiratory movements (prior to stage 33, when mouth opening and gill slits and associated muscles have not yet completely formed), or late in the larval phase when lungs are not yet used as an accessory respiratory organ (De Smet, 1966; Bartsch et al., 1997). Ciliation of the epidermis in the postembryonic stages of many bony fishes and tetrapods (e.g., Assheton, 1896; Whiting, 1980; Kemp, 1996; Nokhbatolfoghahai et al., 2005) plays without doubt an important role in ventilation of respiratory water and removal of particles from the bodysurface. The intensive vascularisation of the embryonic and larval finfold has also to be interpreted in this context and is described for many sarcopterygians and actinopterygian larvae (McElman and Balon, 1979; Picard and Voßwinkel, 1992). The external hyoid gills are the only specialized respiratory organ in the embryonic and postembryonic phase and they serve as an important accessory respiratory organ during the early ontogenesis up to the juvenile period. Their early appearance in embryonic stages (from stage 25 onward) indicates this functional significance. External gills at the branchial arches exist in several sarcopterygian larvae (*Protopterus, Lepidosiren* and Urodela), but these are certainly not the only solution of relatively large larvae of primary aquatic vertebrates if oxygen is to be acquired and endangering exposure at the water surface should be avoided at the same time. Field studies have always demonstrated a variable but often very long persistence and significant size of external gills in juveniles of *Polypterus* of different species (Johnels, 1954; Reitzer et al., 1972), which apparently depends upon the actual water conditions of the nursing site. It is not really possible to generalize about all polypterid species based upon the few that have been investigated in detail, but descriptions of *P. endlicheri* (Azuma, 1995), *Erpetoichthys* (Britz and Bartsch, 1998) and data of particular larvae (e.g., Boulenger, 1902 and Clausen, 1956) indicate that a similarity of general characteristics of polypterid ontogeny is quite probable.

The eyes have differentiated histologically and the lip-folds and teeth are present a short time before these become employed for active feeding.

The development of the olfactory organ and elongation of the preorbital head region obviously coincide with the onset of directed, olfactory-guided search for food by polypterid larvae at the onset of the pterolarval phase. This also correlates with the emergence of a strong, accentuated biting function of the jaw apparatus together with the ability of the larvae to seize bigger prey and the emergence of cannibalism (Bartsch, 1997; Bartsch et al., 1997).

The development of attachment glands, external gills, and pectoral fins demonstrate clear differences in timing of the developmental progress prior to full differentiation. Development of external gills and attachment glands considerably precede pectoral fin development, which correlates well with the time of hatching, general activity and locomotory behaviour of the embryos and larvae. The attachment glands reach their full size and functional stage before hatching. Reduction occurs with formation of mouth and labial folds causing them to vanish without trace by stage 34, immediately before the onset of extrinsic feeding. The external gills show three developmental steps (bud, primary and secondary branching) within only three days and starting during the embryonic phase. The pectoral fin, in contrast, appears only two days after hatching, as a small, flat bud at the dorsal border of the region of the yolksac. Much later this bud differentiates into a peduncle and a distal fan. Only after this condition has been attained are the larvae able to balance and swim by movement of the pectoral fins.

Only fast movements in the free water column, escape reactions, prey capture, turns and fast approaches at the water surface are performed partly or completely by axial undulation (cf. Gemballa and Bartsch, 2002). First, the postembryonic or yolksac larval stage is rather passive. Just hatched by vigorous movements from the decaying egg envelope, it shows long periods of inactivity that are interrupted by short bursts of axial–undulatory swimming upon disturbance, during which they break free from one attachment site to reach their next attachment site.

These interpretations of biological significance and adaptation values of characteristic features are hypothetical. They can be documented by careful observation in aquaria, but must be tested experimentally and in the field. For most embryonic and larval characteristics, this has not yet been done and also might prove to involve a very difficult experimental design. Correlation between emerging ethological traits, supposed function and differentiating structure is consistent in the polypterid species that have been studied. With these demonstrated correlations, a picture of probable functional interdependencies appears and one can look on the early ontogenetic stages of anamnians as living beings in their special environment rather than as incomplete, constructional steps towards the adult organism. With respect to the relatively small size of the eggs and short developmental time of *Polypterus* as compared to other holoblastic anamnian eggs, it seems appropriate to interpret many of the early morphogenetic processes in *Polypterus* in terms of acceleration and miniaturization.

Phylogenetic Conclusions

An analysis under an ontogenetic perspective may provide additional characters. But, it provides mainly an improved character description, a deeper understanding of the developmental course and mechanisms involved in the shaping of an "adult" structure.

With respect to ontogeny, the Polypteriformes or Cladistia (*sensu* Patterson, 1982) as sister group of all other Recent Actinopterygii, show many additional plesiomorphies of Actinopterygians, viz.: derived and primitive ground pattern characters of osteichthyans or Osteognathostomata. **Holoblastic cleavage, mode of gastrulation and embryonic epidermal ciliation** certainly belong to the basic pattern of Actinopterygii and Sarcopterygii. In addition to ampullar electroreceptors, **lungs** and a multitude of skeletal characteristics a persisting spiraculum is found as well. These structural characters form mostly later in the ontogeny by the pterolarval stages and were not described here. The development of the intestine corresponds mainly to an assumed original condition of osteichthyans, with the formation of spiral gut folds as well as paired, ventral foregut diverticles or lungs (Purser, 1926, 1928; De Smet, 1966; Poll, 1967). Likewise, the urogenital system shows few peculiarities in early development that are useful for determining the systematic question of an exact classification of Polypteriformes among the Osteognathostomata. Generally, the connection between opisthonephros and the gonads points to a primitive condition (Kunz, 2004). One paired pronephros and primary urinary ducts (Wolffian ducts; cf. Lebedinsky, 1895; Budgett, 1901a, 1902; De Smet, 1975), and Müllerian ducts equipped with an ostium tubae in the female are present, as is assumed for the ground pattern of gnathostomes. There are no accessory uterus glands forming a secondary egg envelope or adhesive substances. In the adult male, we encounter a more advanced, "almost teleostean" condition, as compared to most basal gnathostomes: there is no close urogenital connection of the testes, but a secondary seminal duct is formed, which is well separated from the mesonephros and primary urinary duct for most of its course (Van den Broek, 1933; Gérard, 1958).

Many features are indeed specialized and autapomorphic for Polypteriformes, e.g., the **pectoral fin skeleton**, the **dorsal filets and their pterygiophores**, formation of the **symmetrical, compound caudal fin**, the

spiracle and series of **spiracular plates** or **structure of the olfactory organ** (see, e.g., Pfeiffer, 1968; Bjerring, 1988; Bartsch and Gemballa, 1992; Bartsch, 1997).

Several synapomorphies of the skeletal system, however, place the Polypteriformes within Actinopterygii: **Ganoin** of the Exoskeleton (e.g., Goodrich, 1928; Schultze, 1977), **Acrodin** of the teeth (s. e.g., Meinke, 1982; Patterson, 1982; Clemen et al., 1998; Wacker et al., 2001), a **dermohyal** at the hyomandibular and a scale shaped **postcleithrum** (Patterson, 1982; Gardiner and Schaeffer, 1989; Bartsch et al., 1997—comp. Bjerring, 1986) also allow for comparison and a classification of diverse fossil actinopterygians (e.g., Lauder and Liem, 1983; Arratia and Cloutier, 1996). The **singular micropyle** of the egg envelope is interpreted as an additional synapomorphy of Actinopterygii (Bartsch and Britz, 1996). The structured, **follicular attachment tufts** of the adhesive envelope may be a very convincing additional, jointly derived character of the ray-finned fishes. The significance of formation of different types of embryonic attachment glands is still not clear and is difficult to reconcile with the present phylogenetic concept of Osteognathostomata. **Rostral** or **upper labial complex adhesive organs, attachment glands** of **entodermal origin** have only been demonstrated in *Polypterus* (and *Erpetoichthys*), *Lepisosteus*, and *Amia*. Whether the attachment glands of sarcopterygians, positioned ventrally or close to the embryonic heart, are homologous, is not clear. Often, the attachment glands of different osteognathostomes have been interpreted as being convergent in origin (Lieberkind, 1937). A similar problem occurs with the **hyoid (or opercular) external gills** of *Polypterus* embryos and larvae. These are well comparable in terms of construction and insertion to a visceral arch with the external gills of Dipnoi and amphibians, but not with respect to their splanchnomere position.

In early ontogeny, Acipenseriformes (Chondrostei) may exhibit some more primitive characteristic features corresponding to concepts of osteichthyan stem group representatives of Sewertzoff (1925) or a "ganoid concept" (Müller, 1844; Pollard, 1892). Rapid development of *Polypterus* is observed in particular in the differentiation of the mesoderm and neural crest. Poorly delimited epithelia or ill-defined cell populations arise before the formation of the definitive functional structures. Also the addition of the somites and the myomere differentiation at the posterior trunk and caudal region in late postembryos are much accelerated. This, however, may only generally point to an advanced reproductive strategy of the production of a high number of offspring within restricted space and time limits.

In the more recent systematic publications based on molecular data sometimes alternative trees of Gnathostomata are favoured (Arnason et al.,

2001), but often the present concept also finds clear (and less spectacular) support (Venkatesch et al., 2001).

It is clear that the classification of Polypteriformes within the Actinopterygii is not questioned. However, the characters of rapid development of purely **subepidermal olfactory** and **auditory placodes**, the existence of a **singular micropyle**, distinct **follicular attachment filaments of the egg envelope, upper labial attachment glands,** could be considered as derived characters of a subgroup of Actinopterygii, originating above Chondrostei. The traditional view of the exact position of Polypteriformes as the sister group to all other Recent Actinopterygii may be questioned and with respect to other basal actinopterygian groups a sister group relationship to Neopterygii (Ginglymodi and Halecomorphi and Teleostei) favored. Based on these characters of the early ontogeny a more plausible scenario of evolutionary change results. This contradicts, however, the present concept of Neopterygii (Gardiner et al., 1997) and some of the well-established character evaluations of the skeleton as well as the significance of lungs.

Acknowledgements

First of all, we wish to thank Ulrich Zeller, who has enabled this study at the facilities of the Museum of Natural History in Berlin, and who has generously and steadily supported the work on polypterid development. Our sincerest thanks are also due to Frank Kirschbaum of the Institute of Freshwater Ecology and Inland Fisheries in Berlin, who often was available for discussions and provided many useful ideas for the project. Nils Hoff, Jutta Zeller, and Gabriele Drescher were of indispensable help with the graphics, histological, and microscopic techniques. John Wourms of Clemson University is gratefully acknowledged for critically reading an earlier version of the manuscript and correcting our style. Last, but not least, Yvette Kunz invested much careful work in critically reading and editing the contribution.

The study was enabled by a doctoral research grant of the DAAD.

References

Agassiz, A. 1879. The development of *Lepidosteus*. *Proceedings of the American Academy of Arts and Sciences* 13: 65-76.
Allis, E.P. 1922. The cranial anatomy of *Polypterus*, with special reference to *Polypterus bichir*. *Journal of Anatomy* 56: 189-294.
Arezo, M.J., L. Pereiro and N. Berois. 2005. Early development in the annual fish *Cynolebias viarius*. *Journal of Biology* 66: 1357-1370.

Arnason, U., A. Gullberg and A. Janke. 2001. Molecular phylogenetics of gnathostomous (jawed) fishes: old bones, new cartilage. *Zoologica Scripta* 30: 249-255.
Arndt, E.A. 1960. Untersuchungen über Eihüllen von Cypriniden. *Zeitschrift für Zellforschung* 52: 315-327.
Arnoult, J. 1964. Comportement et reproduction en captivité de *Polypterus senegalus* C. *Acta Zoologica (Stockholm)* 45: 191-199.
Arratia, G. and R. Cloutier. 1996. Reassessment of the morphology of *Cheirolepis canadensis* (Actinopterygii). In: *Devonian Fishes and Plants of Miguasha, Quebec, Canada*, Schultze, H.P. and R. Cloutier (Eds.), Friedrich Pfeil, München, pp. 165-197.
Assheton, R. 1896. Notes on the ciliation of the ectoderm of the amphibian embryo. *Quarterly Journal of Microscopical Science* 38: 465-484.
Arvedlund, M. and A. Takemara. 2006. The importance of chemical environmental cues for juvenile *Lethrinus nebulosus* Forsskal (Lethrinidae, Teleostei) when settling into their first habitat. *Journal of Experimental Marine Biology and Ecology* 338: 112-122.
Azuma, H. 1995. Breeding *Polypterus endlicheri*. *Tropical Fish Hobbyist* 44: 116-128.
Ballard, W.W. 1986a. Stages and rates of normal development in the holostean fish *Amia calva*. *Journal of Experimental Zoology* 238: 337-354.
Ballard, W.W. 1986b. Morphogenetic movements and a provisional fate map of development in the holostean fish *Amia calva*. *Journal of Experimental Zoology* 238: 355-372.
Ballard, W.W. and R.G. Needham. 1964. Normal embryonic stages of *Polyodon spathula* (Walbaum). *Journal of Morphology* 114: 465-478.
Balon, E.K. 1975. Terminology of intervals in fish development. *Journal of the Fisheries Research Board of Canada* 32: 1663-1670.
Balon, E.K. 1985. The theory of saltatory ontogeny and life history models revisited. Early life histories of fishes. In: *Developments in Environmental Biology of Fishes*, 5, Dr. W. Junk Publishers, Dordrecht, pp. 13-31.
Bartsch, P. 1988. Funktionelle Morphologie und Evolution des Axialskelettes und der Caudalis ursprünglicher Knochenfische. *Palaeontographica (A)* 204: 117-226.
Bartsch, P. 1997. Aspects of craniogenesis and evolutionary biology in polypteriform fishes. *Netherlands Journal of Zoology* 47: 365-381.
Bartsch, P. and R. Britz. 1996. Die Zucht und Entwicklung von *Polypterus ornatipinnis*. *Die Aquarien- und Terrarienzeitschrift* 1: 15-20.
Bartsch, P. and R. Britz. 1997. A single micropyle in the eggs of the most primitive, living actinopterygian fish *Polypterus*. *Journal of Zoology (London)* 241: 589-592.
Bartsch, P. and S. Gemballa. 1992. On the anatomy and development of the vertebral column and pterygiophores in *Polypterus senegalus* Cuvier, 1829 ("Pisces", Polypteriformes). *Zoologische Jahrbücher für Anatomie und Ontogenie der Tiere* 122: 497-529.
Bartsch, P., S. Gemballa and T. Piotrowski. 1997. The embryonic and larval development of *Polypterus senegalus* Cuvier, 1829: its staging with reference to external and skeletal features, behaviour and locomotory habits. *Acta Zoologica (Stockholm)* 78: 309-328.

Bemis, W.E. and L. Grande. 1992. Early development of the actinopterygian head. I External development and staging of the paddlefish *Polyodon spathula*. *Journal of Morphology* 213: 47-83.

Bjerring, H.C. 1986. The question of a dermohyal in brachiopterygian fishes. *Acta Zoologica (Stockholm)* 67: 1-4.

Bjerring, H.C. 1988. The morphology of the *organum olfactus* of a 32 mm embryo of the brachiopterygian fish *Polypterus senegalus*. *Acta Zoologica (Stockholm)* 69: 47-54.

Bjerring, H.C. 1991a. Two intracranial ligaments supporting the brain of the brachiopterygian fish *Polypterus senegalus*. *Acta Zoologica (Stockholm)* 72: 41-47.

Bjerring, H.C. 1991b. The question of a vomer in brachiopterygian fish. *Acta Zoologica (Stockholm)* 72: 223-232.

Boulenger, G.A. 1902. On some characters distinguishing the young of various species of *Polypterus*. *Proceedings of the Zoological Society of London* 1: 121-125.

Braum, E., N. Peters and M. Stolz. 1996. The adhesive organ of larval pike *Esox lucius* L., (Pisces). *Internationale Revue der Gesamten Hydrobiologie* 81: 101-108.

Braus, H. 1909. Die Muskeln und Nerven der Ceratodusflosse. Ein Beitrag zur Vergleichenden Morphologie der freien Gliedmaasse bei niederen Fischen und zur Archipterygiumtheorie. In: *Denkschriften der Medizinisch-Naturwissenshaftlichen Gesellschaft zu Jena IV. Zoologische Forschungsreisen in Australien und dem Malayischen Archipel I*, G. Fischer, Jena, pp. 137-300, pls. 21-29.

Brien, P. and J. Boullion. 1959. Ethologie des larves de *Protopterus dolloi* Blgr. et étude de leurs organes respiratoires. *Annales du Musée Royal du Congo-Belge (Tervuren)* 71: 23-74.

Britz, R. 1997. Egg surface structure and larval cement glands in nandid and badid fishes with remarks on phylogeny and biogeography. *American Museum Novitates* 3195: 1-17.

Britz, R. 2004. *Polypterus teugelsi*, a new species of *bichir* from the Upper Cross River system in Cameroon (Actinopterygii: Cladistia: Polypteridae). *Ichthyological Explorations of Freshwaters* 15: 325-334.

Britz, R., and P. Bartsch. 1998. On the reproduction and early development of *Erpetoichthys calabaricus*, *Polypterus senegalus*, and *P. ornatipinnis* (Actinopterygii: Polypteridae). *Ichthyological Exploration of Freshwaters* 9: 325-334.

Britz, R., F. Kirschbaum and A. Heyd. 2000. Observations on the structure of larval attachment organs in three species of gymnotiforms (Teleostei: Ostariophysi). *Acta Zoologica (Stockholm)* 81: 57-67.

Britz, R., M. Kokoscha and R. Riehl. 1995. The Anabantoid genera *Ctenops*, *Luciocephalus*, *Parasphaerichthys*, and *Sphaerichthys* (Teleostei: Perciformes) as a monophyletic Group: evidence from egg surface structure and reproductive behavior. *Japanese Journal of Ichthyology* 42: 71-79.

Budgett, J.S. 1899. Observations on *Polypterus* and *Protopterus*. *Proceedings of the Cambridge Philosophical Society* 10: 236-240.

Budgett, J.S. 1901a. On some points in the anatomy of *Polypterus*. *Transaction of the Zoological Society of London* 15: 323-338.

Budgett, J.S. 1901b. On the breeding habits of some West-African fishes, with an account of the external features in development of *Protopterus annectens* and

a description of a larva of *Polypterus lapradei*. *Transactions of the Zoological Society of London* 16: 115-136.
Budgett, J.S. 1902. On the structure of the larval *Polypterus*. *Transactions of the Zoological Society of London* 16: 315-340.
Burgess, W.E. 1986. Spawning *Polypterus ornatipinnis*. *Tropical Fish Hobbyist* 1986: 10-23.
Chen, K.C., K.T. Shao and J.S. Yang. 1999. Using micropylar ultrastructure for species identification and phylogenetic inference among four species of Sparidae. *Journal of Fish Biology* 55: 288-300.
Cherr, G.N. and W.H. Clark. 1982. Fine structure of the envelope and micropyles in the eggs of the white sturgeon, *Acipenser transmontanus*. *Development Growth and Differentiation* 24: 341-352.
Cherr, G.N. and W.H. Clark. 1985. Gamete interaction in the white sturgeon *Acipenser transmontanus*: a morphological and physiological review. *Environmental Biology of Fishes* 14: 11-22.
Clausen, H.S. 1956. Larvae of polypterid fish *Erpetoichthys* Smith. *Nature*, London, 178: 932-933.
Clemen, G., P. Bartsch and K. Wacker. 1998. Dentition and dentigerous bones in juveniles and adults of *Polypterus senegalus*. (Cladistia, Actinopterygii). *Annals of Anatomy* 180: 211-221.
Dean, B. 1895a. The early development of the gar-pike and sturgeon. *Journal of Morphology* 11: 1-62.
Dean, B. 1895b. On the spawning habits and early development of *Amia*. *Quarterly Journal of Microscopical Science* 38: 413-444.
Dean, B. 1896. On the larval development of *Amia calva*. *Zoologische Jahrbücher, Systematik, Geographie und Biologie der Tiere* 9: 639-672.
De Smet, W.M.A. 1966. Le développement des sacs aériens des Polyptères. *Acta Zoologica (Stockholm)* 47: 151-183.
De Smet, W.M.A. 1975. Les premiers stades de développement du feuillet mesodermique chez *Polypterus senegalus* Cuvier. *Acta Zoologica et Pathologia Antverpiensia* 62: 3-28.
Dettlaff, T.A. and A.S. Ginsburg. 1954. *Embryonic development of sturgeons (stellate, russian and giant sturgeon), with special reference to the problems of their Breeding*. Izdatel'stvo Akad. Nauk. SSSR, Moscow.
Dogiel, A. 1887. Über den Bau des Geruchsorgans bei Ganoiden, Knochenfischen und Amphibien. *Archiv für mikroskopische Anatomie* 29: 74-139.
Dutheil, D.B. 1999. First articulated fossil cladistian: *Serenoichthys kemkemensis*, gen. et spec. nov., from the Cretaceous of Morocco. *Journal of Vertebrate Paleontology* 19: 243-246.
Eyclesheymer, A.C. and J.M. Wilson. 1906. The gastrulation and embryo formation in *Amia calva*. *American Journal of Anatomy* 5: 132-162.
Eyclesheymer, A.C. and J.M. Wilson. 1908. The adhesive organs of *Amia*. *Biological Bulletin* 14: 134-148.
Gardiner, B.G. and B. Schaeffer. 1989. Interrelationships of lower actinopterygian fishes. *Zoological Journal of the Linnean Society of London* 97: 135-187.
Gardiner, B.G., J.G. Maisey and D.T.J. Littlewood. 1997. Interrelationships of basal Neopterygians. In: *Interrelationships of Fishes*, Stiassny, M.L.J., Parenti, L.R. and G.D. Johnson (Eds.), Academic Press, San Diego, pp. 117-146.

Gawrilenko, A. 1910. Die Entwicklung des Geruchsorgans bei *Salmo salar*. Zur Stammesentwicklung des Jacobson'schen Organs. *Anatomischer Anzeiger* 36: 411-427.

Gayet, M. and F.J. Meunier. 1991. First discovery of Polypteridae (Pisces, Cladistia, Polypteriformes) outside Africa. *Géobios Note brève* 24: 463-466.

Gayet, M. and F.J. Meunier. 1992. Polypteriformes (Pisces, Cladistia) du Maastrichtien et du Paléocène de Bolivie. *Geobios* 14: 159-168.

Gayet, M. and F.J. Meunier. 1996. Nouveaux Polypteriformes du gisement Coniacien-Sénonien d´In Becetem (Niger). *Comptes Rendus de L'Académie des Sciences (Paris)* 322: 701-707.

Gemballa, S. and P. Bartsch. 2002. Architecture of the integument in lower teleostomes: functional morphology and evolutionary implications. *Journal of Morphology* 253: 290-309.

Geoffroy, S.H.E. 1802. Histoire naturelle et description d'un nouveau genre de poisson du Nil nommé Polyptère. *Annales du Muséum d'Histoire Naturelle de Paris* 1: 57-68.

Gérard, P. 1958. Organes Reproducteurs. In: *Traité de Zoologie. Tome XIII. Agnathes et Poissons*, Grassé, P.P. (Ed.), Masson et Cie, Paris, pp. 1565-1583.

Goette, A. 1875. *Die Entwicklungsgeschichte der Unke (Bombinator igneus) als Grundlage einer Vergleichenden Morphologie der Wirbeltiere.* Verlag von Leopold Voss, Leipzig.

Götting, K.J. 1964. Entwicklung Bau und Bedeutung der Eihüllen des Steinpickers (*Agonus cataphractus* L.). *Helgoländer Wissenschaftliche Meeresuntersuchungen* 11: 1-12.

Goodrich, E.S. 1928. *Polypterus* a palaeoniscid. *Palaeobiologica* 1: 87-92.

Grande, L. and W.E. Bemis. 1991. Osteology and phylogenetic relationships of fossil and Recent paddlefishes (Polyodontidae) with comments on the interrelationships of Acipenseriformes. *Journal of Vertebrate Paleontology* 11: 1-121.

Grandel, H. and S. Schulte-Merker. 1998. The development of the paired fins in the zebrafish (*Danio rerio*). *Mechanisms of Development* 79: 99-120.

Groot, E.P. and D.F. Alderdice. 1984. Fine structure of the external egg membrane of the species of pacific salmon and steelhead trout. *Canadian Journal of Zoology* 63: 552-566.

Hagenmaier, H.E. 1974. Zum Schlüpfprozeß bei Fischen. VI. Entwicklung, Struktur und Funktion der Schlüpfdrüsenzellen bei der Regenbogenforelle, *Salmo gairdneri* Rich. *Zeitschrift für Ökologie und Morphologie der Tiere* 79: 233-244.

Hansen, A., P. Bartsch, A.O. Kasumyan and E. Zeiske. 1999. A comparison of developmental processes in the olfactory organs of lower aquatic vertebrates. In: *Proceedings of the Göttingen Conference of the German Neuroscience Society 1999; Volume 127th Göttingen Neurobiology Conference. From Molecular Neurobiology to Clinical Neuroscience*, Elser, N. and U. Eysel (Eds.), Thieme, Stuttgart, p. 16.

Hansen, A. and E. Zeiske. 1993. Development of the olfactory organ in the zebrafish, *Brachydanio rerio*. *Journal of Comparative Neurology* 333: 289-300.

Hansen, A. and E. Zeiske. 1995. Development of the olfactory organ in the zebrafish: an electron microscopic immunocytochemical study of early differentiation and growth. *Biophysics* 40: 159-170.

Harrison, R.G. 1969. Harrison stages and description of the normal development of the spotted salamander, *Amblystoma punctatum* (L.). In: *Organization and development of the embryo*, Harrison, R.G. (Ed.), Yale University Press, New Haven, pp. 44-66.
Horie, N., T. Utoh, Y. Yamyda, A. Okumura, H. Zhang, N. Mikawa, A. Akazawa, S. Tanaka and H. Poka. 2002. Development of the embryos and larvae in the common japanese *Conger myriaster*. *Fisheries science* 68: 972-983.
Ilg, L. 1952. Über larvale Haftorgane bei Teleosteern. *Zoologischer Jahrbücher für Anatomie und Ontogenie der Tiere* 72: 577-600.
Iwamatsu, T., K. Onitake, Y. Yoshimoto and Y. Hiramoto. 1991. Time sequence in the early fertilization in the medaka egg. *Development Growth and Differentiation* 33: 479-490.
Iuchi, U., M. Yamamoto and K. Yamagami. 1982. Presence of active hatching enzyme in secretory granule of prehatching medaka embryos. *Development Growth and Differentiation* 24: 135-143.
Jessen, H.L. 1973. Interrelationships of actinopterygians and brachiopterygians: evidence from pectoral anatomy. *Journal of Linnean Society of London* 53: 227-232.
Johnels, A.G. 1954. Notes on the fishes from the Gambia River. *Arkiv för Zoologi*, 2nd series 6: 327-411.
Johnson, E.Z. and R.G. Werner. 1986. Scanning electron microscopy of the chorion of selected freswater fishes. *Journal of Fish Biology* 29: 257-265.
Jones, S. 1937. On the origin and development of the cement glands in *Etroplus maculatus* (Bloch). *Proceedings of the Indian Academy of Sciences* 6: 79-90.
Keller, R.E. 1980. The cellular basis of epiboly: An SEM study of deep-cell rearrangement during gastrulation in *Xenopus laevis*. *Journal of Embryology and Experimental Morphology* 60: 201-234.
Kemp, A. 1982. The embryological development of the Queensland lungfish, *Neoceratodus forsteri* (Krefft). *Memoirs of the Queensland Museum* 20: 553-597.
Kemp, A. 1996. Role of epidermal cilia in development of the australian lungfish, *Neoceratodus forsteri* (Osteichthyes: Dipnoi). *Journal of Morphology* 228: 203-221.
Kerr, J.G. 1907. The development of *Polypterus senegalus* Cuvier. In: *Budgett memorial*, Kerr, J.G. (Ed.), Cambrige University Press, Cambridge, pp. 195-284.
Kerr, J.G. 1909. Normal plates of the development of *Lepidosiren paradoxa* and *Protopterus annectens*. In: *Normentafeln zur Entwicklungsgeschichte der Wirbeltiere*, Keibel, F. (Ed.), G. Fischer, Jena, pp. 163-194.
Kryzanovsky, S. 1927. Die Entwicklung der paarigen Flossen bei *Acipenser*, *Amia* und *Lepidosteus*. *Acta Zoologica (Stockholm)* 8: 277-352.
Kudo, S. 1980. Sperm penetration and the formation of a fertilization cone in the common carp egg. *Development Growth Differentiation* 22: 403-414.
Kunz, Y.W. 2004. Developmental biology of teleost fishes. Fish and Fisheries Series 28. Springer, Dordrecht.
Kupffer, C. 1868. Beobachtungen über die Entwicklung der Knochenfische. *Archiv für Mikroskopische Anatomie* 4: 209-272.
Laale, H.W. 1980. The perivitelline space and egg envelopes of bony fishes: A Review. *Copeia* 1980: 210-226.

Lauder, G.V. and K.F. Liem. 1983. The evolution of the interrelationships of the actinopterygian fishes. *Bulletin of the Museum of Comparative Zoology* 150: 95-197.

Lebedinsky, J. 1895. Über die Embryonalniere von *Calamoichthys calabaricus* (Smith). *Archiv mikroskopische Anatomie* 44: 216-228.

Lieberkind, I. 1932. Über die Haftorgane bei Jungen von *Pterophyllum eimekei* E. Ahl. *Zoologischer Anzeiger* 97: 55-61.

Lieberkind, I. 1937. *Vergleichende Studien über die Morphologie und Histogenese der larvalen Haftorgane bei den Amphibien*. CA Reitzel, Kopenhagen.

Linhart, S. and S. Kudo. 1997. Surface Ultrastructure of paddle fish eggs before and after fertilization. *Journal of Fish Biology* 52: 573-582.

Lindahl, P.E. 1944. Zur Kenntnis der Entwicklung von Haftorgan und Hypophyse bei *Lepidosteus*. *Acta Zoologica (Stockholm)* 25: 97-133.

Long, W.L. and W.W. Ballard. 2001. Normal embryonic stages of the Longnose Gar, *Lepisosteus osseus*. *Developmental Biology* 1: 6-23.

Long, W. and J.P. Wourms. 1991. Gastrulation in the gar *Lepisosteus osseus*. *American Zoologist* 31: 81 A.

Mark, E.L. 1890. Studies on *Lepidosteus*. *Bulletin of the Museum of Comparative Zoology (Harward)* 19: 1-128.

Martinez-Palacios, C.A. and L.G. Ross. 2004. Post-hatching geotactic behaviour and substrate attachment *Cichlasoma urophthalmus*. *Journal of Applied Ichthyology* 20: 545-547.

McElman, J.F. and E.K. Balon. 1979. Early ontogeny of walleye, *Stizostedion vitreum*, with steps of saltatory development. *Environmental Biology of Fishes* 4: 309-348.

Meijide, F.J. and G.A. Gerrero. 2000. Embryonic and larval development of substrate-brooding cichlid *Cichlasoma dimerus* (Heckel, 1840) under laboratory conditions. *Journal of Zoology* 252: 481-493.

Meinke, D.K. 1982. A histological and histochemical study of the teeth in *Polypterus* (Pisces, Actinopterygii). *Archives of Oral Biology* 27: 197-206.

Meunier, F.J. and M. Gayet. 1996. A new polypteriform from the late Cretaceous and the middle Paleocene of South America. In: *Mesozoic Fishes: Systematics and Paleoecology*, Arratia, G. and G. Viohl (Eds.), Munich, pp. 95-105.

Mooi, R.D. 1990. Egg surface morphology development of Pseudochromoids (Perciformes: Percoidei), with comment on its phylogenetic implications. *Copeia* 1990: 455-475.

Morisawa, S. 1999. Fine structure of micropylar region during late oogenesis in eggs of the hagfish *Eptatretus burgeri* (Agnatha). *Development Growth and Differentiation* 41: 611-618.

Müller, J. 1844. Über den Bau und die Grenzen der Ganoiden und über das natürliche System der Fische. *Abhandlungen der Königlichen Akademie der Wissenschaften (Berlin)* 1846: 117-216.

Nieuwenhuys, R., R. Bauchot and J. Arnoult. 1969. Le développement du télencephale d´un poisson osseux primitif, *Polypterus senegalus* C. *Acta Zoologica (Stockholm)* 50: 101-125.

Nieuwenhuys, R. 1982. An overview of the organisation of the brain in actinopterygian fishes. *American Zoologist* 22: 287-310.

Nokhbatolfoghahai, M. 2005. The surface ciliation of anuran amphibian embryos and early larvae: Patterns, timing differences and functions. *Journal of Natural History* 39: 887-929.

Nelson, J.S. 1994. *Fishes of the world*. 3rd Edition. John Wiley and Sons, New York.

Nelson, J.S. 2006. *Fishes of the world*. 4th Edition. John Wiley and Sons, New York.

Nüesch, H. 1958. Augenentwicklung und Schlüpftermin bei Hecht und Forelle. *Revue Suisse de Zoologie* 65: 396-403.

Patterson, C. 1982. Morphology and interrelationships of primitive actinopterygian fishes. *American Zoologist* 22: 241-259.

Patzner, R.A. and R. Riehl. 1992. Die Eier heimischer Fische 1. Rutte, *Lota lota* L. (1758), (Gadidae). *Österreichische Fischereigesellschaft* 45: 235-238.

Pehrson, T. 1947. Some new interpretations of the skull in *Polypterus*. *Acta Zoologica (Stockholm)* 28: 400-455.

Pehrson, T. 1958. The early ontogeny of the sensory lines and the dermal skull in *Polypterus*. *Acta Zoologica (Stockholm)* 39: 241-258.

Peters, H.M. 1965. Über larvale Haftorgane bei *Tilapia* (Cichlidae, Teleostei) und ihre Rückbildung in der Evolution. *Zoologischer Jahrbücher für, Physiologie und Ökologie der Tiere* 71: 287-300.

Pfeiffer, W. 1968. Das Geruchsorgan der Polypteridae (Pisces, Brachiopterygii). *Zeitschrift für Morphologie der Tiere* 63: 75-110.

Pfeiffer, W. 1969. Der Geruchssinn der Polypteridae (Pisces, Brachiopterygii). *Zeitschrift für vergleichende Physiologie* 63: 151-164.

Picard, F. and R. Voßwinkel. 1992. Normal developmental stages and levels of organogenesis of the rainbow trout *Oncorhynchus mykiss* (Walbaum, 1792). *Zeitschrift für Fischkunde (Solingen)* 3: 127-204.

Picard, J.J. 1975. *Xenopus laevis* cement gland as an experimental model for embryonic differentiation. I. *In vitro* stimulation of differentiation by ammonium chloride. *Journal of Embryology and Experimental Morphology* 3: 957-967.

Piotrowski, T. and G.R. Northcutt. 1996. The cranial nerves of the senegal Bichir, *Polypterus senegalus* (Osteichtyes: Actinopterygii, Cladistia). *Brain Behavior and Evolution* 47: 55-102.

Poll, M. 1954. Zoogéographie des Protoptères et des Polyptères. *Bulletin de la Société Zoologique de France* 79: 282-289.

Poll, M. 1967. Etude systématique des appareils respiratoire et circulatoire des Polypteridae. *Annales du Musee Royal de l'Afrique Centrale* 158: 1-63.

Pollard, H.B. 1892. On the anatomy and phylogenetic position of *Polypterus*. *Zoologische Jahrbücher für Anatomie und Ontogenie der Tiere* 5: 387-425.

Purser, G.L. 1926. *Calamoichthys calabaricus* J.A. Smith. Part I. The alimentary and respiratory systems. *Transactions of the Royal Society of Edinburgh* 54: 767-84.

Purser, G.L. 1928. *Calamoichthys calabaricus* J.A. Smith Part I. The alimentary and respiratory systems. *Transactions of the Royal Society of Edinburgh* 56: 89-101.

Reighard, J. and J. Phelps. 1908. The development of the adhesive organ and head mesoblast of *Amia*. *Journal of Morphology* 19: 469-496.

Reitzer, C., X. Mattei and J.L. Chevalier. 1972. Contribution à l'étude de la faune ichthyologique du bassin du fleuve Sénégal. *Bulletin de l'Institut Fondamental d'Afrique Noire, série A*, 1: 111-125

Riehl, R. 1978. Licht- und elektromikroskopische Untersuchungen an den Oocyten der Süßwasser-Teleosteer *Noemacheilus barbatulus* (L.) und *Gobio gobio* (L.) (Pisces, Teleostei). *Zoologischer Anzeiger* 201: 199-219.

Riehl, R. 1991. Die Struktur der Oocyten und Eihüllen oviparer Knochenfische—eine Übersicht. *Acta Biologica Benrodis* 3: 27-65.

Riehl, R. 1995. Eier und Eihüllen von Knochenfischen. In: *Fortpflanzungsbiologie der Aquarienfische*, Greven, H. and R. Riehl (Eds.), Birgitt Schmettkamp Verlag, Bornheim, pp. 11-25.

Riehl, R. 1996. The ecological significance of the egg envelope in teleosts with special reference to limnic species. *Limnologica* 26: 183-189.

Riehl, R. and R.A. Patzner. 1998. Minireview: The modes of attachment in the eggs of teleost fishes. *The Italian journal of zoology (Modena)* 65: 415-420.

Salensky, W. 1881. Le développement du Sterlet (*Acipenser ruthenus*). *Archives de Biologie* 2: 233-341.

Sawadski, A.M. 1925a. Veränderungen vor der Gastrulation, Bildung der Keimblätter und die damit verbundenen Prozesse. *Zeitschrift für Anatomie und Entwicklungsgeschichte* 78: 26-63.

Sawadski, A.M. 1925b. Untersuchungen zur Entwicklungsgeschichte des Sterlets (*Acipenser ruthenus* L.). *Zeitschrift für Anatomie und Entwicklungsgeschichte* 78: 682-713.

Schindler, O. 1935. Zur Biologie der Larven von Barsch (*Perca fluviatilis* L.) und Hecht (*Esox lucius* L.). *Verhandlungen der deutschen zoologischen Gesellschaft* 37: 141-149.

Schliewen, U.K. and F. Schäfer. 2006. *Polypterus mokelembembe*, a new species of bichir from the Congo River basin (Actinopterygii: Cladistia: Polypteridae). *Zootaxa* 1129: 23-36.

Schmalhausen, J.J. 1913. Bau und Phylogenese der unpaaren Flossen und insbesondere der Schwanzflosse der Fische. *Zeitschrift für wissenschaftliche Zoologie* 104: 1-80.

Schoots, A.F.M., J.J.M. Stikkelbroeck, J.F. Bekhuis and J.M. Denucé. 1982. Hatching in teleostean fishes: fine structural changes in the egg envelope during enzymatic breakdown in vivo and in vitro. *Journal of Ultrastructure Research* 80: 185-196.

Schugardt, C. 1997. *Experimentelle Untersuchung zur exogenen Kontrolle der zyklischen Fortpflanzung afrikanischer Süßwasserfische: Vergleich von Mormyriden und Polypterus*. Dissertationsschrift, Humboldt-Universität, Berlin.

Schultze, H.P. 1977. Ausgangsform und Entwicklung der rhombischen Schuppen der Osteichthyes (Pisces). *Paläontologische Zeitschrift* 51: 152-168.

Semon, R. 1901. Normentafel zur Entwicklungsschichte des *Ceratodus forsteri*. In: *Normentafeln zur Entwicklungsgeschichte der Wirbeltiere*, Keibel, F. (Ed.), Verlag von Gustav Fischer, pp. 1-38.

Senn, D.G. 1976. Brain structure in *Calamoichthys calabaricus* Smith 1865 (Polypteridae, Brachiopterygii). *Acta Zoologica (Stockholm)* 57: 121-128.

Sewertzoff, A.N. 1924. The development of the dorsal fin of *Polypterus delhezi*. *Journal of Morphology* 38: 551-580.

Sewertzoff, A.N. 1925. Zur Morphologie des Schädels von *Polypterus delhezi*. *Anatomischer Anzeiger* 59: 271-278.

Shardo, J.D. 1995. Comparative embryology of teleostean fishes. I. development and staging of the american shad, *Alosa sapidissima* (Wilson, 1811). *Journal of Morphology* 225: 125-165.

Tahara, Y. 1959. Table of the normal developmental stages of the frog, *Rana japonica* I. Early development (Stages 1-25). *Japanese Journal of Experimental Morphology* 13: 49-60.

Theisen, B. 1970. The morphology and vascularisation of the olfactory organ in *Calamoichthys calabaricus* Smith, 1865. (Polypteridae, Brachiopterygii). *Videnskabelige Meddelver av de Dansk Naturhistorisk Foreningen, Koebenhavn* 133: 31-50.

Van den Broek, A.J.P. 1933. Gonaden und Ausführgänge. In: *Vergeichende Anatomie der Wirbeltiere*, Bolk, L., Göppert, E., Kallius, E. and W. Lubosch (Eds.), Urban and Schwarzenberg, Berlin, pp. 1-154.

Venkatesch, B., M.V. Erdmann and S. Brenner. 2001. Molecular synapomorphies resolve evolutionary realtionships of extant jawed vertebrates. *Proceedings of the National Academy of Sciences* 98: 11382-11387.

Verraes, W. 1976. Postembryonic development of the nasal organs, sacs and surrounding skeletal elements in *Salmo gairdneri* (Teleostei: Salmonidae), with some functional interpretations. *Copeia* 1976: 71-75.

Wacker, K., P. Bartsch and G. Clemen. 2001. The development of the tooth pattern and dentigerous bones in *Polypterus senegalus* (Cladistia, Actinopterygii). *Annals of Anatomy* 183: 37-52.

Webb, J.F. and R.G. Northcutt. 1991. Ciliated epidermal cells in non-teleost actinopterygian fish. *Acta Zoologica (Stockholm)* 72: 107-111.

Whiting, H.P. 1980. Ciliary cells in the epidermis of the larval Australian dipnoan, *Neoceratodus*. *Zoological Journal of the Linnean Society* 68: 125-137.

Wickler, W. 1956. Der Haftapparat einiger Cichliden-Eier. *Zeitschrift Zellforschung* 45: 304-327.

Willmess, M.Th.M. and J.M. Denucé. 1973. Hatching glands in Teleosts, *Brachydanio rerio*, *Danio malabaricus*, *Moenkhausia oligolepis* and *Barbus schuberti*. *Development Growth and Differentiation* 15: 169-177.

Wolff, E. 1992. Die Zucht von *Polypterus ornatipinnis* in zweiter Generation. *Jahrbücher Löbbecke Museum und Aquazoo*, Düsseldorf, 1991. p. 58-61.

Wourms, J.P. and P. Bartsch. 1998. Early development and gastrulation of the cladistian fish, *Polypterus senegalus*. *American Zoologist* 38: 187A.

Wunder, W. 1935. Das Verhalten von Knochenfischen beim Ausschlüpfen aus dem Ei und in den ersten Lebenstagen. *Zoologischer Anzeiger* 8: 60-65.

Yamamoto, T.S. and W. Kobayashi. 1992. Closure of the micropyle during embryonic development of some pelagic fish eggs. *Journal of Fish Biology* 40: 225-241.

Yanagimachi, R., G.N. Cherr, M.C. Pillai and J.D. Baldwin. 1992. Factors controlling sperm entry into the micropyles of salmonids and herring eggs. *Development, Growth and Differentiation* 34: 447-462.

Yoshizaki, N. 1979. Induction of the frog hatching gland cell from explanted presumptive ectodermal tissue by LiCl. *Development, Growth and Differentiation* 21: 11-18.

Yoshizaki, N., R.J. Sackers, A.F.M. Schoots and J.M. Denucé. 1980. Isolation of hatching gland cells from the teleost, *Oryzias latipes*, by centrifugation through Percoll. *Journal of Experimental Zoology* 213: 427-429.

Zeiske, E., A.O. Kasumyan and P. Bartsch. 2003. Early development of the olfactory organ in sturgeons of the genus *Acipenser*: a comparative and electron microscopic study. *Anatomy and Embryology* 206: 357-372.

3 | Early Development of Acipenseriformes (Chondrostei: Actinopterygii)

Teresa Ostaszewska[1] and Konrad Dabrowski[2]

[1]Warsaw University of Life Science, Division of Ichthyobiology and Fisheries, Faculty of Animal Science, Ciszewskiego 8, 02-786 Warsaw, Poland.
E-mail: teresa_ostaszewska@sggw.pl

[2]School of Environment and Natural Resources, the Ohio State University, Columbus 43210, OH, USA. *E-mail:* dabrowski.1@osu.edu

INTRODUCTION

The order Acipenseriformes includes two families of ancient ray finned fishes *Acipenseridae* and *Polyodontidae*. The family *Acipenseridae* consists of 25 extant sturgeon species (19 species of *Acipenseridae* and 6 species of *Scaphirhynchinae*). The family *Polyodontidae* includes two genera, *Polyodon* and *Psephurus*, only one species in each (Nelson, 2006).

All species of Acipenseriformes live in the Northern Hemisphere. They reproduce in freshwater and many species migrate to the sea, either living in brackish water (Caspian, Azov, Black and Baltic Seas) or in seawater on the oceanic continental shelf. There are also exclusively freshwater forms. Most species feed on benthic organisms. They reach sexual maturity very late (e.g., *Acipenser transmontanus*, male at 12, female at 16-35 years). However, there are exceptions, such as *Acipenser baerii*, with males maturing at 3-4 years and females 4-5 years. Acipenseriformes do not reproduce annually. Spawning frequency is once in 2-11 years for females, and 1-6 years for males (Billard and Lecointre, 2001). Sturgeon growth is continuous with age, and sexual maturity does not involve a reduction in growth rate. Some species live even 100 years, and reach up to 1000 kg (e.g., Beluga).

Though different in ways of living, sturgeons are very similar in terms of morphology and anatomy. They all show an elongate body with a central flat base, a rostrum, cartilaginous skeleton, notochord, intestinal spiral valve, conus arteriosus, bony dermal plates in longitudinal rows on the body and gill rakers. The *Polyodontidae* do not have rows of shields but are totally covered by small scutes and have a long flattened rostrum, a very perceptive sense organ bearing numerous electroreceptors.

Biogeographic distribution of presently living species reflects ancient relationships among fish of Europe, Asia, and North America (Bemis *et al.*, 1997). Acipenseriformes exist at least since the Lower Jurassic Period (approximately 200 MYBP), and all fossil and recent taxa are from the Holarctic. Phylogenic relationship between the Paleozoic and early Mesozoic actinopterygians is uncertain but most investigators agree that Acipenseriformes are monophyletic and form a sister group of all extant Neopterygii (Bemis *et al.*, 1997). Acipenseriformes karyotypes consist of an extremely large number of chromosomes, which in some sturgeons is estimated to be as high as 500 (Blacklidge and Bidwell, 1993). All Acipenseriformes are polyploids. Polyploidy is a major driving force in evolution and resulted in subsequent radiation of taxa within this group (Birstein and DeSalle, 1998).

The earliest studies of embryonic development in sterlet (*Acipenser ruthenus*) were undertaken in the 19th century by Kowalewsky *et al.* (1870) and Salensky (1881). They were soon followed by similar studies of Atlantic sturgeon, *Acipenser oxyrhynchus oxyrhynchus* (Ryder, 1890; Dean, 1895). Beginning in the sixties of the 20th century, extensive research on the embryonic and post-embryonic development of several species of sturgeons: Russian sturgeon (*Acipenser gueldenstaedti colchicus*), stellate sturgeon (*Acipenser stellatus*) and beluga (*Huso huso*) was carried out in USSR (Dettlaff and Ginsburg, 1954; Ignatieva, 1960, 1961, 1963, 1965; Ginsburg, 1961, 1968; 'Dettlaff, 1962; Ginsburg and Dettlaff, 1991; Schmalhausen, 1991; Dettlaff *et al.*, 1993).

On the North American continent, studies on reproduction and development of paddlefish (*Polyodon spathula*) were started much later (Purkett, 1961, 1963). However, the data concerning early embryonic and larval development of this species are still scarce (Ballard and Needham, 1964; Yeager and Wallus, 1982, 1990). Embryonic development of white sturgeon (*A. transmontanus* [Richardson]), was described by Beer (1981) and Bolker (1993a,b, 2004).

All species of sturgeons in the world are strongly endangered, mainly due to overfishing and environmental degradation, accumulation of pollutants to fish body from contaminated sediments, dredging, embankment and damming of the rivers. Sturgeons are migratory fish, and under such conditions anadromous species cannot reach their upstream spawning grounds, thus fail to reproduce.

Nowadays all species of sturgeons are under protection, according to The Washington Convention (Convention on International Trade in Endangered Species of Wild Fauna and Flora) (Lenhardt et al., 2006), and listed in the IUCN Red Data Books.

The present publication has been completed, in part, on the basis of the experimental work conducted on the embryonic development of three species of sturgeon, of *A. ruthenus*, *A. gueldenstaedti*, *A. baerii*.[1]

EGG AND SPERM

The eggs of sturgeons are slightly elongated, and vary in size: from 1.9-2.5 mm in diameter in the sterlet (*A. ruthenus* L.) (Berg et al., 1949), and *A. oxyrhynchus* (Mitchell) (Valadykov and Greeley, 1963), to 3.5-4 mm in *H. huso*, kaluga (*Huso dauricus*) (Georgi) and *A. transmontanus* (Dettlaff and Ginsburg, 1954; Berg et al., 1949; Cherr and Clark, 1982). Other species of sturgeons have mid-size eggs, from 2.5 to 3.8 mm (Sokolov, 1965; Berg et al., 1949; Dettlaff and Ginsburg, 1954; Jones et al., 1978).

Unfertilized mature eggs of *H. huso*, *A. gueldenstaedti colchicus* and *A. stellatus* are brownish-grey (Dettlaff et al., 1993), with lighter animal pole, and darker vegetal pole. In the centre of the animal pole there is a light polar spot surrounded by a dark central ring. Egg pigmentation may differ even among the spawn of various females of the same species. The eggs of the *A. transmontanus* (Richardson) are heavily pigmented, except for light rings surrounding the micropylar region (Cherr and Clark, 1982). The main difference between immature oocytes, and mature eggs is the lack of a germinal vesicle, which breaks down during maturation. Complex architecture and structure of the egg-shell is a taxonomic as well as a phylogenetic feature. The eggs of Acipenseriformes, e.g., of the Russian sturgeon *A. gueldenstaedti*, are encased by a large egg envelope which consists of three distinct layers (Ginsburg, 1968). According to Cherr and

---

[1]The specimens of *A. ruthenus*, *A. gueldenstaedti*, *A. baerii* respectively, were obtained from the Fishing Farm RYBA in Olecenica, Poland, and were reared from artificially inseminated eggs. The eggs were kept in Zuger glasses until hatching and then transferred to recirculatory glass aquaria of 25 dm^3 volume at a stocking density of 150 fish per tank. Rearing larvae and all histological procedures were carried out in the Warsaw University of Life Science, Division of Ichthyobiology and Fisheries. The larvae were fed from 9th day after hatching with A*rtemia naupli ad libitum*. The average water temperature was 15°C±0.8°C. Samples (n = 10) of the successive developmental stages were taken from fertilization to day 30 post hatching and then anesthetized, preserved in Bouin's solution, embedded in Paraplast and then subjected to standard histological procedures (Ostaszewska et al., 2007; Wegner et al., 2008).

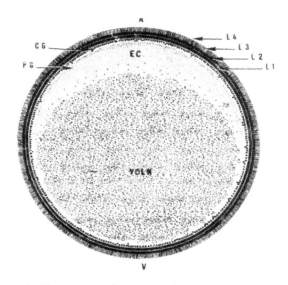

Fig. 1. A schematic illustration of a mature white sturgeon egg, showing the four distinct envelope layers (L1-L4); the animal pole (A); vegetal pole (V); egg cytoplasm (EC); cortical granules (CG); pigment granules (PG). (Cherr and Clark (1982), with kind permission of the Japanese Society of Developmental Biology).

Clark (1982) in *A. transmontanus* the egg-shell of 50 µm thickness consists of four layers. The innermost (first) layer is closely apposed to the oolemma, the second layer contains screw like projections that anchor it on the first layer. The third (interior) layer contains numerous pores, or ductules, and the fourth (the outermost) layer is amorphous (jelly) (Fig. 1).

In contrast, the embryos of bony fish are enclosed in a chorion (*zona radiata*) consisting of two layers: the outer (*zona radiata externa*), and the inner (*zona radiata interna*) in contact with the perivitelline fluid. Teleost eggs contain a single funnel-shaped micropyle and the degree of tapering varies among the different fish (Kobayashi and Yamamoto, 1981; Kunz, 2004).

The number of micropyles markedly varies in various acipenserid species and within the limits of the species, in different females, as well as in regard to the eggs of one female. For instance, in *A. stellatus* from 1 to 13 micropyles were observed, in *H. huso* up to 33 (Dettlaff and Ginsburg, 1954), while in *A. ruthenus* from 5 to 13 micropyles are present (Kowalewsky *et al.*, 1870).

The outer opening of these micropyles is 15 µm in diameter. The micropylar canal tapers twice, eventually terminating at the oolemma with its inner opening of 1.2 µm. Such a high number of micropyles in sturgeons

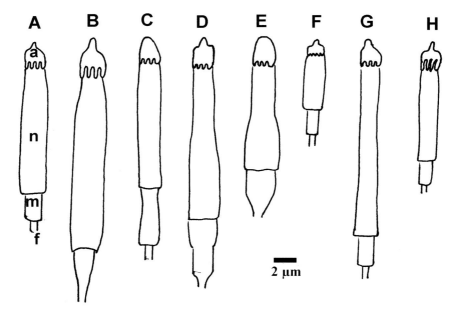

Fig. 2. Drawings to scale of the various sperm cells of several different sturgeon species (redrawn from DiLauro *et al.*, 2001. With kind permission of NRC Research Press, Canada). (A) Lake sturgeon, *A. fulvescens*, (B) White sturgeon, *A. transmontanus*, (C) Russian sturgeon, *A. gueldenstaedti colchicus*, (D) Stellate sturgeon, *A. stellatus*, (E) Chinese sturgeon, *A. sinensis*, (F) Atlantic sturgeon, *A. oxyrhynchus oxyrhynchus*, (G) Shortnose sturgeon, *A. brevirostrum*, (H) Pallid sturgeon, *S. albus*; a—acrosome, f—flagellum, m—midpiece, n—nucleus.

probably increase the possibility of polyspermia which is, however reduced by a double tapering of the canal (Cherr and Clark, 1982).

The spermatozoa of various species differ in total length (Fig. 2). The shortest spermatozoa (38.7 µm) were observed in the Chinese sturgeon (*Acipenser sinensis*) (Wei *et al.*, 2007), while the longest (51.05-81.05 µm) was found in *A. stellatus* (Ginsburg, 1977).

The spermatozoa of the sturgeons are ancient in organization, with a cell having almost a radial symmetry (Fig. 3). The spermatozoon shows a rod-like elongated head with an acrosome and its subacrosome, a short columnar midpiece with the centriolar complex and a long flagellum with the 9+2 microtubular structure (Ginsburg, 1968; Cherr and Clark, 1984; DiLauro *et al.*, 2001; Wei *et al.*, 2007) (Fig. 3). The nucleus occupies the major part of the head, and usually contains three endonucelar canals leading from the implantation fossa to the acrosome. Such a morphology was confirmed for the spermatozoa of *A. transmontanus* (Cherr and Clark, 1984),

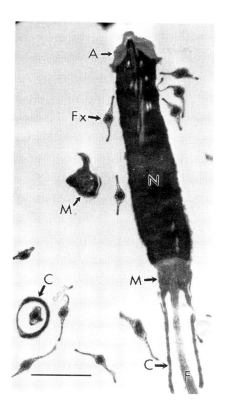

Fig. 3. Longitudinal section of entire pallid sturgeon sperm cell (acrosome, nucleus and mid piece) as revealed by TEM (A) acrosome, (C) cytoplasmic sheath, (F) flagellum, (Fx) cross section of flagellum, (M) midpiece, (N) nucleus, nuclear region. Scale bar 2.0 μm. (DiLauro *et al.*, 2001. With kind permission of NRC Research Press, Canada).

the lake sturgeon (*Acipenser fulvescens*) (DiLauro *et al.*, 2000), *A. stellatus* (Ginsburg, 1977), *A. sinensis* (Xu and Xiong, 1988), *A. gueldenstaedti colchicus* (Ginsburg, 1968), shortnose sturgeon (*Acipenser brevirostrum*) (DiLauro *et al.*, 1999), and pallid sturgeon (*Scaphirhynchus albus*) (DiLauro *et al.*, 2001). In these species, except in *A. oxyrhynchus*, a gradual tapering of the nuclear diameter from posterior to anterior occurs. In the latter species the nucleus is shorter and gradually tapered in the opposite direction (DiLauro *et al.*, 1998).

The acrosome often forms several posteriolateral projections, which were observed in *A. oxyrhynchus, A. fulvescens,* and *S. albus* (DiLauro *et al.*, 1998, 2000, 2001). The sperm cell of *A. stellatus* (Ginsburg, 1977), *A. transmontanus* (Cherr and Clark, 1984), *A. brevirostrum* (DiLauro *et al.*, 1999), and *S. albus* (DiLauro *et al.*, 2001) had three membrane-bound, interweaving endonuclear canals that spanned the nucleus from the acrosome to the implantation fossa. *A. oxyrhynchus* is the only species of sturgeon that has two endonuclear canals (DiLauro *et al.*, 1998). The endonuclear canals probably help in transferring the centriole into the egg after sperm-egg interaction during fertilization (DiLauro *et al.*, 1999). According to Cherr and Clark (1984) and Jamieson (1991) material in the nuclear canals plays a perforatorial role in penetration of the sperm into the egg.

The midpiece of sturgeon sperm cells contains the mitochondria, proximal and distal centrioles and the ovoidal vacuoles (Ginsburg, 1968; Cherr and Clark, 1984; DiLauro *et al.*, 1998, 1999, 2000, 2001). The flagellum of all sturgeon spermatozoa exhibits the classical 9+2 microtubule arrangement (DeRobertis and DeRobertis, 1980), with the axonome originating from the near distal centriole and the flagellum emanating from the axonome. *A. stellatus, A. transmontanus, A. oxyrhynchus, A. brevirostrum* and the Siberian sturgeon (*A. baerii*) show similar structure and length of the flagellum, and gradually assume a distal paddle- or fin-like shape upon adaptation for locomotion during fertilization (Ginsburg, 1977; Cherr and Clark, 1984; DiLauro *et al.*, 1998, 1999; Psenicka *et al.*, 2007). Paddlefish spermatozoa show a very similar structure to sturgeon sperm cells. Their ultrastructure and architectural features were described by Zarnescu (2005). Sperm motility is greatest immediately after activation. All sperm motility parameters (motion frequency, velocity and wave amplitude) decrease rapidly during the period after activation, and the percentage of motile cells also gradually decreases (Cosson *et al.*, 2000). During the earliest period of motility, spermatozoa of sturgeons and paddlefish move at velocities of 175-250 mm^{-1} (Cosson *et al.*, 2000) and then the forward motility is gradually reduced to between 50 and 100 mm s^{-1} at 3-6 min after activation. According to Ginsburg (1968), in 5-10 min. after addition of water, spermatozoa of *A. gueldenstaedti* cease motion, but small part of the spermatozoa swim for a long time, up to 20-60 min., or even several hours.

The results of various studies (Cherr and Clark, 1982, 1985a,b) revealed considerable differences between the gametes of Acipenseriformes and the gametes of bony fish (teleostei). It was suggested that the eggs of white sturgeon (*A. transmontanus*) can remain in freshwater for hours without activation. Spermatozoa of sturgeons that undergo exocytosis and filament formation show an acrosome while the eggs have numerous micropyles. The egg-shell of *A. transmontanus* releases a soluble factor that induces the acrosome reaction in homologous sperm (Cherr and Clark, 1985a). Gamete interaction in *A. transmontanus* was described by Cherr and Clark (1985b). Sturgeon spermatozoa have to pass a 50 µm thick egg shell to reach the oolemma. Therefore, the acrosomal reaction takes place in the micropylar canal, at the level of the third layer (L3), where the inducer, 66KD glycoprotein, is located (Cherr and Clark, 1985a).

The penetration of spermatozoon into the oocyte induces many changes leading to the continuation of the second meiotic division, which was interrupted in metaphase. The cortical reaction acts as a reaction blockage of polyspermy, and the egg cytoplasm shrinkage is resulting in development of the perivitelline space between the egg-shell and the ooplasm. The block to polyspermy in sturgeon is apparently the extremely fast cortical reaction that is triggered when spermatozoa contact the oolemma. This reaction involves the exocytosis of numerous cortical granules that lie beneath the oolemma so that the plasma membrane of the egg is altered, and the cortical granule material is rapidly sealing the inner micropylar openings. This cortical reaction appears quick enough to block supernumerary spermatozoa from penetrating the oolemma (Cherr and Clark, 1985b).

Dettlaff and Ginsburg (1954), and Ginsburg (1968) observed that the somewhat oval, unfertilized egg of the sturgeon always lies with its animal-vegetal axis parallel to the substrate to which it is attached. Immediately after fertilization and the cortical reaction the zygote is released from the tight grip of the envelope, becomes spherical and rotates by 90° so that the animal pole is brought to the highest point. According to Bemis and Grande (1992), unfertilized eggs of paddlefish, *P. spathula,* are situated on the substrate with the animal pole upwards. Similarly to sturgeons, within a few minutes post-fertilization an abrupt expansion of the

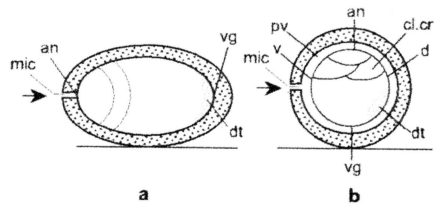

Fig. 4. The radially symmetrical egg (a) prior to fertilization, b) reorganization of the cytoplasm into a bilateral symmetric pattern following the rotation of the egg after fertilization. (an) animal pole, (cl. cr) clear crescent, (d) dorsal, (dt) vegetal determinations (yellow), (vg) vegetal pole, (mic) microopyle, (v) ventral, (pv) perivitelline space. (Eyal-Giladi (1997). With kind permission of The Company of Biologists Ltd. Cambridge, England).

perivitelline space and rearrangement of pigment granules in the ooplasm take place. This results in the development of a concentric ring of cortical granules around the animal pole. Dark pigment distributed under the cell membrane relocates together with the cortical cytoplasm. This results in a clear region (called a clear crescent, similar to amphibians) at the boundary between the animal and vegetal hemisphere. A rearrangement of the previously radially distributed cytoplasm follows, and a clear crescent appears on the descending side of the vegetal pole. The clear crescent denotes also the dorsal and rear part of the embryo. If the zygote is not disturbed, the middle point of the clear crescent, together with the animal and vegetal poles, will indicate the plane of bilateral symmetry (Eyal-Giladi, 1997) (Figs. 4a and b).

CLEAVAGE

In Acipenseridae and Polyodontidae cleavage is holoblastic, and its course is similar in both cases. Developmental stages of sturgeons, from the unfertilized egg (stage 1; Fig. 5a) and Fig. 5 (b-u) until the prolarval stage (stage 45), were described by Dettlaff and Ginsburg (1954) and Dettlaff *et al.* (1993). This description was applied by various authors also for other acipenserid species, i.e. *A. stellatus, A. gueldenstaedti* and *H. huso* (Ballard and Needham, 1964; Bemis and Grande, 1992; Bolker, 1993a). The present contribution has been completed, in part, on the basis of the experimental work conducted on the embryonic development of three species of sturgeons, *A. ruthenus, A. gueldenstaedti* and *A. baerii*. The developmental stages[1] correspond to the classification by Dettlaff and Ginsburg (1954) and Dettlaff *et al.* (1993).

The first cleavage in sturgeons divides the cytoplasm into two halves (stage 4; Fig. 5b). The second cleavage furrow passes at right angles to the first one, across the centre of animal pole, and divides the egg into four more or less even parts (stage 5). Two clefts of the third division are parallel to the first cleft, and in the animal pole eight blastomeres develop (stage 6; Fig. 5c,d). The depth of the clefts is limited. Due to the abundance of yolk in the vegetal pole, separation depth in this part of the egg is lowered compared to the animal pole. The fourth division results in an equatorial cleft, which is irregular and situated near the animal pole (stage 7; Fig. 5e,f). Concentration of the yolk and delay in divisions of the vegetal pole result in a difference in blastomere size at the end of the cleavage phase (Fig. 5g,h). In the animal hemisphere, blastomeres are much smaller compared to the yolk-loaded blastomeres of the vegetal part. Bolker (2004) observed in *A. transmontanus* at the end of cleavage in the animal pole, that the cytoplasm subdivided into cells too small to be distinguished under a dissecting microscope (Fig. 5i,j). At stage 11 the animal hemisphere is lighter

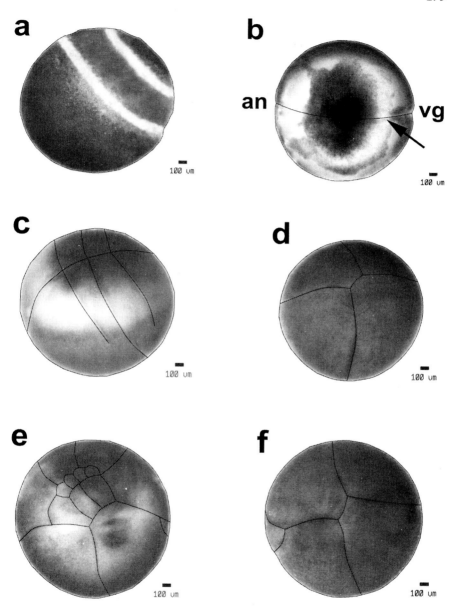

Fig. 5a-f. See page 182 for caption.

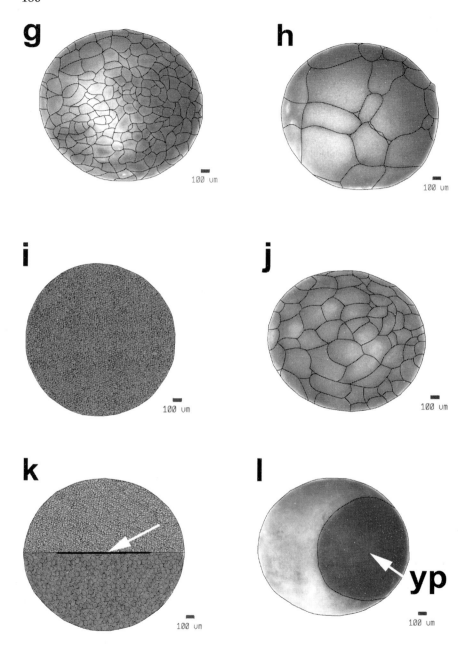

Fig. 5g-l. See page 182 for caption.

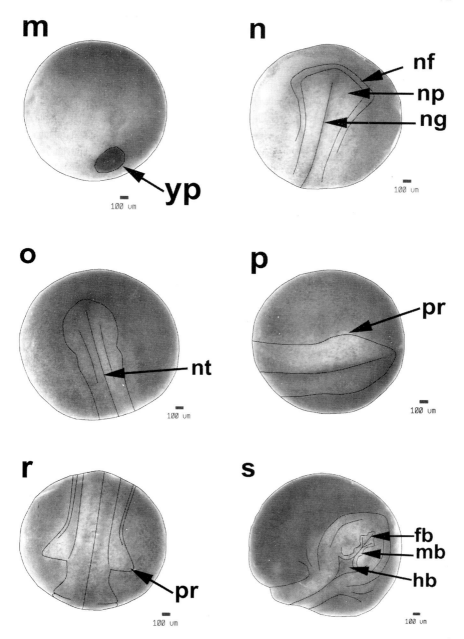

Fig. 5m-s. See page 182 for caption.

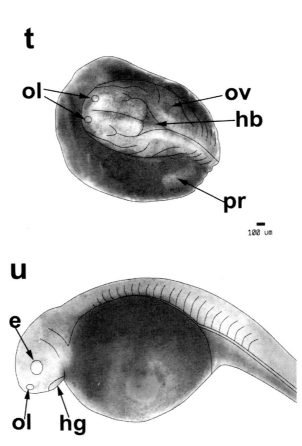

Fig. 5t-u. Selected stages of embryonic development of *A. ruthenus*. (a) Unfertilized oocyte (stage 1); (b) First cleavage (arrow) (stage 4) an—animal pole, vg—vegetal pole; (c) Eight blastomere animal view (stage 6); (d) Vegetal view (stage 6); (e) 16-cells animal view (stage 7); (f) Vegetal view (stage 7); (g) Animal hemisphere (stage 9); h) Vegetal hemisphere (stage 9); (i) Animal hemisphere (stage 11); (j) Vegetal hemisphere (stage 11); (k) Equatorial blastopore (arrow) formation (stage 14); (l) Large yolk plug (yp) (stage 16); (m) Small yolk plug (yp) (stage 17); (n) Neural plate head view (stage 20/21) nf—neural folds, ng—neural groove, np—neural plate; (o) Neural plate head view (stage 23) nt—neural tube; (p) Pronephros (pr) wing like (stage 24); (r) Excretory system (stage 25) pr—pronephros; (s) Brain development (stage 26) fb—forebrain, hb—hindbrain, mb—midbrain; (t) Sensory organs (stage 28) hb—hindbrain, ol—olfactory placode, ov—otic placode/vesicle; pr—pronephros; (u) Differentiation of sensory organs (stage 32-35) e—optic placode, hg—hatching glands, ol—olfactory placode.

compared to the vegetal part. The marginal zone between the animal (unpigmented), and vegetal (pigmented) hemisphere is placed in the region corresponding to the clear crescent of the egg before the cleavage, and the cells of this zone show medium size.

In sturgeons, at the stage of 16-32 blastomeres, accumulation of the fluid and development of small cavities of irregular shape between the blastomeres were observed. Later on, a single cavity appeared as a result of the fusion of the small cavities, and the embryo showed a structure typical for the blastula stage (Dettlaff and Ginsburg, 1954). The blastula cavity is located at the centre of the animal pole, at the boundary with blastomeres which contain a large amount of the coarse-granular yolk. The size of the cavity gradually increases, while its roof becomes thinner (Dettlaff *et al.*, 1993). The blastula is covered by a superficial layer of flattened epithelial cells, surrounding blockier internal cells that increase in size from the animal to the vegetal pole, reflecting the gradient in yolk distribution. In *A. transmontanus* (Bolker, 1993a), in the late blastula stage (stage 12), cells in the equatorial marginal zone are 0.25-0.5 mm in diameter, while vegetal macromeres are 0.5-1.0 mm across. The floor of the blastocoel is about 20° above the equator of the egg. The thick lateral walls of the blastocoel are composed of about 10 cell layers at the beginning of gastrulation, and the blastocoel roof is three to four cells thick in the white sturgeon (Bolker, 1993a).

GASTRULATION

Gastrulation is the process in which blastomeres of similar developmental characteristics develop groups of cells called germ layers. Gastrulation is a complex process involving a complete alteration of embryonic metabolism, first processes of embryonic induction, activation of embryonic genome, determination and the onset of differentiation of the cells.

The first visible sign of gastrulation in sturgeon is a small transversal pigmented line between the clear crescent, and the vegetal part (stage 13). The pigmented line formed by the apices of the bottle cells denotes the onset of development of the dorsal lip of blastopore (Fig. 5k). The bottle cells show the ability of active movement, and in sturgeon may act to anchor the deep layer of the marginal zone to the epithelium during involution, as they do in *Xenopus* (Bolker, 1993a.) In various species of sturgeons the dorsal lip develops at various distances above the egg equator. In the *A. transmontanus*, gastrulation starts in the equatorial marginal zone, mid way between the animal and vegetal parts (Bolker, 1993a), while in A. *stellatus* bottle cells develop closer to the animal part (Ignatieva, 1965).

The apical surface of the bottle cells is primarily a part of the marginal zone or upper vegetal region epithelium. The dorsal lip of the blastopore develops in *A. transmontanus* at stage 14, when the bottle cells migrate from the pigmented equatorial line inwards. The movement of dorsal superficial material into the interior of the embryo begins as an invagination at the side of the dorsal lip, and continues by a process of involution as marginal zone material rolls around the lip, which simultaneously advances toward the vegetal pole during epiboly (Bolker, 1993a). During the gastrulation, material gradually involutes into the depth of blastula cavity, which decreases, and finally completely disappears at the moment of blastopore closure (stage 18). At the same time this is the end of involution. The archenteron appears, and opens outwards through the blastopore.

Epiboly (expansion) of animal cap tissue is a process beginning early during gastrulation (stage 14) It transfers marginal zone vegetally, so that its construction is mechanically effective in production of involution (Bolker, 2004). Bolker (1993a) monitored by time-lapse filming the movement of points (pigment concentration) occurring at the surface of the dorsal side of the gastrula. She observed that at the stage 14 the surface of the animal hemisphere moves downwards to the dorsal blastopore lip. During gastrulation, the blastocoel roof shows gradual thinning—a decrease in number of cell layers. Initially, at stage 13 it consists of about 4-5 layers of the cells. At stage 14, when the invagination that initiates archenteron formation begins, the blastocoel roof thins to 2-3 cell layers. By stage 15-16 it consists of 1-2 cell layers (Bolker, 1993a). Blastopore dorsal lip extends (stage 15) laterally, and migrates ventrally towards the vegetal pole, forming a complete ring around the large yolk plug (stage 16; Fig. 5l). It then moves downwards from the equator to the latitude of about $50°$ above the vegetal pole. At the early gastrula stage, the dorsal body portion of the embryo extends. Convergence of the dorsal part begins after extension of the lip to $90°$, when its lateral ends migrate dorsally. Further extension towards the vegetal pole takes place after the stage 15, and still continues during neurulation.

During gastrulation, a distinct zone (cleft of Brachet) appears between the pre- and post- involuted material. The tissues at each side of this zone migrate in opposite directions, the outer vegetally toward the lip, and the inner up inside the embryo toward the animal pole (Fig. 6a,b). The involuting and non-involuting components of the dorsal marginal zone with the cleft of Brachet are situated outside of the developing archenteron. The involuted layer will be transformed into the mesodermal mantle and archenteron roof lining. The non-involuted layer will develop into the outer ectoderm, and the neural plate. Endoderm develops from macromeres, and yolky endodermal cells give rise to liver, pancreas, and gut. This process in sturgeon bears similarity to amphibians as described in *Xenopus* (Balinsky, 1965). Following involution, the prospective axial mesoderm

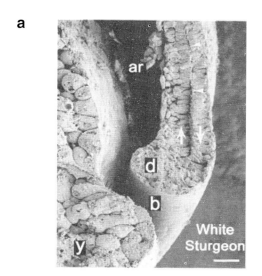

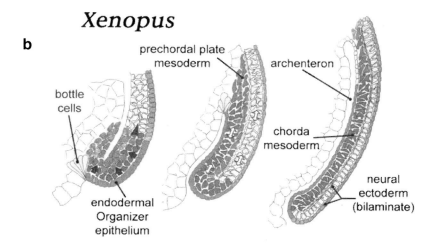

Fig. 6. (a) Gastrulation in the white sturgeon (*A. transmontanus*), a basal neopterygian fish, is similar to gastrulation in *Xenopus*. The Organizer of the white sturgeon (chondrostean) has an Organizer epithelium that involutes into a large archenteron, similar to that of the amphibian *Xenopus* (reprinted from (Bolker, 1993a) with permission of Wiley-Liss, a subsidiary of John Wiley and Sons, Inc.). Scale bar 75 μm. (b) Morphogenetic domains in Spemann Organizer of a *Xenopus* frog gastrula. The endodermal Organizer Epithelium involutes and becomes the roof the archenteron (reprinted from (Cooper and Virta (2007). With permission of Wiley-Liss, a subsidiary of John Wiley and Sons, Inc.). ar—archenteron, b—blastopore, d—dorsal blastopore lip, y—yolky endoderm.

located on the dorsal surface of the late blastula ingresses from a central zone in the posterior archenteron roof surface (Bolker, 1993a).

At the end of gastrulation, the blastopore edges are still closer together (stage 17 *A. ruthenus*; Fig. 5m). The lateral lips of the blastopore approach each other, the yolk plug is drawn inside, and the blastopore edges are closed at stage 18. It is a transitory stage between the gastrulation and the neurulation stage. At the end of gastrulation, the archenteron considerably increases in size thus displacing the centre of gravity of the embryo. Therefore, at this time, the embryo turns its dorsal side upwards inside the egg.

The goal of gastrulation is involution of the outer layers of the embryo (containing the future endoderm and mesoderm) through the blastopore. In experiments with vital dye on *H. huso*, and three other sturgeon species, Ballard and Ginsburg (1980) developed a fate map of late blastula. Staining of small groups of cells allowed observation of the same cells at later stages of development. The results obtained in sturgeon confirmed the situation of the presumptive regions earlier determined in amphibian blastula. They observed that most cells of the animal hemisphere (micromeres) would develop into ectoderm, with the dorsalmost ectoderm developing into neural plate and tube. The cells of the marginal zone would produce mesoderm, and further differentiate into the notochord cells located near the dorsal midline. The somatic mesoderm will transform into two symmetric regions laterally from the notochord, and lateral and ventral plate mesoderm farther around the embryo. Large cells of the vegetal hemisphere (macromeres), filled with yolk, would participate in the development of yolky endoderm, the outer layer of which, situated dorsally, will line the pre-intestine or presumptive gut.

NEURULATION

According to the observations made by Ballard and Ginsburg (1980), the central neural system develops from the superficial dorsal ectoderm, initially as the neural plate and neural tube, being a primordial neural system. At the end of gastrulation, the inner ectoderm layer starts to thicken on the dorsal side of the embryo, above the region of the archenteron roof. Initially, on the dorsal side, the outline of the neural plate is surrounded by neural folds. The neural folds first appear around the anterior margin of the thickened neural plate (a region of the columnar ectodermal cells on the dorsal side), and progress posteriorly parallel to the midline, where a faint neural groove is often visible (Bolker, 2004). The folds grow and lift up, then fuse dorsally into the neural tube, which differentiates later into the brain and spinal cord.

Neurulation in sturgeons was divided by Dettlaff et al. (1993) into five stages. During the first, in early gastrula, the outline of head neural plate and neural folds become visible (*A. ruthenus*; stage 19). It is followed by the second stage characterized by a wide neural plate, all contours of which are well visible, and the neural folds in the head region are distinct (stage 20; Fig. 5n). The third stage begins when the folds begin to draw together, and a primordial secretory system develops (stage 21). The fourth stage of the late neurula, is characterized with abdominal folds which are drawn together in the region of the neural plate (stage 22). The last stage can be distinguished by closure of the neural tube (stage 23; Fig. 5o) (Fig. 7). The beginning of neural plate closure takes place in the anteriormost part, in the region of the future forebrain (*prosencephalon*). In the posterior part, closure proceeds from the rear towards the head. The closure of the neural plate ends in the widest part, in the region of the future midbrain (*mesencephalon*). A similar course of neurulation was described by Bemis and Grande (1992) in paddlefish.

During neurulation two parallel groups of neuroectodermal cells, the neural crest, at the border between the neural folds and ectoderm develop. Some of these cells probably migrate forward and backward, in a similar way as it takes place in other vertebrates, and develop into the pigment cells, neuro-sensory cells, into the cells of autonomic neural system and mesenchyme associated with the neural system and head mesenchyme.

In the *A. transmontanus* (Bolker, 2004) a separate population of neural crest cells migrate around the developing placodes and invaginates into the central neural system. Bemis and Grande (1992) distinguished In the paddlefish the following populations of neural crest cells: mandibular, hyoid, glossopharyngeal. They also found an additional population located behind the midbrain, and described it as a forebrain crest. They additionally explained that two other populations of cranial neural crest cells may simply be part of the mesencephalic crest that is migrating anteriorly and dorsally to the eyes.

During the neurulation also mesoderm starts to divide. The clear crescent material that moved over the pre-intestine roof during gastrulation transforms into the axial mesoderm, from which a primordial notochord and chondromesoderm develop. A cluster of cells from which the notochord will develop migrates within the centre of the pre-intestine roof towards the notochord under the neural plate. The notochord is underlaid by the hypochord plate, which is an endodermally derived tissue.

The notochord is bordered at both sides by adaxial mesoderm that divides into segments, somites, developing during neurulation. At the end of this process, eight to nine somites are visible in the anterior part of the abdomen (*A. ruthenus*; stage 23-24; Fig. 5p). Segmentation proceeds from

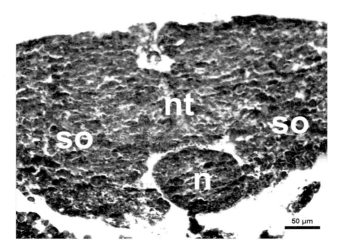

Fig. 7. Transverse section through the trunk regions. Closing of neural tube (stage 22) *A. ruthenus*; n—notochord; nt—neural tube; so—somite.

the anterior to the posterior part. At the sides of the somites, a thin mesoderm layer, called lateral mesoderm, is situated. It splits into two lamellae separated with a slit. The one adjacent to the intestine is called a visceral lamella (*splanchnopleura*), while another, adjacent to the ectoderm, is called a wall lamella (*somatopleura*). The slit between both lamellae is a primordium of the secondary body cavity. Part of material between the rows of somites and lateral mesoderm develops into the secretory system, the primordium of which was observed in several species of sturgeons at the stage 21, as clear ribbons (Dettlaff *et al.*, 1993). In paddlefish, also a hairpin-shaped pronephros was visible at this stage (Ballard and Needham, 1964; Bemis and Grande, 1992) (stage 25; Fig. 5r).

During neurulation, the cleft opens outwards in the posterior part of neural plate, loses its contact with the external environment after neural tube closure, and opens up to the neural tube cavity. This results in a transformation of the blastopore into a neuro-intestinal channel connecting gut cavity with the lumen of neural tube. During neurulation, the growth of endoderm towards the dorsal part of the embryo (which started already during gastrulation) is completed and the archenteron transforms into a gut cavity. The large content of yolk in sturgeon eggs results in a large size of gut tube. Intestinal-stomach cavity is a small area located in the upper part of the gut tube.

ORGANOGENESIS

Brain Development

At the end of neurulation, the widened end of the head elongates and divides into three main brain vesicles: forebrain, midbrain, and hindbrain. Then, secondary division separates the forebrain into telencephalon with olfactory placodes and diencephalon with optic placodes. The midbrain does not divide, while the hindbrain develops into medulla oblongata and the primordium of cerebellum (*A. ruthenus*; stage 26; Fig. 5s, Fig. 8) The placodes are ectodermal thickenings from which optic lenses, otic vesicles and sensory ganglia of the sensory neurons develop (Webb and Noden, 1993).

Development of Sensory Organs

The details of the development of the olfactory organ in *A. ruthenus* and *A. baerii* were described by Zeiske *et al.* (2003). Early during embryonic development (stage 25), the beginning of differentiation of olfactory pits may be observed between the optic placodes, and a depression is visible in the anterior part of the head (Rathke's pouch). Olfactory placodes consist of two superficial epidermal and subepidermal layers. In the olfactory organ three different types of receptor cells occur: ciliated, microvillous and crypt cells. Ciliated and microvillous receptor cells develop from the subepidermal

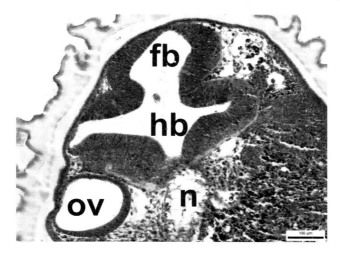

Fig. 8. Sagittal section of the brain of the *A. ruthenus* embryo (stage 29) showing the secondary division of brain in to 5 portions; fb—forebrain; hb—hindbrain; n—notochord; ov—otic placode/vesicle.

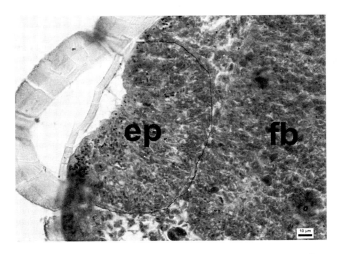

Fig. 9. Sagittal section of the *A. ruthenus* embryo (stage 27). Differentiation of olfactory organ; ep—epidermal layer; fb—forebrain.

placode layer. Non-sensory cells of the olfactory epithelium, such as ciliated non-sensory cells, and supporting cells develop from the superficial epidermal layer. Differentiation of the olfactory placode towards the olfactory epithelium in *A. ruthenus* is shown in Figs. 5t,u and Fig. 9.

At stage 39, a deeply invaginated cavity becomes visible, followed by the development of superficial extensions at each side of the cavity. They grow and finally meet over the ventral cavity, forming a bridge that separates the anterior and posterior opening. The division of the primary olfactory opening into anterior (incurrent) and posterior (excurrent) opening in front of the eye is completed in embryos six days post hatching (*A. ruthenus*; 45 stage; Fig. 10). Both nasal openings show a typical shape, in both juvenile and adult individuals (Zeiske *et al.*, 2003) (Fig. 11). Olfactory placode and optic vesicles are recognizable in *A. gueldenstaedti* (Dettlaff *et al.*, 1993) and in paddlefish at stage 24 (Bemis and Grande, 1992).

Development of the eye begins with the appearance of two involuted structures of nervous tissue, the optic vesicles, at both sides of anterior part of the forebrain. They grow towards the sides and upwards and finally reach the ectoderm. The ectoderm forms a sensory placode. Together with the adjacent vesicle wall, it invaginates inside the embryo, which results in an optic cup formation. Then, the placode separates from the ectoderm, and closes the vesicle, which becomes a primordium of the optic lens (*A. ruthenus*, stage 28; Fig. 12). The optic vesicle transforms into a retina connected with the brain via the optic nerve. The lens, adjacent to the ectoderm

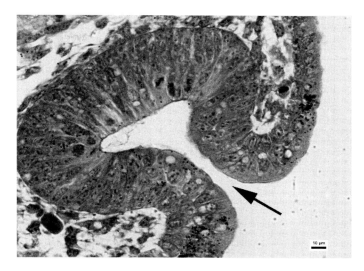

Fig. 10. Longitudinal section of the olfactory organ. Olfactory pit (arrow) is visible. The day of hatching (stage 45) *A. ruthenus*.

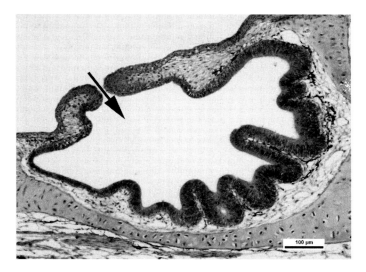

Fig. 11. Longitudinal section of the totally formed olfactory organ of the *A. ruthenus* (15 dph).

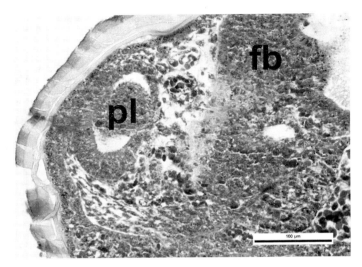

Fig. 12. Transverse section through the optic cup of the *A. ruthenus* embryo (stage 28). Placode separated from the ectoderm becomes optic lens; fb—forebrain; pl—primordium of the optic lens.

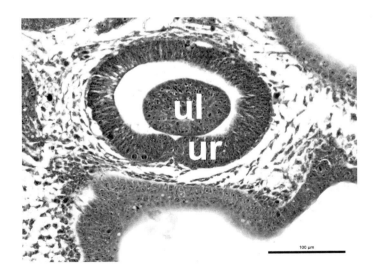

Fig. 13. Longitudinal section of the eye on the day of hatching *A. ruthenus*. Undifferentiated lens (ul) and retina (ur).

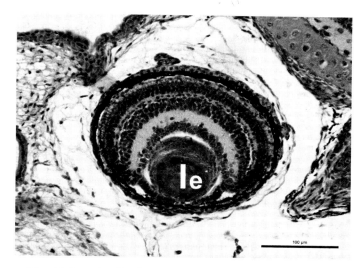

Fig. 14. Longitudinal section of the completely developed optic system in *A. ruthenus* at 7 dph; le—lens.

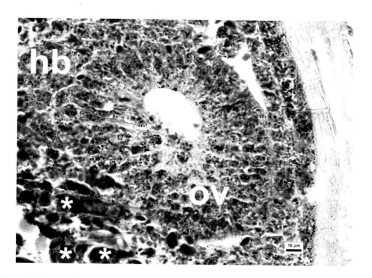

Fig. 15. Sagittal section of the otic vesicle (ov) showing the beginning of differentiation of otic vesicle in the *A. ruthenus* embryo (stage 24); stars—yolk platelets; hb—hindbrain.

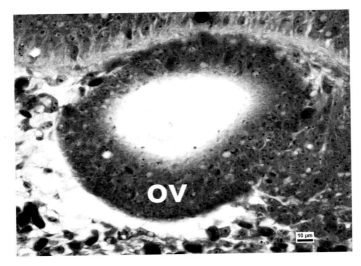

Fig. 16. Longitudinal section of the otic vesicle (ov) on the day of hatching *A. ruthenus*.

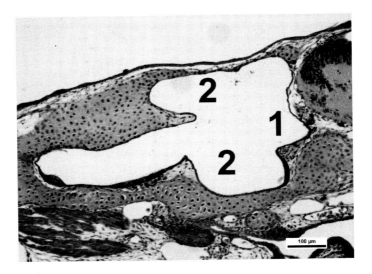

Fig. 17. Longitudinal section of the totally formed otic cavity in *A. ruthenus* at 8 dph. 1-2 semicircular canals are visible.

and mesoderm, proceeds towards differentiation of a colourless cornea.

In the larvae (the newly hatched eleuthero-embryos or postembryos) of sturgeons the eyes have dark patches of pigment and lack well-differentiated photoreceptors (Dettlaff et al., 1993; Loew and Sillman, 1993; Boglione et al., 1999). According to Rodríguez and Gisbert (2002) larvae of *A. baerii* and *A. ruthenus* at hatching show an undifferentiated lens and retina (Figs. 5u, 13). In 3 days posthatching (dph), the following layers may be identified: an external layer of cylindrical epithelium, a rudimentary retina with some melanosomes and tissue, which constitutes crystalline lens. Between the 5th-6th dph the retina of *A. baerii* consisted of 7 layers: the pigment epithelium, the outer nuclear layer (consisting of two types of photoreceptor cells, rods and single cones), the outer plexiform layer and the inner nuclear layer (consisting of horizontal, amacrine and bipolar cells). The inner plexiform layer, the ganglion cell layer and the optic nerve were also observed. The visual system of *A. baerii* and *A. ruthenus* was completely developed on 7 dph (Rodríguez and Gisbert, 2002) (Fig. 14).

Simultaneously with the development of gill arches, otic placodes form at the level of the second arch. Initially they are plates of columnar ectodermal epithelium, then the cells proliferate and invaginate into the mesoderm. This results in the formation of an otic cavity followed by closure of the otic vesicle. In both, sturgeon (Dettlaff et al., 1993) and paddlefish (Bemis and Grande, 1992) otic vesicles develop at the stage 24 (Fig. 5t, 15). A further differentiation of ear opening took place after hatching in *A. ruthenus* (Fig. 16) and a completion of the process was observed at 8 dph (Fig. 17).

Development of Pituitary and Pineal Gland

The pituitary gland, which controls the activity of other endocrine glands, develops from two separate ectodermal primordia, one being a projection of the epithelium of the buccal cavity, and the other a midbrain projection. It appears in the anterior part of the head of the sturgeon and paddlefish at stage 24 as a hypophysial invagination (homologue of Rathke's pouch) (Bemis and Grande, 1992). In the epidermal layer, rostral to the hypophysial invagination, is situated a cluster of cells, from which hatching glands would develop (*A. ruthenus*; Fig. 18) (Ostaszewska, 2002a). When the embryo raises its head over the yolksac (stage 27-32), the pituitary cells migrate towards the ventral surface of the head (Fig.19). In *A. transmontanus* these cells appear about 2.5 days before hatching as an oval mass in a narrow central depression (hypophysial cavity), near the ventral diencephalon boundary (Grandi and Chicca, 2004). The ventral region of the pituitary gland is adjacent to the epithelium of the buccal cavity. According to the

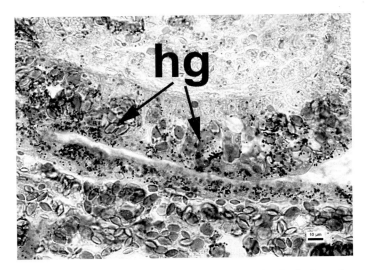

Fig. 18. Longitudinal section of the hatching glands (hg) on the ventral side of the head *A. ruthenus*.

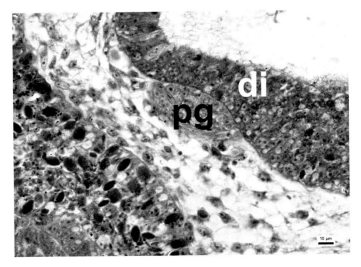

Fig. 19. Longitudinal section of the pituitary gland (pg) on the day of hatching in *A. ruthenus*; di—diencephalon.

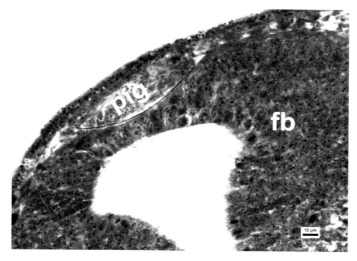

Fig. 20. Sagittal section of the pineal gland in the *A. ruthenus* embryo (stage 29-30). In the rostral end of the forebrain (fb) pineal gland (pig) is visible.

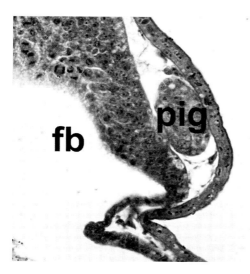

Fig. 21. Longitudinal section of the pineal gland (pig) on the day of hatching in *A. ruthenus*; fb—forebrain.

results of immunocytochemical studies carried out by Grandi and Chicca (2004) a structural division of the pituitary gland into *pars intermedia* and *pars distalis* did not yet occur at this time, and the gland shows exclusively growth hormone immunoreactive cells. One day before hatching, immunoreactivity of thyreotropic and adrenocorticotropic cells was observed. On the 8th dph, the developing gland shows two regions of adenohypophysis: posterior (presumptive *pars intermedia*) and an anterior part (presumptive *pars distalis*). In the larvae between 30 and 35 dph, the adenohypophysis elongates and remains close to the diencephalon, and no nerves were observed to reach it from the hypothalamus (Grandi and Chicca, 2004). The process of development and differentiation of the cells of the anterior pituitary (adenohypophysis) in Acipenseriformes is similar to the one observed in teleost fishes (Saga *et al.*, 1999; Pandolfi *et al.*, 2001).

At this time of hatching, also the pineal gland primordium becomes visible (*A. ruthenus*; stage 24) in the rostral end of the prosencephalon as a small thickening extending dorsally (Figs. 20-21). The pineal gland is a sensory and endocrine organ. Light inhibits its secretory activity, while darkness stimulates melatonin production.

Muscle Development

Development of muscle in sturgeon embryos begins after formation of the spinal chord, notochord, and tailbud. It takes place at the stage of 20 somites, at 3 (dpf).

Somites develop from undifferentiated cells of adaxial mesoderm. Like in teleosts, within the sturgeon myotome slow muscle fibers, derived from the adaxial precursor cells, are located laterally and fast fibers are located medially (Flood *et al.*, 1987; Daczewska and Saczko, 2005; Steinbacher *et al.*, 2006). Similarly to the teleosts (Devoto *et al.*, 1996; Stoiber *et al.*, 1998), the area of adaxial cells is first extended dorsally and ventrally along the medial surface of the myotomes. The adaxial cells are gradually transferred laterally through the rest of the myotome until they eventually form the superficial slow muscle fiber layer (Steinbacher *et al.*, 2006). Somitogenesis proceeds from the head towards the tail. Presumptive myoblasts contain large centrally located nuclei (often in the stage of mitosis) and numerous yolk platelets (Fig. 22). Developing somites are covered with embryonic epithelium which does not show segmentation. Myoblasts become spindle-shaped, with large heterochromatic nuclei. Before fusion to form multi-nucleated muscle lamellae, mononucleated myoblasts of the somites are situated linearly in rows between the adjacent presumptive myosepts. Synthesis of myofibrils begins just before hatching (*A. ruthenus*; stage 44; Fig. 23), and myogenesis continues after hatching. The myosept tissue differentiates and the somites become V-shaped. Differentiation of the outer

199

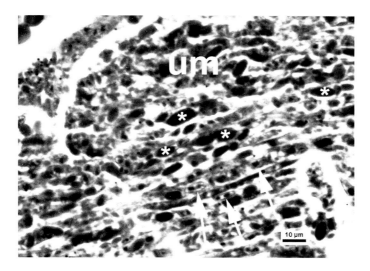

Fig. 22. Sagittal section through the embryo. Unsegmented mesoderm (um). Presumptive cells of myoblasts (white arrows) in *A. ruthenus* 3 dpf; yolk platelets (stars).

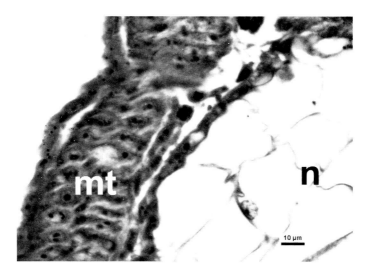

Fig. 23. Transverse section of the myotome (mt) in *A. ruthenus* (stage 44). The beginning of myofibril synthesis; n—notochord.

muscle layer was observed on the 2nd dph, and on the 3rd day two distinct layers were visible: an inner layer consisting of lamellar white muscle fibers, and an outer single layer of red muscle fibers (*A. ruthenus*; 3 dph, Fig. 24).

The outer layer of red fibers is accompanied by undifferentiated cells situated between the red fibers and the embryonic epithelium, and at the boundary between the red and white fibers. According to Groves *et al.* (2005), Devoto *et al.* (2006) and Steinbacher *et al.* (2006) these undifferentiated external cells on the lateral surface of the embryonic myotome are myogenic, and in fact constitute the fish dermomyotome. At hatching, the PCNA-positive cells were observed in the muscle tissue, and the most intensely dividing cells occurred near the notochord, while on the 4th day post hatching the PCNA-positive cells were present in the red muscle layer, and on the top of the apical myotomes (both ventral and dorsal) of white muscles. Recruitment of new fibers of smaller diameter was also observed in the lateral parts of the myotomes, at the boundary with the red fiber layer (*A. ruthenus*; Fig. 25) and at the apices of the ventral and dorsal myotomes. The recruitment of new fibers in sturgeons is similar to teleosts, and reflects stratified growth of the fast muscle (Johnston *et al.*, 1998). Fast fibers in sturgeons arise *de novo* from myogenic precursor cells within growth zones lateral to the original fast fiber stock, and at the layer of the myotomal apices (Steinbacher *et al.*, 2006).

During the myogenesis in *A. ruthenus* no differentiation of horizontal connective tissue, septum, dividing dorsal and ventral myomeres was observed (such a septum occurs in teleost fishes until the 14th dph), neither a mid-lateral insertion of the red muscle layer is present (Steinbacher *et al.*, 2006; own data).

Mesenchymal cells adjacent to the apical and central myotomes were present at the boundary with white muscle fibers, and in myosepts between the myotomes. Transversal sections of the white muscle fibers showed nuclei of various shape: from columnarly elongated to more round ones. The PCNA-staining revealed mitotic activity of the cells of both myosepts and myotomes between the myotubes. These mesenchymal cells probably migrate from the myosepts to the myotomes. The PCNA-positive nuclei were observed at the surface of muscle fibers and inside the myotubes. Darker elongated nuclei of mesenchymal cells were present between the myotubes. The ultrastructural study by Daczewska and Saczko (2005) revealed that in embryos of *A. baerii* cells comprising intermyotomal areas differentiate into fibroblasts, while the cells that have migrated into the myotomes differentiate into secondary myoblasts and contribute to the hypertrophic growth. During myogenesis in sturgeon multinuclear myotubes develop as a result of fusion of myoblasts derived from somite

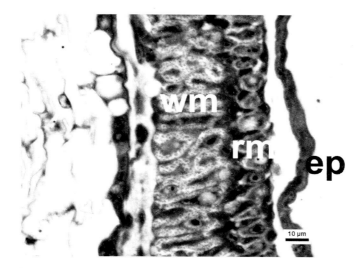

Fig. 24. Transverse section of the red (rm) and white (wm) muscle *A. ruthenus* 3dph; ep—epidermis.

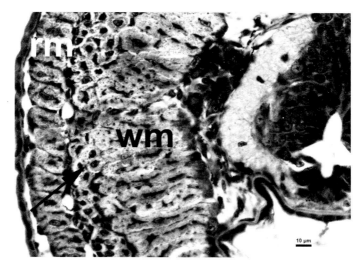

Fig. 25. Transverse section of the muscle of *A. ruthenus* (6 dph). Recruitment of new fibers (arrow) in lateral parts of myotomes; rm—red muscle; wm—white muscle.

cells. Muscle growth in early embryonic development results mainly from the hypertrophy of somite cells. After hatching both hypertrophy and hyperplasia occur simultaneously and contribute to muscle growth. Mesenchymal cells that have migrated from the intermyotomal space participate in both processes (Daczewska and Saczko, 2005).

Excretory System and Gonad Development

The excretory organs develop from the intermediate mesoderm, which becomes arranged into segmental rows of nephrotome cells connecting the somite with lateral mesoderm. The beginning of the development of the excretory system in sturgeons and paddlefish takes place at the end of neurulation (Ballard and Needham, 1964; Bemis and Grande, 1992; Dettlaff *et al.*, 1993). The primordial excretory organ (pronephros) forms a loop at the level of the 4th-7th somite (bends outwards and returns) (Fig. 5p,r). It gradually grows posteriorly in length (*A. ruthenus*; stage 24). The anterior part concentrates and undergoes secondary division becoming a primordial pronephros consisting of six pronephric tubules. The remaining part of the primordium develops into the majority of the pronephric duct (Dettlaff *et al.*, 1993). The ducts reach endodermal cloacal invagination (stage 27-28; Fig. 26) and open into it (Ballard and Needham, 1964). At the time when somites change their shape, pronephric ducts migrate towards the notochord.

Further kidney differentiation takes place already during the post-embryonic development of sturgeons. Initially, the primordia of opisthonephric placodes appear in the ventral part and form regularly distributed nephrons (one per segment). The larger caudal kidney portion of *Acipenser* is not divided into a meso- and metanephros, but represents a continuous structure, the opistonephros (Wrobel *et al.*, 2002). According to these authors in the first days after hatching opisthonephric nephrons consist of three parts: an individual nephric corpuscle (Malpighian body; Fig. 27), the tortuous, non-ciliated opistonephric tubule and the developing nephrostomial tubule (Fig. 28). The opistonephric tubule begins in the vestibulum, turns from dorsal to the ventral direction, and then opens from the dorsal side into the Wolffian duct. In *A. ruthenus, A. baerii* and *A. gueldenstaedti* the nephrostomial tubule opens into the body cavity between the 9th and 10th day after hatching (Fig. 29). At this time the largest nephric corpuscles are situated between the 18th and 25th myotome and show a completely developed Malpighian body, urine space and tubules opening into the Wolffian duct. At the level of 10th-14th myotome the opisthonephros gradually passes over to the cranial pronephros that consists of reticular tissue with solitary tubules and renal bodies (Lepilina, 2007).

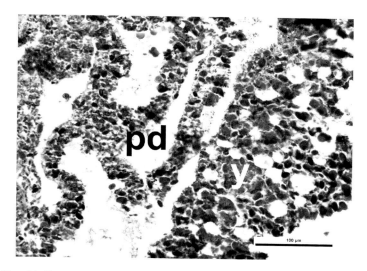

Fig. 26. Transverse section of the *A. ruthenus* embryo (stage 27-28). Opening of the pronephric duct (pd) into endodermal cloacal invagination; y—yolksac.

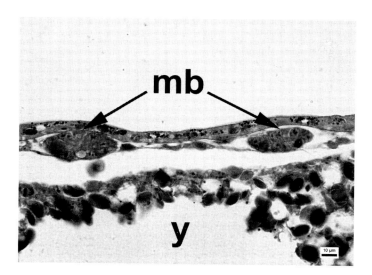

Fig. 27. Longitudinal section of the kidney—Malpighian body (mb) in *A. ruthenus* on the day of hatching; y—yolksac.

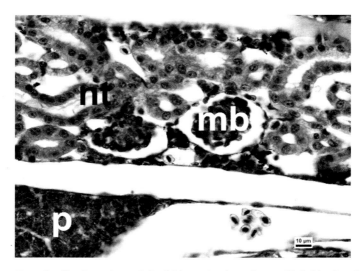

Fig. 28. Longitudinal section of the kidney in *A. ruthenus* (7 dph); developing nephrostomial tubule (nt); mb—Malpighian body; p—pancreas.

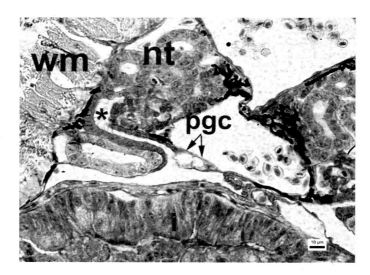

Fig. 29. Transverse section of the kidney. Nephrostomial tubule (nt) opening into the body cavity (star) in *A. ruthenus* (10 dph); i—intestine; pgc—primordial germ cells; wm—white muscle.

The excretory and sexual organs in sturgeons develop from the intermediate mesoderm and their primordial germ cells (PGCs) are located near each other. Nephrostomes open into the body cavity with nephrostomial tubules that widen into a nephrostomial funnel. The cells of the nephrostomial funnel and the lip region proliferate and spread over the body cavity where they replace the original flat mesothelial cells of the lateral plate mesoderm. Proliferating cells of the central part of the nephrostome give rise to the gonad primordium. Somatic cells of the mid-lip origin grow and begin to surround the adjacent PGCs (*A. ruthenus*; 8 dph, Fig. 30). This results in the development of the gonadal crest situated at the level of the stomach, and spiral valve, and spreading in the caudal direction (Fig. 31; *A. ruthenus*, 21 dph). The migrating primordial germ cells were observed along the dorsal wall of the peritoneal cavity and in the presumptive gonad area before formation of the gonadal crest at 6.5 dpf in the Adriatic sturgeon (*Acipenser naccarii*) (Grandi et al., 2007). These observations suggest active migration of primordial germ cells (PGCs) through the mesenchyme (Yoshizaki et al., 2002). PGCs reach their final destination in the gonad crest at about the 10th dph (Grandi et al., 2007; *A. ruthenus*; Fig. 31).

Proliferating cells of the lateral nephrostome side spread towards the body surface above the Wolffian duct as an unilayered cubical epithelium. The area occupied by this special surface epithelium is called the infundibular field (Wrobel, 2003). Development of the infundibular field by the migrating cells begins at about the 14th dph. Proliferating infundibular field cells give rise to the Müllerian infundibulum and Müllerian duct (*A. ruthenus*; Fig. 32). The Müllerian duct grows towards the posterior, in parallel to the Wolffian duct. In *Acipenser* Müllerian ducts unite with Wolffian ducts before the latter fuse to form the urogenital sinus. The infundibular field and Müllerian duct arise in the same manner as described for Amniota (Wrobel, 2003). This model represents the oldest vertebrate on the evolutionary scale with the Müllerian duct development typical for birds and mammals (Wrobel, 2003).

The Development of the Heart

The heart develops from the visceral mesoderm (stage 26-27; *A. ruthenus*, Fig. 33). At the site where the visceral mesoderm contacts the yolk endoderm, at the opposite side of the notochord, a heart primordium develops. It is accompanied by a pair of the yolksac vessels (ducts of Cuvier) that carry the blood back to the sinus venosus. The development of the ducts of Cuvier is simultaneous with the development of the heart primordium. Hepatic veins form ventrally to the primordium of the sinus venosus. At the same time when the cardiac primordium develops as a

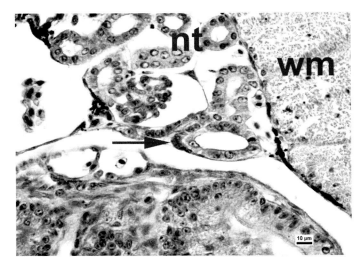

Fig. 30. Transverse section of the *A. ruthenus* larvae (8 dph). Development of gonad primordium from the somatic cells of the mid lip (arrow) of nephrostome; nt—nephrostomial tubule; wm—white muscle.

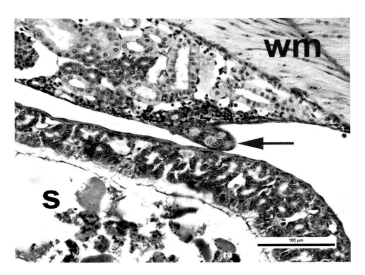

Fig. 31. Transverse section of the *A. ruthenus* larvae (21 dph). Pgc cells in side gonadal crest (arrow) are visible; s—stomach; wm—white muscle.

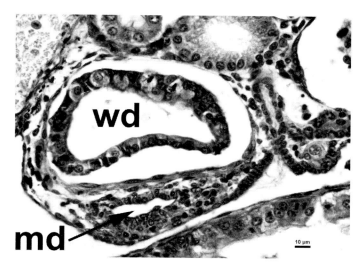

Fig. 32. Transverse section of the Müllerian duct (md) and Wolffian duct (wd) *A. ruthenus* (28 dph).

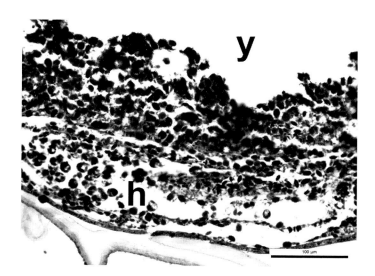

Fig. 33. Transverse section of the yolksac (y) of the *A. ruthenus* embryo (stage 26-27). Differentiation of the heart (h).

straight tube, blood vessels are formed from the visceral mesoderm adjacent to the yolksac. Blood vessels develop simultaneously with the red blood cells in regions called haematopoietic islets. Endothelial walls surround clusters of red blood cells, and then the islets fuse into the blood vessel network. The heart axis is aligned with the axis of the developing head. Then the heart elongates and becomes C-shaped, while its axis becomes perpendicular to the head axis (Fig. 34; *A. ruthenus*). Migration of the heart and pericardial cavity towards the head takes place gradually after hatching, together with yolk material utilization. The heart loop disappears at hatching. In *A. naccarii* at hatching the heart is a straight tube in which the following parts may be identified: outflow tract, ventricle, atrium and small *sinus venosus* (Icardo *et al.*, 2004). During the first four days after hatching the heart becomes C-shaped, starts looping again and undergoes a counterclockwise movement. These processes result in a transfer of the atrium left of the outflow tract while the ventricle moves to a caudal position. At the same time, a *bulbus arteriosus* and *conus* start to differentiate from the outflow tract (Guerrero *et al.*, 2004). The conus arteriosus and the atrium open at the base of the ventricle trough separate right and left orifices. The sinus venosus is connected with the liver primordium and receives two yolksac veins. Morphological differentiation on the 7th-9th dph is already similar to the adult sturgeons (Fig. 35). Development of the cone valves begins on the 8th dph. The primordium of the atrio-ventricular valve appears at the 4th dph and the system of the *chorda tendinae* at 22-24 dph (Icardo *et al.*, 2004). A similar course of embryonic heart development was described for the paddlefish (Ballard and Needham, 1964; Bemis and Grande, 1992).

The primary cardiac loop in sturgeon appears to be an ancient developmental feature that has been lost at some point in evolution. The relationship between the *sinus venosus* and the developing liver of the sturgeon seems particularly interesting. In the sturgeon the developing liver is connected to the heart through a large opening in such a way that the liver sinusoids drain directly into the *sinus venosus*. The studies on *Xenopus* have shown that the part of the embryonic heart (*sinus venosus*) has a close relationship with the developing liver also in amphibians (Mohun *et al.*, 2000).

Development of Digestive System

Development of the alimentary tract is similar in sturgeons and paddlefish (Ballard and Needham, 1964; Weisel, 1973; Ballard and Ginsburg, 1980; Buddington and Doroshov, 1986; Dettlaff *et al.*, 1993; Gawlicka *et al.*, 1995; Gisbert *et al.*, 1998, 1999; Gisbert and Doroshov, 2003). However, there are some differences concerning the time of differentiation of various structures,

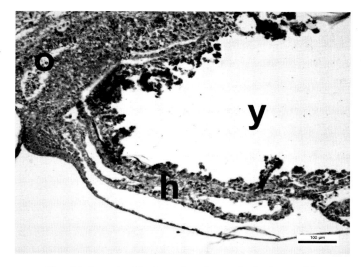

Fig. 34. Longitudinal section through the heart (h).
The heart in the shape of straight tube on the day of hatching
in *A. ruthenus*; o—oesophagus; y—yolksac.

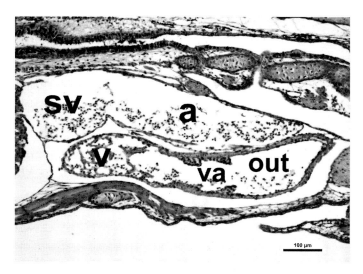

Fig. 35. Longitudinal section through the heart of *A. ruthenus* (8 dph);
a—atrium; out—outflow tract; sv—sinus venosus; v—ventricle; va—valves.

and thus also the moment of functional ability of food utilization by the digestive tract. Embryos of sturgeon differ from those of *Lepisosteus*, *Amia* and teleost fishes in that they show intraembryonic yolk and their alimentary tract develops from the yolksac endoderm (Dettlaff *et al.*, 1993). This is a result of holoblastic cleavage in sturgeon, contrary to the teleost fish in which the egg is subjected to meroblastic cleavage, and in which the alimentary tract develops independently from the extraembryonic yolksac. During development of fish species, the eggs of which undergo meroblastic cleavage, proteolysis of yolk proteins has been considered to occur gradually with a participation of the yolk syncytial layer surrounding the embryonic yolk mass (Sire *et al.*, 1994; Ostaszewska, 2002b; Kunz, 2004).

Immediately after hatching the alimentary tract of the sturgeon is divided into two regions that have no contact with the external environment, because both mouth and anus are closed. The anterior section consists of the closed buccal cavity, occluded oesophagus and the future anterior part of the gut filled with yolk mass, and surrounded by the endodermal cells which will develop into the stomach walls and intestine. The buccal cavity is closed and its lumen filled with yolk platelets (Fig. 36). The posterior section of the alimentary tract consists of the posterior intestine, which will develop into the spiral intestine and the anus. The postembryonic development of the alimentary tract of the sturgeon takes place at the time of endogenous feeding. Differentiation proceeds from the posterior towards the anterior part. The spiral intestine is the first to develop and the stomach (gastric stomach) the last (Buddington, 1985).

The embryos of *A. baerii* (Gisbert *et al.*, 1998), *A. naccarii* (Cataldi *et al.*, 2002) and the green sturgeon (*Acipenser medirostris*) (Gisbert and Doroshov, 2003) at hatching show a closed mouth and anus. The entire alimentary tract is lined with a squamous epithelium. The buccal cavity in *A. medirostris* opens between the first and the second dph and the mouth opening is surrounded by buccal folds covered with multilayered squamous epithelium. Epithelial cells lining the buccal cavity show vacuoles filled with remains of acidophilous yolk and pigment granules (Gisbert *et al.*, 1998). The first mucous cells (goblet cells) appear in *A. medirostris* on the 6th dph, but already on the 7th-8th dph in *A. baerii* (Fig. 37). They show the presence of neutral and acidic carbohydrate compounds. Basophilic taste buds in the epithelium of the buccal cavity differentiate about on 8-9 dph (*A. gueldenstaedti*, Fig. 38) and are completely developed about 10-11 dph in *A. medirostris* (Gisbert and Doroshov, 2003). In *A. baerii*, the development of taste buds takes place much earlier and they are visible already in fish at 3 dph (Gisbert *et al.*, 1998). In *A. naccarii* the taste buds appear in the buccopharynx, on the 4th dph (Boglione *et al.*, 1999). The buccal cavity of sturgeon larvae shows temporary teeth, which in various species develop

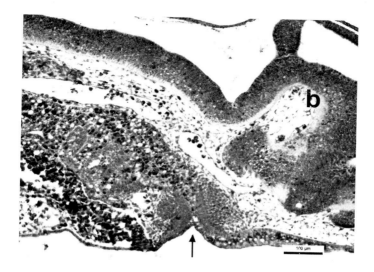

Fig. 36. Longitudinal section of the head showing the buccal cavity still closed (arrow) on the day of hatching in *A. ruthenus*; b—brain.

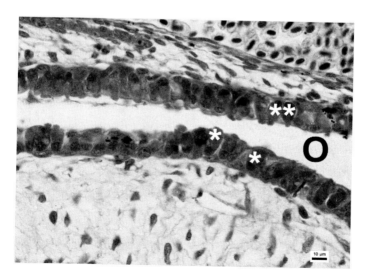

Fig. 37. Longitudinal section of the oesophagus (o) showing mucous cells (stars) *A. baerii* (8 dph).

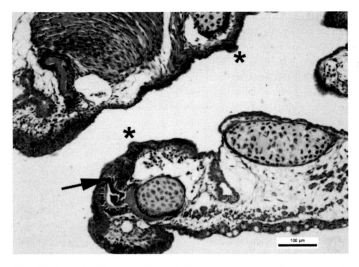

Fig. 38. Longitudinal section of the buccal cavity showing taste buds (stars) and teeth (arrow) *A. gueldenstaedti* (8 dph).

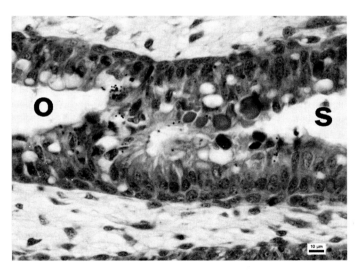

Fig. 39. Longitudinal section of the oesophagus (o) separated from the presumptive stomach (s) by a thin diaphragm in *A. gueldenstaedti* (6 dph).

at different times after hatching (*A. gueldenstaedti,* Fig. 38). Larvae of *A. naccarii* show tiny teeth already on the 4th dph (Cataldi *et al.,* 2002) while in *A. baerii* the teeth appear in the buccal cavity at the time of the shift to exogenous feeding, between 9 and 10 dph (Gisbert *et al.,* 1998). In *A. medirostris* larvae the teeth develop even later, at about 11-12 dph (Gisbert and Doroshov, 2003). According to Balfour (1885) sturgeons lose these teeth at the age of about 3 months. Boglione *et al.* (1999) observed in *A. nacarii* that the teeth on the upper lip disappear after 36 and the palatal ones after 180 days, when the tongue teeth are still present. Adult sturgeons are toothless.

The oesophagus in 1 day old larvae of *A. baerii* is undifferentiated and on the 3rd dph it is covered with a layer of columnar epithelium. The oesophageal cells show the presence of acidophilic yolk and pigment granules in the supranuclear vacuoles. At this time, the oesophagus is not yet connected with the presumptive stomach (Gisbert *et al.,* 1998) (see also *A. gueldenstaedti,* Fig. 39). In *A. medirostris* the oesophagus remains undifferentiated until the 6th dph and the yolk remains are present in the posterior part of the buccopharynx. Yolk particles and melanin granules disappear between the 8th and 9th dph (Gisbert and Doroshov, 2003). At this time in larvae of *A. baerii* and *A. ruthenus* (8-9 dph, Fig. 40) gradual yolksac material utilization results in establishing a connection between the oesophagus and stomach. The oesophagus elongates and differentiates into two functional regions: the anterior part for the secretory function, and the posterior part participating in food storage and transport (Fig. 40).

At hatching sturgeon and paddlefish show an undifferentiated stomach, which is filled with the yolk mass. In two days post hatching, the hind dorsal wall of the yolksac forms a cleft (*A. ruthenus,* Fig. 41), which will divide stomach and the intestine (Buddington and Christofferson, 1985; Buddington and Doroshov, 1986). The process of differentiation of the non-glandular stomach in *A. medirostris* larvae begins on the 6th dph in the anterior ventral region of the yolksac and is completed between the 11th and 12th dph (see also *A. ruthenus,* Fig. 42). At the same time, the stomach shows well developed mucosal folds lined with ciliated columnar epithelium with mucous cells producing acidic and neutral mucins. A smooth muscle layer occurs under the epithelium (Gisbert and Doroshov, 2003). Gisbert *et al.* (1998) reported that the non-glandular stomach with several mucosal folds appears already in 5 days old *A. baerii* larvae. At this time the non-glandular stomach is separated from the intestine with epithelial folds that will develop into a sphincter. The non-glandular stomach of *A. transmontanus* starts to differentiate at 9 dph as a region of intensive proliferation of smooth muscles fiber. At the same time glands begin to develop in the glandular stomach (Buddington and Doroshov, 1986). In *A. baerii, A. naccarii* and *A. gueldenstaedti* gastric glands develop

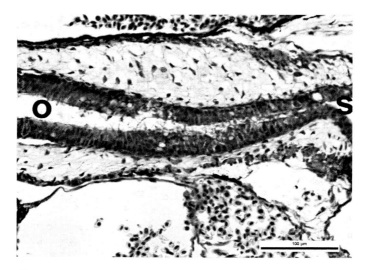

Fig. 40. Longitudinal section of the oesophagus (o) connected with presumptive stomach (s) in *A. baerii* (8 dph).

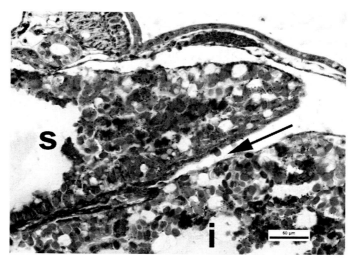

Fig. 41. Transverse section of the *A. ruthenus* larvae (2 dph). A cleft (arrow) dividing the stomach (s) from the intestine (i).

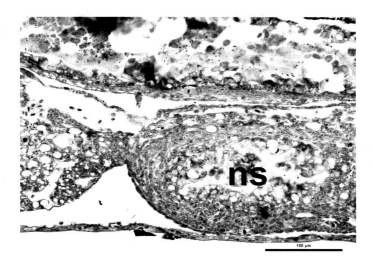

Fig. 42. Longitudinal section of the gastric region of the *A. ruthenus* larvae (5 dph). The beginning of differentiation of non-glandular stomach (ns).

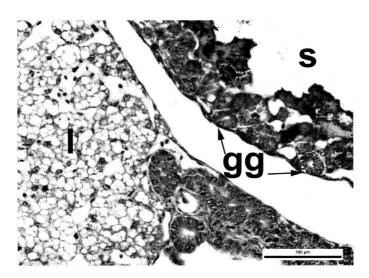

Fig. 43. Longitudinal section of the gastric region of the *A. gueldenstaedti* larvae (8 dph). Differentiating gastric glands (gg) in the stomach (s); l—liver.

between the 8th and 9th dph (*A. gueldenstaedti*, Fig. 43) (Gisbert *et al.*, 1998; Cataldi *et al.*, 2002). At this time, in *A. naccarii* the glandular stomach is still almost entirely filled with yolk (Cataldi *et al.*, 2002). The endodermal epithelium of the yolksac of *A. medirostris* on the 8th dph transforms from the squamous epithelium into columnar epithelium which is the beginning of glandular stomach development. Multicellular gastric glands are distinctly visible on the 12th dph (Gisbert and Doroshov, 2003).

The absorptive surface area of the sturgeon intestine increases during development due to the elongation of the intestine and formation of pyloric caeca (*A. baerii*, 11 dph) Fig. 44). In the adult fish pyloric caeca are the flattened triangular structures situated between the stomach and intestine (Buddington and Christofferson, 1985). A multilobed pyloric caecum is associated with the alimentary canal of all chondrosteans. Apparently this is a structure resulting from a fusion of numerous caeca (Weisel, 1979). In *A. naccarii* organogenesis, pyloric caeca differentiate on the 3rd dph (Boglione *et al.*, 1999). Cataldi *et al.* (2002) observed that in *A. naccarii* on the 8th dph pyloric caeca consisted of two or three cavities while their mucosa was still differentiating. After two days more cavities appeared and the mucosa was completely developed, similarly to the intestinal mucosa. On the 12th dph all pyloric caeca are already developed (Boglione *et al.*, 1999).

The primordial intestine of 1 day old larvae is lined with a differentiating simple columnar ciliated epithelium (Gisbert *et al.*, 1998) (see also *A. ruthenus*, Fig. 45). Yolk platelets fill the supranuclear regions of enterocytes. Such a cellular organization and low concentration of enzymes indicate enhanced endocytosis of yolk material (Buddington, 1985; Heming and Buddington, 1988). At that time lipids produced as a result of yolk digestion are gradually accumulated in the ciliated anterior intestine epithelium and in the liver (Dettlaff *et al.*, 1993). Also accumulation of large lipid droplets in the cytoplasm of enterocytes is considered temporary storage (Watanabe and Sawada, 1985). Numerous lipid vacuoles disappear from the supranuclear regions of the enterocytes of the anterior intestine between the 12th and 13th dph, shortly before the onset of exogenous feeding (Gisbert *et al.*, 1998).

In *A. medirostris* the spiral intestine (posterior intestine section) starts to differentiate as does the first gut segment between the first and second day post hatching. The central part of the intestine develops between the 2nd and 3rd dph, while the anterior section does so on the 7th dph (Gisbert and Doroshov, 2003). Intestinal mucous cells appear for the first time between the 6th and 9th dph.

Organs such as the liver and pancreas are absent at hatching (Buddington and Doroshov, 1986; Gisbert and Doroshov, 2003). The liver

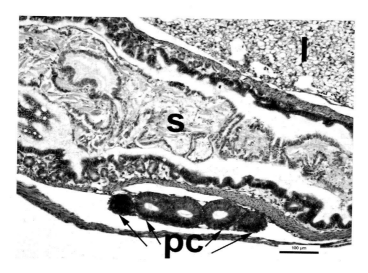

Fig. 44. Longitudinal section of the gastric region of *A. baerii* larvae (11 dph). Formation of the pyloric caeca (pc); l—liver; s—stomach.

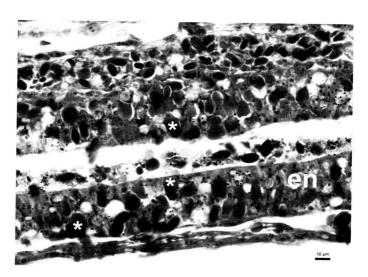

Fig. 45. Longitudinal section of the primordial intestine on the day of hatching *A. ruthenus*; stars—yolk plateles; en—enterocytes.

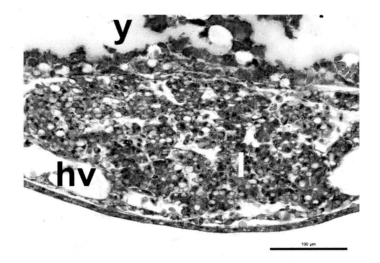

Fig. 46. Transverse section of the liver (l) of the *A. ruthenus* (2 dph). Differentiating hepatocytes; hv—hepatic veins; y—yolksac.

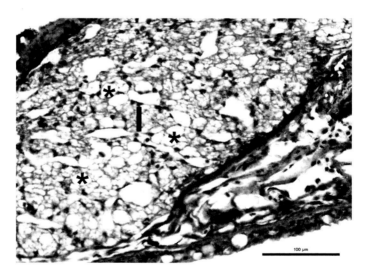

Fig. 47. Longitudinal section of the liver (l) of the *A. gueldenstaedti* larvae (4 dph). Hepatocytes cytoplasm containing lipid vacuoles (stars), and glycogen.

219

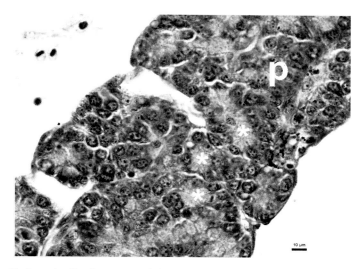

Fig. 48. Longitudinal section of the pancreas (p) of *A. gueldenstaedti* larvae (8 dph). The zymogen granules (stars) present in the pancreas acinar cells.

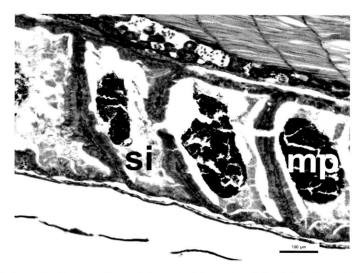

Fig. 49. Longitudinal section of the spiral intestine (si) of the *A. gueldenstaedti* larvae (7 dph). A melanin plug (mp) expelled from the spiral intestine.

primordium in *A. naccarii* and *A. ruthenus* appears already on the first day post hatching, and in 3 days old larvae two hepatic head and lateral lobes portions are visible (Boglione *et al.*, 1999) (Fig. 46). In *A. medirostris* the liver starts to develop at 2 dph in the ventral part of the yolksac. At the same time a pancreas primordium appears dorsally from the cleft dividing the yolksac into a part giving rise to the stomach and into another part from which the intestine will develop (Fig. 41, *A. ruthenus*). On the 4th day, hepatocyte cytoplasm contained lipid vacuoles and glycogen (Fig. 47, *A. gueldenstaedti*). On the 4th-5th dph the exocrine pancreatic cells showed basally situated nuclei and basophilic cytoplasm and they were clustered around a small intercellular space. The zymogen granules were present in the acinar cells before the onset of the exogenous feeding (Fig. 48, *A. gueldenstaedti*). However, Gawlicka *et al.* (1995) observed the presence of proenzyme granules in the exocrine pancreatic cells in *A. transmontanus* larvae only on the 11th dph, before the beginning of exogenous feeding. The digestive tract gradually develops, and at about 2 dph the mouth opens. At the first exogenous feeding the digestive tract consists of a well developed buccal cavity, oesophagus, glandular and non-glandular stomach, anterior intestine, spiral intestine and anus. After completion of the yolk resorption, a melanin plug is expelled from the spiral intestine lumen and the anus opens (Fig. 49, *A. gueldenstaedti*).

DISCUSSION AND SUMMARY

Investigation of the development of sturgeons and paddlefish, both belonging to the order Acipenseriformes, is extremely interesting since their early stages resemble some amphibians rather than teleost fishes. Their phylogenetic position as an early out-group to teleost fishes, together with their amphibian-like embryonic development, makes them a uniquely valuable point of comparison for evolutionary systematics of both teleost and amphibian studies.

According to Cooper and Virta (2007) gastrulation of chondrostean fishes and bichirs (*Polypterus*) proceeds in a manner similar to that of many amphibians and the lamprey, suggesting that this conserved mode of gastrulation stems from a common ancestor that predates the Devonian-Paleozoic period in which amphibians diverged from the common ancestor of fish.

Contrary to Acipenseriformes and many amphibians (with holoblastic cleavage), teleosts undergo meroblastic cleavage with blastodisc formation on top of the uncleaved yolk. In Acipenseriformes and amphibians yolk-material is present in platelets within the cells. Fertilization is followed by a cortical reaction, which is similar in both groups. Cortical rotation results in the formation of a clear crescent on the dorsal side of the egg (Detlaff and Ginsburg, 1954). In the sturgeon (*Acipenser*) the relatively large amount of yolk causes cleavage to be slower in the vegetal than in the animal half. As a result, the more vegetally situated cells remain 'open' for a longer time, which may allow more time for the vegetal determinants to continue their gravity-directed movement towards the future dorsoposterior side.

Cleavage results in a blastula, which consists of small cells called called micomeres (at the animal pole) and large cells called macromeres (at the vegetal pole), with a marginal zone of middle-sized cells between them. The blastocoel is located in the animal hemisphere. The main difference between sturgeons and the clawed frog (*Xenopus*) is the marginal zone location, equatorial in sturgeons, and in the vegetal hemisphere in *Xenopus*. Thus bottle cells develop at the equator in sturgeons, while in *Xenopus* they do near the vegetal pole (Bolker, 1994). Bottle cells initiate involution of superficial material, which develops into the mesodermal and endodermal lining of the gastrocoel. Involution correlates with, and may in part be driven by, the extension of the dorsal side (involuted prospective axial mesoderm, and non-involuted prospective neural tissues) (Bolker, 1993b). The mechanism causing extension in *Xenopus* are radial intercalation of cells within a tissue and mediolateral intercalation in which the tissue converges and extends (Keller *et al.*, 1985). The difference concerns only timing of this process. In *Xenopus* it occurs during epibolic expansion of the animal cap before gastrulation, while in sturgeon it does so through early gastrulation (Bolker, 1994). The early phase of dorsal thinning and extension is longer and more important in sturgeons, where the marginal zone must be moved below the equator before convergence begins (Bolker, 1994).

In sturgeons and *Xenopus laevis* the differences occur in the process of axial mesoderm morphogenesis. In sturgeons the presence of axial mesoderm in the surface layer of the late blastula and its ingression from the archenteron roof into the deep layer after involution was observed. In *Xenopus laevis* all prospective mesoderm lies in the deep layer of the gastrula before involution and there is no ingression of the material from the surface of the ancherteron roof (Bolker, 1993a).

The process of gastrulation in teleost fishes, which is a result of meroblastic cleavage, is quite different from gastrulation in Acipenseriformes. During teleost cleavage multilayered blastoderm develops and overlies an acellular yolk syncytial layer (Sire *et al.*, 1994; Ostaszewska, 2002b; Kunz, 2004).

Also organogenesis is different in Acipenseriformes and teleosts. In general, teleost larvae after hatching have a straight intestine, located on the yolk sac and formed by simple epithelia that differentiate during the absorption of yolk sac and lipid globule. At this time the intestine can be differentiated into foregut, midgut and hindgut (Sarasquete et al., 1995; Ostaszewska and Wegiel, 2002; Ostaszewska et al., 2003; Sysa et al., 2006). On the contrary, unlike in other fish, but similarly to amphibians, cleavage of Acipenseriformes eggs is holoblastic and results in the formation of a yolk endoderm, which participates in the formation of the digestive system (Buddington and Christofferson, 1985; Buddington and Doroshov, 1986; Gawlicka et al., 1995; Wegner et al., 2008).

The primary cardiac loop in sturgeons and paddlefish appears to be an ancient developmental feature that has been lost at some point in evolution. Relationship between the venous sinus (*sinus venosus*) and the liver of developing sturgeons and paddlefish seems particularly interesting. In sturgeon the developing liver is connected to the heart through a large opening in such a way that the liver sinusoids drain directly into the *sinus venosus*. The studies on the clawed frog (*Xenopus*) have shown that in amphibians the embryonic heart also has a close relationship with the developing liver (Mohun et al., 2000).

Steinbacher et al. (2006) confirmed that initial muscle development and associated innervation and vascularisation of the sturgeon follow the well-known teleost pattern, but that this occurs without the presence of slow muscle cells equivalent to the teleost-type 'muscle pioneers' (MPs). MPs form cellular divisions between the dorsal and ventral myotomes at the site of the future collageneous horizontal septum in teleosts (Barresi et al., 2000). Lack of MPs cells in sturgeons may be related to late differentiation (14th day of development) of the collageneous horizontal septum in these fishes, compared to teleosts (in which it is present already in embryos) (Ostaszewska et al., 2008).

Moreover, in sturgeons, the myotome consists of stacked lamellae of the fast muscle fibers covered laterally by a layer of slow muscle fibers (Ostaszewska et al., 2007) while in teleosts the fast muscle fibers are tube-like structures with an isometric diameter (Ostaszewska et al., 2008).

Acknowledgements

The present study was financed from the funds of the Ministry of Science and Higher Education (grant No N311 030 32/2256). The authors would like to express their thanks to Dr. Eugeniusz Bogdan from the Fish Farm RYBA in Olesnica, Poland for supplying fish, and to Mr. M. Karaszewski M.Sc. for assistance in preparation of micrographs. We thank Yvette Kunz for discussions and comments on the final version of the manuscript.

References

Balfour, F.M. 1885. A Treatise on Comparative Embryology. Macmillan Publishers Ltd., London.

Balinsky, B.I. 1965. An Introduction To Embryology. 2nd Edition. W.B. Saunder Company, Philadelphia, pp. 673.

Ballard, W.W. and R.G. Needham. 1964. Normal embryonic stages of *Polyodon spathula* (Walbaum). *Journal of Morphology* 114: 465-477.

Ballard, W.W. and A.S. Ginsburg. 1980. Morphogenetic movements in acipenserid embryos. *Journal of Experimental Zoology* 213: 69-103.

Barresi, M.J.F., H.L. Stickney and S.H. Devoto. 2000. The zebrafish slow muscle-omitted gene product is required for hedgehog signal transduction and the development of slow muscle identity. *Development* 127: 2189-2199.

Beer, K.E. 1981. Embryonic and larval development of the white sturgeon (*Acipenser transmontanus*). Master's thesis. University of California, Davis.

Bemis, W.E. and L. Grande. 1992. Early development of the actinopterygian head. I. General observations and comments on staging of the paddlefish, *Polyodon spathula*. *Journal of Morphology* 213: 47-83.

Bemis, W.E., E.K. Findeis and L. Grande. 1997. An overview of Acipenseriformes. *Environmental Biology of Fishes* 48: 25-71.

Berg, L.S., A.S. Bogdanov, N.I. Kozhin and T.S. Rass. (Eds). 1949. Commercial Fishes of the USSR. Fishes Description. Pishchepromizdat, Moscow. (In Russian).

Billard, R. and G. Lecointre. 2001. Biology and conservation of sturgeon and paddlefish. *Reviews in Fish Biology and Fisheries* 10: 355-392.

Birstein, B.J. and R. DeSalle. 1998. Molecular phylogeny of Acipenserinae. *Molecular Phylogeny and Evolution* 9: 141-155.

Blacklidge, K.H. and C.A. Bidwell. 1993. Three ploidy levels indicated by genome quantification in Acipenseriformes of North America. *Journal of Heredity* 84: 427-431.

Boglione, C., P. Bronzi, E. Cataldi, S. Serra, F. Gagliardi and S. Cataudella. 1999. Aspects of early development in the Adriatic sturgeon *Acipenser nacarii*. *Journal of Applied Ichthyology* 15: 207-213.

Bolker, J.A. 1993a. Gastrulation and mesoderm morphogenesis in the white sturgeon. *Journal of Experimental Zoology* 266: 116-131.

Bolker, J.A. 1993b. The mechanism of gastrulation in the white sturgeon. *Journal of Experimental Zoology* 266: 132-145.

Bolker, J.A. 1994. Comparison of gastrulation in frogs and fish. *American Zoologist* 34: 313-322.

Bolker, J.A. 2004. Embryology. In: *Sturgeons and Paddlefish of North America*, G. LeBreton, F. Beamish and R. McKinley (Eds.). Kluwer Academic Publishers, Dordrecht, pp. 134-146.

Buddington, R.K. 1985. Digestive secretion of lake sturgeon (*Acipenser fulvescens*) during early development. *Journal of Fish Biology* 26: 715-723.

Buddington, R.K. and J.P Christofferson. 1985. Digestive and feeding characteristics of the chondrosteans. *Environmental Biology of Fishes* 14: 31-41.

Buddington, R.K. and S.I. Doroshov. 1986. Structural and functional relations of white sturgeon alimentary canal (*Acipenser transmontaneus*). *Journal of*

Morphology 190: 201-213.
Cataldi, E., Albano, C. Boglione, L. Dini, G. Monaco, P. Bronzi and S. Cataudella. 2002. *Acipenser naccarii*: fine structure of the alimentary canal with references to its ontogenesis. *Journal of Applied Ichthyology* 18: 329-337.
Cherr, G.N. and W.H. Clark. 1982: Fine structure of the envelope and micropyles in the eggs of the white sturgeon *Acipenser transmontanus* Richardson. *Development, Growth and Differentiation* 24: 341-352.
Cherr, G.N. and W.H. Clark. 1984. An acrosome reaction in sperm from the white sturgeon *Acipenser transmontanus*. *Journal of Experimental Zoology* 232: 129-39.
Cherr, G.N. and W.H. Clark, Jr. 1985a. An egg envelope component induces the acrosome reaction in sturgeon sperm. *Journal of Experimental Zoology* 234: 75-85.
Cherr, G.N. and W.H. Clark. 1985b. Gamete interaction in the white sturgeon *Acipenser transmontanus*: a morphological and physiological review. *Environmental Biology of Fishes* 14: 11-12.
Cooper, M.S. and V.C. Virta. 2007. Evolution of gastrulation in the ray-finned (Actinopterygian) fishes. *Journal of Experimental Zoology* 308: 591-608.
Cosson, J., O. Linhart, S.D. Mims, W.L. Shelton and M. Rodina. 2000. Analysis of motility parameters from paddlefish (*Polyodon spathula*) and shovelnose sturgeon (*Scaphirhynchus platorynchus*) spermatozoa. *Journal of Fish Biology* 56: 1348-1367.
Daczewska, M. and J. Saczko. 2005. Myotomal myogenesis of axial muscle in the sturgeon *Acipenser baeri* (Chondrostei, Acipenseriformes). *Folia Biologica (Krakow)* 53: 29-38.
Dean, B. 1895. The early development of gar-pike and sturgeon. *Journal of Morphology* 11: 1-62.
DeRobertis, E.D.P. and E.M.F. DeRobertis. 1980. *Cell and Molecular Biology*, 7th Edition, Saunders College-Holt, Rinehart and Winston, Philadelphia.
Dettlaff, T.A. 1962. Cortical changes in acipenserid eggs during fertilization and artificial activation. *Journal of Embryology and Experimental Morphology* 10: 1-26.
Dettlaff, T.A. and A.S. Ginsburg. 1954. Embryonic development of sturgeon (in connection with problems of artificial propagation). In: *Biology*. J.F. Muir & R.J. Roberts (Eds.). Westview Press. Vol. 2, pp. 251-275.
Dettlaff, T.A., A.S. Ginsburg and O.I. Schmalhausen. 1993. Sturgeon fishes: developmental biology and aquaculture. Springer-Verlag, New York.
Devoto, S.H., E. Melancon, J.S. Eisen and M. Westerfield. 1996. Identification of separate slow and fast muscle precursor cells in vivo, prior to somite formation. *Development* 122: 3371-3380.
Devoto, S.H., W. Stoiber, C.L. Hammond, P. Steinbacher, J.R. Haslett, M.J.F. Barresi, S.E. Patterson, E. Adiarte and S.M. Hughes. 2006. Generality of vertebrate developmental patterns: evidence for a dermomyotome in fish. *Evolution and Development* 8: 101-110.
DiLauro, M.N., W. Kaboord and R.A. Walsh. 1998. Sperm-cell ultrastructure of North American sturgeons. I. The Atlantic sturgeon (*Acipenser oxyrhynchus*). *Canadian Journal of Zoology* 76: 1822-1836.
DiLauro, M.N., W.S. Kaboord and R.A. Walsh. 1999. Sperm-cell ultrastructure of

North American sturgeons. II. The shortnose sturgeon (*Acipenser brevirostrum*, Lesueur, 1818). *Canadian Journal of Zoology* 77: 321-330.

DiLauro, M.N., W.S. Kaboord and R.A. Walsh. 2000. Sperm-cell ultrastructure of North American sturgeon. III. The Atlantic sturgeon (*Acipenser oxyrhynchus*). *Canadian Journal of Zoology* 78: 438-447.

DiLauro, M.N., R.A. Walsh and M. Peiffer. 2001. Sperm-cell ultrastructure of North American sturgeons. IV. The pallid sturgeon (*Scaphirhynchus albus* Forbes and Richardson, 1905). *Canadian Journal of Zoology* 79: 802-808.

Eyal-Giladi, H. 1997. Establishment of the axis in chordates: facts and speculations. *Development* 124: 2285-2296.

Flood, P.R., D. Gulyaev and H. Kryvi. 1987. Origin and differentiation of muscle fibre types in the trunk of the sturgeon, *Acipenser stellatus*. *Sarsia* 72: 343-344.

Gawlicka, A., S.J. Teh, S.S.O. Hung, D.E. Hinton and J. de la Noüe. 1995. Histological and histochemical changes in the digestive tract of white sturgeon larvae during ontogeny. *Fish Physiology and Biochemistry* 14: 357-371.

Ginsburg, A.S. 1961. The block to polyspermy in sturgeon and trout with special reference to the role of cortical granules (alveoli). *Journal of Embryology and Experimental Morphology* 9: 173-190.

Ginsburg A.S. 1968. *Fertilization in Fishes and the Problem of Polyspermy*. Nauka Press, Moscow.

Ginsburg, A.S. 1977. Fine structure of the spermatozoan and acrosome reaction in *Acipenser stellatus*. In: *Problemy Eksperimentalnioj Biology*, D.K. Beijaev (Ed). Nauka, Moscow, pp. 246-256.

Gisbert, E. and S.I. Doroshov. 2003. Histology of developing digestive system and the effect of food deprivation in larval green sturgeon (*Acipenser medirostris* Ayres). *Aquatic Living Resources* 16: 77-89.

Ginsburg, A.S. and T.A. Dettlaff. 1991. The Russian sturgeon *Acipenser güldenstädti*. I. Gametes and early development up to time of hatching. In: *Animal Species for Developmental studies*. Dettlaff, T.A. and S.G. Vassetzky (Eds). Plenum, NY. Vol. 2. Vertebrates, pp. 15-65.

Gisbert, E., A. Rodriguez, F. Castelló-Orvay and P. Williot. 1998. A histological study of the development of the digestive tract of Siberian sturgeon (*Acipenser baeri*) during early ontogeny. *Aquaculture* 167: 195-209.

Gisbert, E., M.C. Sarasquete, P. Williot and F. Castelló-Orvay. 1999. Histochemistry of the development of the digestive system of Siberian sturgeon during early ontogeny. *Journal of Fish Biology* 55: 596-616.

Grandi, G. and M. Chicca. 2004. Early development of the pituitary gland in *Acipenser naccarii* (Chondrostei, Acipenseriformes): an immunocytochemical study. *Anatomy and Embryology* 208: 311-321.

Grandi, G., S. Giovannini and M. Chicca. 2007. Gonadogenesis in early developmental stages of *Acipenser naccarii* and influence of estrogen immersion on feminization. *Journal of Applied Ichthyology* 23: 3-8.

Groves, J.A., C.L. Hammond and S.M. Hughes. 2005. Fgf8 drives myogenic progression of a novel lateral fast muscle fibre population in zebrafish. *Development* 132: 4211-4222.

Guerrero, A., J.M. Icardo, A.C. Durán, A. Gallego, A. Domezain, E. Colvee and V. Sans-Coma. 2004. Differentiation of the cardiac outflow tract components in alevins of the sturgeon *Acipenser naccarii* (Osteichthyes, Acipenseriformes). Implications for heart evolution. *Journal of Morphology* 260: 172-183.

Heming, T.A. and R.K. Buddington. 1988. Yolk sac absorption in embryonic and larval fishes. In: *Fish Physiology*, W.S. Hoar and D.J Randall (Eds.). Academic Press, London, Vol. 11A, pp. 407-446.

Icardo, J.M., A. Guerrero, A.C. Durán, A. Domezain, E. Colvee and V. Sans-Coma. 2004. The development of the sturgeon heart. *Anatomy and Embryology* 208: 439-449.

Ignatieva, G.M. 1960. The regional nature of the inductive action of chordamesoderm in embryos of acipenserid fish. *Doklady Akademii Nauk SSSR* 134: 233-236. (In Russian).

Ignatieva, G.M. 1961. Inductive properties of the chordamesodermal rudiment prior to invagination and the regulation of its defects in sturgeon embryos. *Doklady Akademii Nauk SSSR* 139: 503-505. (In Russian).

Ignatieva, G.M. 1963. A comparison of the dynamics of invagination of chordamesoderm material in embryos of *Acipenser stellatus*, *A. güldenstädti,* and axolotl. *Doklady Akademii Nauk SSSR* 151: 1466-1469. (In Russian).

Ignatieva, G.M. 1965. Relationship between epiboly and invagination in sturgeon embryos during the period of gastrulation. *Proceedings of the Academy of Sciences of the USSR-Biology Section* 165: 970-973.

Jamieson, B.G.M. 1991. *Fish Evolution and Systematics: Evidence from Spermatozoa.* Cambridge University Press, Cambridge.

Jones, P.W., F.D. Martin and J.D. Hardy, Jr. 1978. Development of fishes of the Middle Atlantic Bight. *U.S. Fish and Wildlife Service, Biological Services Program.* Vol. 1. Acipenseridae through Ictaluridae. FWS/OB5-78/12. pp. 366

Johnston, I.A., N.J. Cole, M. Abercromby and V.L.A. Vieira. 1998. Embryonic temperature modulates muscle growth characteristics in larval and juvenile herring. *Journal of Experimental Biology* 201: 623-646.

Keller, R., M. Danilchik, R. Gimlich and J. Shih. 1985. The function and mechanism of convergent extension during gastrulation in *Xenopus laevis*. *Journal of Embryology and Experimental Morphology* 89 (Suppl.): 185-209.

Kobayashi, W. and T.S. Yamamoto. 1981. Fine structure of the micropylar apparatus of the chum salmon egg, with a discussion of the mechanism for blocking polyspermy. *Journal of Experimental Zoology* 217: 265-275.

Kowalewsky, A., P. Owsjannikow and N. Wagner. 1870. Die Entwicklungsgeschichte der Störe. *Bulletin de L'Académie Impériale des Sciences de St. Petersburg.* 14: 317-325, pl. 17.

Kunz, Y.W. 2004. *Developmental Biology of Teleost Fishes.* Springer. Dordrecht, The Netherlands.

Lenhardt, M., I. Jaric, A. Kalauzi and G. Cvijanovic. 2006. Assessment of extinction risk and reasons for decline in sturgeon. *Biodiversity and Conservation* 15: 1967-1976.

Lepilina, I.N. 2007. Development of mesonephros in prolarvae of Acipenseriformes (Acipenseridae). *Journal of Ichthyology* 47: 81-86.

Loew, E. and A.J. Sillman. 1993. Age-related changes in the visual pigments of the white sturgeon (*Acipenser transmontanus*). *Canadian Journal of Zoology* 71: 1552-1557.

Mohun, T.J., L.M. Leong, W.J. Weninger and D.B. Sparrow. 2000. The morphology of heart development in *Xenopus laevis*. *Developmental Biology* 218: 74-88.

Nelson, J.S. 2006: *Fishes of the World*. John Wiley and Sons, Inc. Hoboken, New Jersey. 4th Edition.

Ostaszewska, T. 2002a. Ultrastructural changes in the unicellular hatching glands of common carp (*Cyprinus carpio* L.) incubated at low or high water pH. *Annales of Warsaw Agricultural University-SGGW, Animal Science* 2, 79-86.

Ostaszewska, T. 2002b. The morphological and histological development of digestive tract and swim bladder in early organogenesis of pike-perch larval (*Stizostedion lucioperca* L.) in different rearing environments. Treatises and Monographs, Agricultural University Press, Warsaw.

Ostaszewska, T. and M. Wegiel. 2002. Differentiation of alimentary tract during organogenesis in larval asp (*Aspius aspius* L.). *Acta Academiae Scientiarum Polonarum Piscaria* 1: 23-34.

Ostaszewska, T., A. Wegner and M. Wegiel. 2003. Development of the digestive tract of ide, *Leuciscus idus* (L.) during the larval stage. *Archives of Polish Fisheries* 11, 2, 181-185.

Ostaszewska, T., A. Wegner and K. Dabrowski. 2007.The effect of feeding on muscle growth rate during larval development of sterlet (*Acipenser ruthenus* L.). *Aquaculture Europe 2007: competing claims*, Istanbul, Turkey, 405.

Ostaszewska, T., K. Dabrowski, A. Wegner and M. Krawiec. 2008. The effects of feeding on muscle growth dynamics, and proliferation of myogenic progenitor cells during pike-perch (*Sander lucioperca*). development. *Journal of World Aquaculture Society* 39: 184-195.

Pandolfi, M., A.A. Paz, C. Maggese, M. Ravaglia and P. Vissio. 2001. Ontogeny of immunoreactive somatolactin, prolactin and growth hormone secretory cells in the developing pituitary gland of *Cichlasoma dimerus* (Teleostei, Perciformes). *Anatomy and Embryology* 203: 461-468.

Psenicka, M., S.M.H. Alavi, M. Rodina, D. Gela, J. Nebesarova and O. Linhart. 2007. Morphology and ultrastructure of Siberian sturgeon, *Acipenser baerii*, spermatozoa using scanning and transmission electron microscopy. *Biology of the Cell* 99: 103-115.

Purkett, C.A. 1961. Reproduction and early development of the paddlefish. *Transactions of the American Fisheries Society* 90: 125-129.

Purkett, C.A. 1963. Artificial propagation of paddlefish. *Progressive Fish-Culturist* 25: 31-33.

Rodríguez, A. and E. Gisbert. 2002. Eye development and the role of vision during Siberian sturgeon early ontogeny. *Journal of Applied Ichthyology* 18: 280-285.

Ryder, J.A. 1890. The sturgeons and sturgeon industries of the eastern coast of the United States, with an account of experiments bearing upon sturgeon culture. *Bulletin of the United States Fish Commission* 8: 231-328.

Saga, T., K. Yamaki, Y. Doi and M. Yoshizuka. 1999. Chronological study of the appearance of adenohypophysial cells in the ayu (*Plecoglossus altivelis*). *Anatomy and Embryology* 200: 469-475.

Salensky, V.V. 1881. Recherches sur le développement du sterlet (*Acipenser ruthenus*). *Archive de Biologie* 2: 233-341.

Sarasquete, M.C., A. Polo and M. Yúfera. 1995. Histology and histochemistry of the development of the digestive system of larval gilthead seabream, *Sparus aurata* L. *Aquaculture* 130: 79-92.

Schmalhausen, O.I. 1991. The Russian sturgeon *Acipenser güldenstädti*. Part II. Later prelarval development. Ch. 3. In: *Animal species for developmental studies. Vol. 2. Vertebrates*, T.A. Dettlaff and S.G. Vassetzky (Eds.), Consultants Bureau, New York and London. pp. 67-88.

Sire, M.F., P.J. Babin and J.M. Vernier. 1994. Involvement of the lysosomal system in yolk protein deposit and degradation during vitellogenesis and embryonic development in trout. *Journal of Experimental Zoology* 269: 69-83.

Sokolov, L.I. 1965. Maturation and fertility of the Siberian sturgeon *Acipenser baeri* Br. From the Lena River. *Voprosy Ikhtiologii* 5: 70-81. (In Russian).

Steinbacher, P., J.R. Haslett, A.M. Sanger and W. Stoiber. 2006. Evolution of myogenesis in fish: a sturgeon view of the mechanisms of muscle development. *Anatomy and Embryology* 211: 311-322.

Stoiber, W., J.R. Haslett, A. Goldschmid and A.M. Sänger. 1998. Patterns of superficial fibre formation in the European pearlfish (*Rutilus frisii meidingeri*) provide a general template for slow muscle development in teleost fish. *Anatomy and Embryology* 197: 485-496.

Sysa, P., T. Ostaszewska, and M. Olejniczak. 2006. Development of digestive system and swim bladder of larval nase (*Chondrostoma nasus* L.). *Aquaculture Nutrition* 12, 5: 331-339.

Vladykov, V. and J.R. Greeley. 1963. Order Acipenseroidei. In: *Fishes of the western North Atlantic*, H.B. Bigelow and W.C. Schroeder (Eds.). Sears Foundation for Marine Research. Yale University, New Haven, pp. 24-60.

Watanabe, Y. and N. Sawada. 1985. Larval development of digestive organs and intestinal absorptive functions in the freshwater goby *Chaenogobius annularis*. *Bulletin of Tohoku National Fisheries Research Laboratory* 47: 1-10.

Webb, J.F. and D.M. Noden 1993. Ectodermal placodes: Contributions to the development of the vertebrate head. *American Zoologist* 33: 434-447.

Wegner, A., T. Ostaszewska and M. Karaszewski. 2008. The morphological changes in digestive tract of starlet (*Acipenser ruthenus*, L.) larvae during endogenous stage of feeding. *Aquaculture Europe 2008*, Krakow, Poland.

Wei, Q., P. Li, M. Psenicka, S.M. Hadi Alavi, L. Shen, J. Liu, J. Peknicova and O. Linhart. 2007. Ultrastructure and morphology of spermatozoa in Chinese sturgeon (*Acipenser sinensis* Gray 1835) using scanning and transmission electron microscopy. *Theriogenology* 67: 1269-1278.

Weisel, G.F. 1973. Anatomy and histology of the digestive organs of the paddlefish (*Polyodon spathula*). *Journal of. Morphology* 140: 243-256.

Weisel, G.F. 1979. Histology of the feeding and digestive organs of the shovelnose sturgeon, *Scaphirhyncus platorynchus*. *Copeia* 508-515.

Wrobel, KH., I. Hees, M. Schimmel and E. Stauber. 2002. The genus *Acipenser* as a model system for vertebrate urogenital development: nephrostomial tubules and their significance for theorigin of the gonad. *Anatomy and Embryology* 205: 67-80.

Wrobel, K.H. 2003. The genus *Acipenser* as a model for vertebrate urogenital development: the Müllerian duct. *Anatomy and Embryology* 206: 255-271.

Xu, Y. and Q. Xiong. 1988. The process of fertilization of *Acipenser sinensis* Gray observed by SEM. *Acta Zoologia Sinica* 34: 325-328.

Yeager, B. and R. Wallus. 1982. Development of larval *Polyodon spathula* (Walbaum) from the Cumberland River in Tennessee. In: *Proceedings of the Fifth Annual*

Larval Fish Conference, E.F. Bryan, J.V. Conner and F.M. Truesdale (Eds.). Louisiana State University, Baton Rogue, pp. 73-77.

Yeager, B.L. and R. Wallus. 1990. Family Polyodontidae. In: *Reproductive Biology and Early Life History of Fishes in the Ohio River Drainage,* R. Wallus, T.P. Simson and B.L. Yeager (Eds.), Vol. 1. Acipenseridae Through Esocidae. Chattanooga: Tennessee Valley Authority, pp. 47-56.

Yoshizaki, G., Y. Takeuchi, T. Kobayashi, S. Ihara and T. Takeuchi. 2002. Primordial germ cells: the blueprint for piscine life. *Fish Physiology and Biochemistry* 26: 3-12.

Zarnescu, O. 2005. Ultrastructural study of spermatozoa of the paddlefish, *Polyodon spathula*. *Zygote* 13: 241-247.

Zeiske, E., A. Kasumyan, P. Bartsch and A. Hansen. 2003. Early development of the olfactory organ in sturgeons of the genus *Acipenser*: a comparative and electron microscopic study. *Anatomy and Embryology* 206: 357-372.

4

Early Ontogeny of Semionotiformes and Amiiformes (Neopterygii: Actinopterygii)

Marta Jaroszewska[1*] and Konrad Dabrowski[1**]

[1]School of Environment and Natural Resources, the Ohio State University, Columbus 43210, OH, USA.

*present address: Nicolaus Copernicus University, Laboratory of Histology and Embryology of Vertebrates, Gagarina 9, 87-100 Toruń, Poland.
E-mail: marja@biol.uni.torun.pl
**E-mail:* dabrowski.1@osu.edu

INTRODUCTION

The phylogenetic analysis, based on the morphological and molecular studies, reassembled lepisosteids (Order Semionotiformes), amiids (Order Amiiformes) and teleosts into the subclass Neopterygii of class Actinopterygii (ray-finned fish) (Nelson, 2006; Hurley et al., 2007). The gars (Genera: *Lepisosteus* and *Atractosteus* spp.) and the bowfin (*Amia calva*) occupy an important position in fish phylogeny because they are surviving representatives of ancient ray-finned fish, and they are thought to have shared some common ancestors with modern-day bony fish, the teleostei (Groff and Youson, 1997). Semionotiformes diverted earlier in evolutionary history, whereas Amiiformes have a more distinct link with teleosts according to Groff and Youson (1997) and Nelson (2006). In other words, lepisosteids are regarded as the most ancient sister group of Amiiformes and teleosts (Nelson, 2006). However, the last author emphasized that the current monophylogeny of Neopterygii is based on some conflicting hypotheses and for this reason we do not have a convincing evidence to accept Amiiformes as the sister group of teleosts (Nelson, 1994). The data of molecular markers related to garfish, bowfin and teleosts provided more support to the statement that, among living neopterygians, amiids and lepisosteids may form a monophyletic clade, formerly called holostei, and separate from teleosts (Hurley *et al.*, 2007).

The genus *Lepisosteus* has four species which possess 14-33 small, pear-shaped gill rakers, whereas *Atractosteus* has three species with 59-81, large, and laterally compressed, gill rakers. The family Amiidae includes only a single extant species, the bowfin *Amia calva*. The "hard and soft morphology", as well as nuclear and mitochondrial genetic data, provide important independent evidence regarding the comparison of living teleosts with *Amia, Lepisosteus, Acipenser* (chondrosteans) and *Polypterus* (De Pinna, 1996; Hurley *et al.*, 2007). Long and Ballard (2001) have emphasized that the gaps existing in understanding developmental patterns in living groups of actinopterygian fish, do not allow to describe a comprehensible system. The embryological literature dealing with the gars, of which both genera (*Lepisosteus* and *Atractosteus*) are the representatives, is to be found in an older work published over a century ago. Previous studies on the early development of *Lepisosteus* and *Amia* were carried out by Agassiz (1879a,b), Balfour and Parker (1882), Dean (1895a,b), Eycleshymer (1899, 1903) and Lanzi (1909). These findings were revised, in some aspects, most recently by Long and Ballard (2001). The latest work on the type of cleavage in *Lepisosteus osseus* allowed, for instance, the clarification of some discrepancies which have existed in the literature for years, and resulted from the discussion between scientists from the end of the 19th and beginning of the 20th century. The present publication has been completed, in part, on the basis of the experimental work conducted on the embryonic development of the tropical gar (*Atractosteus tropicus*)[1]. This, in our opinion, validates or questions the earlier findings presented in the span of almost 140 years. However, a review of the literature remains essential to our arguments.

The development of gars differs from that of the paddlefish, sturgeon, bichir, bowfin, as well as of teleosts (Colazzo *et al.*, 1994; Long and Ballard, 2001; Kunz, 2004). Among the developmental characters that separate lepisosteids from amiids, there are the following:

---

[1]Fish were obtained from the Department of Biology, Universidad Juárez Autónoma de Tabasco, Mexico (courtesy of W. Contreras and A. Hernandez-Franyutti) where breeding under laboratory conditions was carried out. Tropical gars were implanted with GnRH-a (Sigma Chemical, St Louis, MO). They spawned after 7 hours in outdoor tanks. Eggs of one female and sperm from 3 males were used. Hatching occurred at 42 hours at 28°C incubation temperature or 49°D. Degree days (°D) are calculated as mean water temperature multiplied by the number of hours, and divided by 24. Sixteen samples of 8 specimens each were collected from 3hrs postfertilisation (hpf) (3.5D°) to hatching (49°D) every three hours, and 4 samples of 8 specimens each were collected every 12 hours from hatching to 54 hours posthatching (hph) (112D°).

1) meroblastic cleavage, 2) a well-defined yolk syncytial layer (YSL) (periblast)[2], and 3) involution of surface cells to the interior during gastrulation (Long and Ballard, 2001).

GAMETES

The spermatozoa of *Lepisosteus* and *Amia*, as the neopterygian fish, agree in the main design with those of teleosts because they lack an acrosome.

The spermatozoa of *Lepisosteus* have a short ring-shaped midpiece containing the mitochondria (Afzelius, 1978; Jamieson, 1991). However, the bullet-shaped nucleus, 2.5 μm long and 1.1 μm wide, presents a slight departure from the more rounded form of the nucleus found in most of the teleostean anacrosomal sperms. The nucleus is covered by the nuclear and cell membranes. Besides mitochondria the sperm midpiece contains two centrioles and a narrow cytoplasmic sleeve which surrounds the proximal part of the sperm tail. The proximal centriole is joined to a fibrous body, which sends an extension to a posterior central fossa of the nucleus (Afzelius, 1978). The distal centriole extends along most of the midpiece region. As is the case in the proximal centriole, the main portion of the distal centriole contains microtubular triplets, but distally there are doublets with a hook-like extension. They are surrounded by a single ring of C-shaped mitochondria, which are not unusual for neopterygians and are known only in five families of teleosts (Jamieson, 1991). The sperm tail has two lateral folds or ridges, and this feature is not unique, as it is present in many teleosts.

Some structures of the bowfin spermatozoa are unique in comparison with those of gars and teleosts. Afzelius and Mims (1995) listed these features as follows: annulate lamellae, mitochondrial matrix granules and vesicles that appear to open at the cell surface. The spermatozoon of *A. calva* differs from that of *Lepisosteus* in the round shape of the head, spheroidal nucleus, absence of a connection between centrioles and the nucleus, axonemal A- and B-subtubules with an electron lucid lumen and a tail lacking lateral folds (Afzelius, 1978; Jamieson, 1991; Afzelius and

---

[2]The review of the literature reveals that there is no consensus on the use of terms *yolk syncytial layer* and *periblast* with reference to the embryonic stage when this structure functions. Betchaku and Trinkaus (1978) suggested to use *ysl* as a proper term, arguing against "periblast" as an inadequate one. Some authors use both these terms as synonymous (Fishelson, 1995). However, in the recent literature very often just the term *ysl* is used (Trinkaus, 1984; Ninhaus-Silveira *et al.*, 2006). Following these descriptions the authors of the present chapter adopted the term of *ysl* for the superficial, multinuclear, and extra-embryonic layer of the yolk.

Mims, 1995). To add to the general description of the bowfin spermatozoa it could be mentioned that in the midpiece (2 µm wide at the base, 1 µm long) there are 12-16 mitochondria arranged in two rings around the two centrioles (Afzelius and Mims, 1995). From the evolutionary perspective spermatozoa structure is regarded as being illustrative of amiids and teleosts as sister groups, within the subclass Neopterygii (Afzelius and Mims, 1995).

In *L. osseus* the eggs are spherical, about 3 mm in diameter (Colazzo *et al.*, 1994). There is some variation in egg size within a range of 2.8-3.1 mm for batches from various females (Long and Ballard, 2001). The egg envelope is composed of two parts: an external layer of pyriform bodies and the internal *zona radiata* (chorion). Beard (1889) stated that the inner egg membrane is composed just of one layer, and that the division into two layers (Balfour and Parker, 1882) is due to the optical effect of thick sections. There is only one micropyle in gar eggs (Dean, 1895a).

The egg envelope enlarges slowly over a few hours after water activation (fertilization), revealing a narrow perivitelline space, about 0.1 mm wide. A cushioning, non-adhesive jelly fills this space. The principal envelope that surrounds the egg is the chorion derived from the vitelline envelope of the unfertilized ovum. The second layer of jelly surrounds the chorion. It is of variable thickness, up to 0.3 mm, and it provides an adhesive characteristic to the eggs (Long and Ballard, 2001). According to Beard (1889) the modified cells of the outer layer of the egg envelope disintegrate into a gluey substance, which causes the excessive stickiness of the newly laid eggs. In his opinion, such a layer, less developed than in the gars, is characteristic of the eggs of sturgeon and lamprey (*Petromyzon*) (Beard, 1889).

The micropyle is the point of sperm entry and lies immediately adjacent to the blastodisc. The blastodisc is about 2.3 mm in diameter and a pit occurs at its apex, which marks the animal pole (Long and Ballard, 2001).

The *Amia* egg is strongly telolecithal, assumes an elongated form, with dimensions of 2.2 × 2.8 mm along the minor and major axes. It possesses a membrane which consists of the *zona radiata*, which is more compact in structure than in other ganoids, and a villous layer with enormously elongated villi at one side of the egg. These villi serve to attach the egg to structures under water (Dean, 1896).

CLEAVAGE AND BLASTULA

Colazzo *et al.* (1994) expressed the view that the pattern of cleavage correlates with the quantity of yolk. Eggs with a small amount of yolk display holoblastic cleavage, while those with more yolk cleave meroblastically. The cleavage of *Amia* is clearly holoblastic, which is an advancement from the condition found in the sturgeon, but it also displays unique features which associate this type of division with the meroblastic mode of the typical teleost (Ballard, 1986a; Whitman and Eycleshymer, 1897 and Eycleshymer and Wilson, 1906). Cleavage in this species produces about a dozen large yolky macroblastomeres (giant blastomeres), upon which the smaller cells of the blastoderm rest (Long and Ballard, 2001). In the gars the type of cleavage is meroblastic, i.e. the furrows divide the blastoderm in a pattern similar to that of the teleosts (*Lepisosteus:* Dean, 1895a; Long and Ballard, 2001; *Atractosteus*: our data). However, in the teleost egg most of the yolk remains uncleaved from the beginning of the division (Colazzo *et al.*, 1994; Kunz, 2004), and this feature constitutes a significant difference between the early blastula of the gars and the teleosts.

In the gars the first furrow divides the germ disc into segments of unequal size. The second cleavage (yielding the 4 cell stage) places the furrow perpendicularly to the first one (Fig. 1a). The furrows yielding the 8 cell stage (third cleavage) are parallel to the first one (Fig. 1b). The fourth cleavage yields the 16 cell stage with the first cleavage groove reaching halfway between the equator and the vegetal pole (Fig. 1c) (Long and Ballard, 2001). The stage of 32 cells is the result of the fifth cleavage (Fig. 1d). Dean (1895a) emphasized that the mode of the fifth cleavage corresponds to the teleost type of segmentation. The cleavage furrows continue past the blastoderm margin as grooves in the yolk cell surface (Beard, 1889; Long and Ballard, 2001). When the animal pole is divided into small segments, the vegetal part, which subsequently forms a large yolksac, is divided by a few vertical superficial furrows (Beard, 1896) (Fig. 1e). Some of them nearly meet at the pole opposite to the blastoderm (Fig. 1f). The majority of the vertical furrows extend only a short way from the edge of the small animal pole, and are partially intercepted by imperfect equatorial furrows (Balfour and Parker, 1882; Dean, 1895a). There is no evidence that some furrows extend significantly deeply into the yolk and it was stated that the yolk furrows gradually regress until they are no longer visible (Long and Ballard, 2001). For this reason, it is thought that the gars have a single yolk cell similar to the teleosts. In the gars (Long and Ballard, 2001) this type of embryonic divisions sets up the stage for the appearance of the yolk syncytial layer (*ysl*), which is a characteristic feature observed in elasmobranchs (Hamlett *et al.*, 1987) and teleosts (Kimmel and Law, 1985; Cooper and Virta, 2007). From the phylogenetic point of view, the *ysl* of gars marks the first evolutionary appearance of this structure in the actinopterygian fish (Long and Ballard, 2001). *Amia* does not have a *ysl*[3]

---

[3]*Ysl* was called also 'periblast syncytium' (Dean, 1895b).

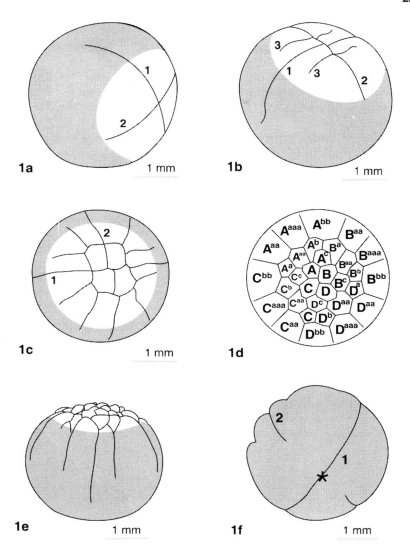

Fig. 1. Schematic cleavage of eggs in *L. osseus*; (a) stage of 4 cells; (b) stage of 8 cells; (c) stage of 16 cells, (d) stage of 32 cells, (e) blastula stage of 64 cells with about 30 cells at the surface of the animal pole surrounded by marginal cells which are continuous with the yolk cell. (f) Furrows continue along the yolk cell towards the vegetal pole; 1, 2, 3—first, second and third cleavage furrow; (*) vegetal pole; A, B, C, D—primordial blastomeres and their derivatives (a, b, c in upper index); (Fig. 1a-c and e-f—reprinted from Long and Ballard (2001) with permission of the corresponding author; from http://www.biomedcentral.com/1471-213X/1/6; 1d according to Dean (1895a).

(Ballard, 1984). Cooper and Virta (2007) concluded, after Dean (1895b), that in the gastrula of *Amia* the layer of primary hypoblast[4] cells is an ancestral progenitor tissue for *ysl* in teleostean embryos. However, a more elaborate description of the affinity between these two structures was given by Eycleshymer and Wilson (1906). These authors noticed that the upper part of the yolk, with the nuclei bordering the blastodisc, should be considered a homologue of the *ysl* of teleosts. Eycleshymer and Wilson (1906) were in opposition to Dean (1895b), who described the meroblastic cleavage in *Amia*. There is also some disagreement between these authors as to whether *Amia* has a segmentation cavity (blastocoel) or not. This question is addressed below.

According to Long and Ballard (2001) cleavages in the gar embryo carry the blastula from 64 to 512 cells. Approximately 30 complete blastomeres appear on the surface of the embryo at the stage of 64-cells. They are surrounded by marginal cells that are continuous with the 'yolk cell' (Fig. 1e). At this stage tangential cleavage divisions separate central blastomeres from the yolk cell. Deep blastoderm cells are still confluent with the yolk, and appear to grow from it (Long and Ballard, 2001). It is believed that some of these cells may split off nuclei into the yolk below (Dean, 1895a). Long and Ballard (2001) determined that the smooth-surfaced blastula is the final stage before epiboly. They described the superficial blastomeres that are tiny and the blastoderm surface consisting of a thin enveloping layer (EVL), a similar structure to that found in teleosts. The blastoderm has an irregular rim and a flat bottom where it joins the yolk cell. A cross section at this stage of *Lepisosteus* development reveals a continuing presence of deep central cells and their connection with the yolk; in many cases they join the yolk via a narrow stalk (Long and Ballard, 2001).

In *Amia*, at the beginning of cleavage the first, second and third furrows divide yolky cytoplasm on the animal pole, into two, four and eight 'cells' respectively (Ballard, 1986a). These cells are still incompletely separated from each other on the ventral part. The cell divisions are generally symmetrical. At the stage of 16-cells there are central and marginal blastomeres, which are connected to the yolk. From this stage onward a dozen or more furrows extend down across the vegetal hemisphere of the yolk, but only some of them extend as far as the vegetal pole (Fig. 2a,b). During the reduced holoblastic cleavage in the yolky endoderm, these furrows, which reach the vegetal pole, separate yolk into twelve mammoth blastomeres, which will be used during development as nutrient reserves (Ballard, 1986a; Cooper and Virta, 2007). As the result of further divisions,

---

[4]Hypoblast is a deep cell layer in the germ ring (Colazzo et al., 1994), also called 'presumptive endodermal cells' (Kunz, 2004).

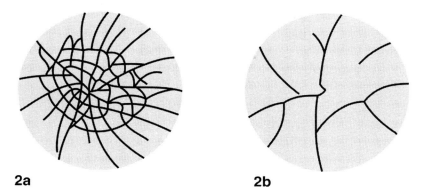

Fig. 2. View of the furrows of the animal (a) and vegetal hemisphere (b) of the egg in the bowfin. As a result of the cleavage the whole yolk is divided into giant blastomeres (Whitman and Eycleshymer, 1897).

internal cells of blastodisc are formed. The smaller and more numerous internal cells originate from the superficial blastomers and the larger ones bud off from the apex of the giant blastomeres (Ballard, 1986a). This process takes place around the margin of the blastodisc and in the central portion of the yolk (Eycleshymer and Wilson, 1906).

Ballard (1986a) reported the presence of a subgerminal cavity in *Amia*, which is defined in teleostei as the space between the cellular blastoderm and *ysl*. However, in the case of bowfin the location is on the upper surface of the giant blastomeres (Fig. 3). The smaller dimension of the subgerminal cavity in the gar and bowfin as opposed to the one in the sturgeon could be assumed as a decisively teleostean feature (Dean, 1896).

The presence of a blastocoel (segmentation cavity) in *Amia* is discussed in length by Eycleshymer, who contradicts himself without convincing evidence for its existence (Whitman and Eycleshymer, 1897; Eycleshymer and Wilson, 1906). According to Dean (1986) the blastula in the bowfin is compact; there is no blastocoel, just a centrally located space between loosely aggregated cells. In the most recent literature regarding teleostean embryonic development, there is no evidence of a true blastocoel, archenteron and neuroenteric canal (Nilsson and Cloud, 1992). As Cooper and Virta (2007) concluded, these three structures were greatly reduced during the transition of the gastrula from ancient actinopterygian to teleost type when embryonic germ layers were compacted into new juxtapositions. They further emphasized the common lack of the blastocoel in the teleostean embryos which favours the opinion that this structure existed in common ancestors before the elaborate radiation of teleosts occurred in the early Mesozoic era.

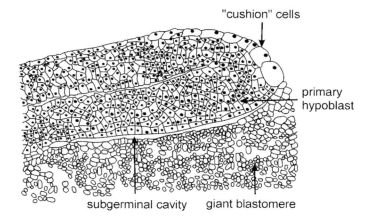

Fig. 3. Sagittal section through dorsal midline of blastoderm rim of the bowfin embryo showing the presence of "cushion" cells of the external epithelium (modified from Ballard, 1986a with permission of Wiley Liss Inc., a subsidiary of John Willey and Sons, Inc.).

GASTRULATION

Gastrulation of *Lepisosteus*

In the lepisosteids transition to gastrulation takes place from the moment when the entire periphery of the blastodisc is clearly separated from the yolk (Dean, 1985a). When epiboly starts, two structures appear at the surface of the embryo: a slight bulge at the dorsal blastoderm margin, which is the first external expression of the embryonic shield, and the germ ring in the internal position. An extensive subgerminal cavity has a floor paved by large yolky cells which are adherent to the yolk syncytial layer (*ysl*) (Long and Ballard, 2001). In the opinion of these authors the subgerminal cavity is not a blastocoel because it is a product of morphogenetic cell movement, not of cleavage. This opinion is in agreement with the information mentioned above regarding the meroblastic egg blastula. When epiboly reaches the equator, the blastoderm overhangs its margin at the dorsal midline. It has a crinkled edge, where it attaches to the yolk cell. This structure was named dorsal blastopore lip by Long and Ballard (2001) because of its similarity to the structure in the amphibians (Fig. 4). At this stage the exposed yolk cell surface beneath the dorsal lip is the first external appearance of the *ysl*. Among the ray-finned fishes, also bichirs, sturgeon and paddlefish display a dorsal lip around which surface cells migrate to their interior. Long and Ballard (2001) emphasized that the presence of the dorsal lip in *Lepisosteus* raises the possibility that invagination and

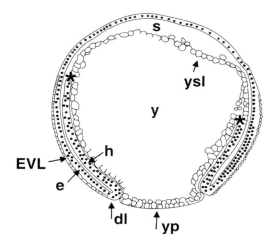

Fig. 4. Sagittal section through the late gastrula (40 hrs) of *L. osseus* before closing of blastopore; dl—dorsal lip; lips are separated from the underlying yolk; e—epiblast; future ectoderm with partial separation of the enveloping layer EVL; (*)—the point where the cells of the hypoblast (h) adjoin the yolk; s—subgerminal cavity; y—yolk; yp—yolk plug; ysl—yolk syncytial layer; (Dean, 1895a), labeling was modified).

involution of surface cells to the interior happens in this fish. However, there is a question regarding the homology of the involution in the gars, chondrostean and teleost fishes. This problem is discussed below in regard to the extra-embryonic sheath of epithelial cells (EVL). In the gar and amphibia alike, the yolk-plug projects from the blastopore (Beard, 1889) (Fig. 4). At the time of the blastopore closure the appearance of the mesoderm is noted which can be traced to its origin from the hypoblast (Fig. 5).

In *L. osseus*, the division of the epiblast, the outermost cells of the early blastoderm, into outer and inner layers is accomplished very early during embryonic development. The structure and function of the outer, or covering layer, described first by Beard (1889), corresponds to the enveloping layer of cells (EVL)[5] in teleostei, where the inner layer represents the future ectoderm (Long and Ballard, 2001; Kunz, 2004) (Figs. 4, 5). In both,

---

[5]In the literature also known as 'periderm', 'outer developing layer'. Ballard (1984; 1986a) described the presence in *Amia* of the thin outermost 'epithelium', also called "cellular envelope', 'Deckschicht', 'couche enveloppante', enveloping layer under which a thick layer of inner cells is present. However, he did not provide any arguments as to its relation to a similar structure in lepisosteids or teleosts.

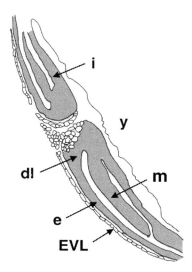

Fig. 5. Sagittal section of the region of blastopore at the late gastrula stage (42 hrs) of *L. osseus* showing the closing of blastopore and the origin of the middle germ layer; dl—dorsal lip of blastopore; e—future ectoderm; EVL—enveloping layer; i—inner germ layer (hypoblast); m—middle germ layer (future mesoderm); y—yolk (Dean, 1895a; labeling was modified).

lepisosteids and teleosts the EVL serves as an extra-embryonic layer of epithelial cells, which does not form any organs (Long and Ballard, 2001; Kunz, 2004). It has been suggested that in teleostei it assumes a protective function over the embryo, and maintains a high electrical resistance by providing an effective barrier to movement of even small ions. There are pluripotent deep cells below the EVL that construct the embryo and form the adult tissues. In the garfish this inner layer of epiblast was named by Beard (1889) the 'formative epiblast'. In teleostei it is known as deep cells (DC[6]), a group of loosely packed cells with irregular shape. It is believed, that in the gars, like in teleosts (Kunz, 2004), this part of epiblast represents the future ectoderm (epidermis and neuroectoderm).

In the newest review by Cooper and Virta (2007) it was stated that cellular identity changes resulted in the appearance of the novel extra-embryonic tissues (*ysl* and EVL), in the embryos of teleostean fish. However, molecular methods are needed to provide evidence of homology between outer and inner layer of blastoderm in lepisosteids and teleosts. In case of

---

[6]Cooper and Virta (2007) described in detail the extra-embryonic layer of cells (EVL-NEM) and the group of underlying cells (deep-NEM cells) that form the Organizer Epithelium (NEM) in zebrafish.

the homology between extra-embryonic tissue of the ray-finned fish would be proven, the hypothesis suggested by Cooper and Virta (2007) could be questioned. Our studies confirmed suggestions by Long *et al.* (1982) that both the *ysl* and EVL could have developed already in Semionotiformes, before the appearance in teleostei. Moreover, the question regarding the presence of the Organizer[7], a superficial marginal zone of the blastula epithelium, and the form of gastrulation in the gars should be directly addressed in further comparative studies of more ancestral, and most modern ray-finned fish.

Gastrulation of *Amia*

In *Amia* the presence of a giant yolky endodermal blastopore was postulated (Dean, 1896; Eycleshymer and Wilson, 1906). This also was most recently emphasized in the review by Cooper and Virta (2007). However, Ballard (1986a) contradicted those conclusions and stated that there is not "an amphibian-like" blastopore in the bowfin. This author described in detail the structure of the "cushion" cells[8], which are present over the advancing blastodisc rim and the yolk plug, pronounced from the moment when the blastoderm is at the distance of two thirds from the equator to the vegetal pole, and then in the later stages. The cap or "cushion" is composed of one layer of 10-15 enlarged squamous or cuboidal superficial cells (Fig. 3), called also 'the cushion of cells of Ballard' (Cooper and Virta, 2007). These cells form the external epithelium of the 'terminal node'[9]. The enlarged "cushion cells" in the bowfin fail to involute during gastrulation (Ballard, 1986a). The results of the comparative studies suggest that "cushion cells" in the superficial organizer epithelium of the bowfin display similarities with the non-involuting epiblastic marginal cells (NEM) in teleosts (zebrafish) (Cooper and Virta, 2007). This means that *Amia*, an ancient neopterygian fish, shows similarities to teleostei, and differences to chondrostean fishes and amphibia, with respect to an intermediate mode

---

[7]The definition of the Organizer, given by Solnica-Krezel (2005), is the following: 'a signaling center that patterns the germ layers and regulates gastrulation movements'. This author concluded that both, 'the pattern of gastrulation cell movement and Organizer' are similar in vertebrates, from fish to mammals.

[8]It seems that Eycleshymer and Wilson (1906) when they described this structure, came short of giving it this name. As they explained: 'at the blastoporic margin, this layer of hypoblastic cells changes in character from the small elongated cells with deeply staining granules to larger cuboidal cells and these in turn shade off into the smaller elongated cells of the superficial ectoblast' (Fig. 27-29 of Plate III in Eycleshymer and Wilson, 1906).

[9]Ballard (1986a) used the term 'terminal node' instead of 'dorsal lip'.

of cleavage, epiboly and involution. The description of the Organizer of bowfin and zebrafish revealed similarities in the morphogenetic behaviour of presumptive germ layers (Cooper and Virta, 2007). In addition to this characteristic of bowfin gastrulation, the loss of 'bottle cells in the superficial marginal zone of epithelium' in bowfin should be emphasized as a similarity to teleostei (Cooper and Virta, 2007). However, in the work of Ballard (1986a) we find the description of some hypoblastic cells, which lie next to the "cushion", and were named by the author "bottle cells". They are elongated cells in a form that is the reverse of the amphibian bottle cells. In the bowfin, these cells have attenuated internal tails, not external ones like in amphibian embryos and in effect, in the opinion of Ballard (1986b) they could be associated with different developmental functions.

The description of the gastrulation in *Amia*, given in this paragraph, is based on the detailed studies of Ballard (1986a,b) on the internalization of germ layers and morphogenetic movements of the labeled cells. Only Ballard provided the most documented account of the gastrulation process in *Amia* in spite of the fact that Eycleshymer and Wilson (1906) called into question the presence of the involution in *Amia* embryos. Cooper and Virta (2007) provide the conclusion that in the bowfin there is evidence of 'a novel mechanical detachment of the involuting deep marginal zone of the meso-endoderm from the endodermal yolk mass'. That means that there is no involution in the superficial marginal zone of the epithelium in the bowfin, or in other words, the epithelial endoderm is not derived from cells of the original surface layer by involution like in sturgeon or in the amphibia (Ballard, 1986b; Cooper and Virta, 2007).

The beginning of gastrulation is signified by the moment when the yolky internal cells, proliferating from the earlier superficial blastomeres, become grouped into about 12 rows of cells (epiblast) under the animal pole. The deeper cells integrate with four-five rows of bigger yolky cells budded off from the top of the giant blastomeres, and form the hypoblast. The next stage is the differentiation of the epiblast and hypoblast and 'separation from each other by the Brachet cleft'. Hypoblast cells migrate outwards from the area below the animal pole and concentrate towards the future germ ring, mainly in its dorsal side. At the same time the number of epiblast cells at the animal pole is limited to three-four rows whereas about nine rows are present close to the dorsal part of the rim. This is the commencement of the embryonic shield arising. When the blastoderm rim is about $10°$ above the equator, the many antero-ventral hypoblast cells are translocated from the animal pole towards the germ ring, which becomes thick on the dorsal side. When the rim of the blastodisc is spread around the equator of the embryo, the dorsal side forms a slight bulge, called a terminal node (Ballard, 1986a). As it was described earlier, the distal cells of the external epithelium of the terminal node constitute a "cushion". At this stage the hypoblast cells, that are similar to bottle cells of amphibian embryos, are well pronounced. During the later stages of the epiboly of bowfin, the bulging of the dorsal rim progresses, and simultaneously the evacuation zone is broadening. The extended process of the movement of hypoblast cells results in the rapid enlargement of the embryonic shield and germ ring. When the hypoblast is commensurated with epiblast in the shield area, a sheet of endoderm cells begins to separate from the deep surface of the hypoblasts. In the bowfin, the primary hypoblast forms the roof of the subgerminal cavity, which is not recognized as a true archenteron (Ballard, 1986a; Cooper and Virta, 2007). Mesoderm becomes visible on both sides of the notochord, when the neural plate in the embryonic shield is well recognized. In some individuals it is possible to observe the prostomal thickening which is formed by 'bottle cells'. They project to the surface below the dorsal lip. At this stage these cells follow the development of elongated cells lying in the middle line of the endoderm (Ballard, 1986a). The epiblast of the bowfin embryo is composed of the enveloping layer and the inner layer of cells which becomes the ectoderm of the embryo (Ballard, 1984).

The remnant of the evolutionarily reduced posterior archenteron and neuroenteric canal has the form of the ciliated epithelial sac called Kupffer's vesicle (KV), which plays a role in establishing left-right asymmetry in the teleost embryo (Essner *et al.*, 2005). In more advanced teleosts, such as the cyprinid zebrafish, Kupffer's vesicle is formed in the tailbud at the end of gastrulation from a group of approximately two-dozen cells, known as dorsal forerunner cells (DFCs) (Cooper and Virta, 2007). In contrast to other cells in this region DFCs do not involute during gastrulation, but remain at the leading edge of the line of epibolic overgrowth. At the end of gastrulation, DFCs migrate deeply into the embryo and are reorganized to form the KV (Cooper and D'Amico, 1996; Melby *et al.*, 1996). During subsequent developmental stages of somite differentiation, KV is found ventrally to the forming notochord in the tailbud and adjacent to the yolk cell. This transient spherical organ (KV) was first described by Kupffer in 1868 and is known as a conservative structure among teleost fishes. Its presence and function in *Amia* could only be presumed in connection with the reduced archenteron. The information given by Dean (1896), who described this organ below the dorsal lip, and by Eycleshymer and Wilson (1906) who postulated its digestive functions, is doubtful and leads us to conclude that further studies are needed which could address the question if KV is the organ characteristic only for teleosts.

Neural System

There is some remarkable resemblance between lepisosteids, amiids and teleostei in the development of the neural system. In *Lepisosteus* and *Amia*, and as it was stated, among teleostei, in sea bass, *Serranus atrarius*, zebrafish *Danio rerio*, but also in the lamprey, *Petromyzon planeri*, neurulation takes place by cavitation (Balfour, 1885; Hall, 1999). This process begins when the three germ layers are fully established (in *L. osseus* on the 5th day after fertilization; Balfour, 1885). Along the axial line there is a solid keel-like thickening of the neuro-ectoderm, the primordium of the spinal cord, which projects towards the hypoblast (Fig. 6) (Balfour, 1885; Dean, 1895a). In the region of the brain the spinal cord is thick, but its posterior part is flatter (Balfour, 1885). The mesoblast in the trunk has the form of two plates, which thin out laterally. The hypoblast is a single layer of cells separated from the neural rod[10] by the notochord (Balfour, 1885). When the length of the embryo encompasses half of the yolk circumference, the three vesicles can be distinguished in the brain. At the same stage the spinal cord is separated from the neuro-ectoderm and a lumen is already formed (Balfour, 1885). The nervous system becomes hollowed out by the separation of its walls from each other, the process named cavitation (Beard, 1889). To put it in perspective: in teleostei the process of neurulation has some specific features in common with the neurulation of the gar, but different from other vertebrates. The neural plate of teleosts thickens but does not fold. According to Kunz (2004) the solid neural keel projects ventrally towards the yolk, later becoming a hollow neural tube by the process mentioned above, called cavitation or 'secondary neurulation'.

According to Long and Ballard (2001) in a very early gar embryo, when the size of the yolk plug is half or less of the diameter at the equator, there is a hint of a neural groove on its surface and at the later stage the neural folds are visible. The authors reported that the neural folds close in a posterio-anterior sequence at this stage, remaining last open in the anterior end. They interpreted this finding that at stage 16 when the axis of the embryo occupies from 150° to 180° of the yolk circumference, the neural folds disappear, intruding the subgerminal cavity (Long and Ballard, 2001). However, it should be emphasized in this place that the groove which appears as the blastoderm thickening, marking the axis of the gar embryo, could not be the groove between neural folds. As it was mentioned above, in the gar the primordium of the central neural system arises as a solid keel (Balfour and Parker, 1882; Beard, 1889). In our opinion, by analogy

---

[10] In the original work called 'neural cord' (Balfour, 1885). According to Kunz (2004): 'neural keel, carina'.

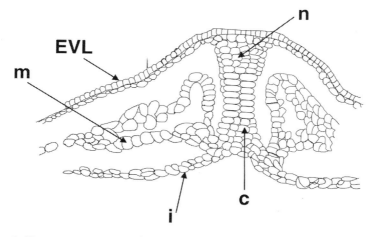

Fig. 6. Transverse section of posterior part of body (in the original text: the hinder trunk) of embryo of *L. osseus* at the 7 somites stage (80 hpf): c—notochord; EVL—enveloping layer; i—inner germ layer; m—middle germ layer; n—neural rod (Dean, 1895a; labeling was modified).

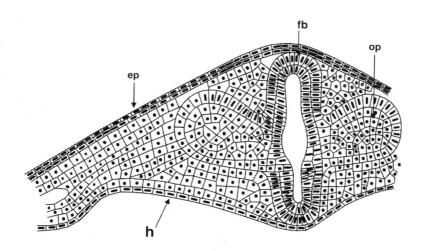

Fig. 7. Transverse section of the forebrain and optical primorida of the embryo of *L. osseus* days after fertilization (Balfour and Parker, 1882; redrawn, modified and with modified labeling); ep—epiblast; fb—forebrain; h—hypoblast; op—optical primordia.

with teleostei, the dorsal groove described by Long and Ballard (2001) is a transient longitudinal indentation and is not homologous with the neural groove of the higher vertebrates, which later folds into a neural tube (Kunz, 2004).

Beard (1889) claimed that the optic primordia in the gar arise as a pair of hollow evaginations of the forebrain. However, Balfour and Parker (1882) illustrated the origin of the optical primordia as solid structures (Fig. 7). Balfour (1885) mentioned that at the stage when the first visceral clefts are visible, but not yet perforated, the optic primordia are very prominent outgrowths of the brain, but they are still solid. This observation was confirmed by Dean (1895a) who stated, for instance that "the pronephric duct occurs before even the appearance of the lumen in the optic vesicle". It should be emphasized that this feature is analogous with the teleostean embryo, in which, the eye cups and ears arise as solid thickenings of the brain primordium and only subsequently acquire a lumen.

In the gar the olfactory bulbs arise in the form of a pair of thickened patches of the nervous layer of the epiblast (Balfour and Parker, 1882; Beard, 1889). The olfactory nerve is formed very early and arises at a stage prior to the first differentiation of an olfactory bulb as a special lobe of the brain (Balfour and Parker, 1882). The olfactory pits became visible in the gar at the stage when the head and the tail are free from the yolk (Balfour, 1885). The 'ear' (auditory organ) and the lateral sense organ arise similar to the olfactory organ, as a product of the neural layer of the epiblast, the openings of which are covered by the epidermis (Balfour, 1885). Immediately after hatching (about 200 hrs after fertilization at 20°C), the larvae of *L. osseus* possess olfactory pits, eyes and auditory primordia. In this species the auditory invaginations appear about 10 h later than the optic primordia (Dean, 1895b). The nasal apparatus is finally separated into incurrent and excurrent openings when the yolk mass is much diminished (Long and Ballard, 2001).

In *Amia*, as in the gar and teleosts, the central nervous system is formed as a solid rod from the deeper ectoblast (Dean, 1896). The neural canal (cavity) is formed secondarily in the solid keel, after the appearance of the optic primordia (Ballard, 1986a; Eycleshymer and Wilson, 1906). The sequence of the appearance of the sensory organs in the bowfin is as follows: optic primordia, ear primordia, olfactory organs and lateral line system. The optic and auditory primordia appear initially as solid buds and become hollow structures during further embryonic development. The olfactory organs appear as proliferation of the deep ectoderm, forming well defined pits as the result of invagination (Eycleshymer and Wilson, 1906).

KIDNEYS

In *L. osseus* the mesoderm, which forms, among other organs, kidneys, arises very early, and before the closure of the blastopore (Beard, 1889). The first part of the excretory system is formed as a pronephric duct[11] (Balfour and Parker, 1882). These ducts extend posteriorly to the 4th somite[12], at the time when the central nervous system is already recognizable, the prospective diencephalon has slightly enlarged, and the optic lobes are formed (Fig. 8) (Dean, 1895a; Long and Ballard, 2001). The pronephric duct (Wolffian duct, primary ureter, primary nephric duct) arises, as described in teleostei and amphibia, by the constriction of a hollow ridge of the nephrotome, along the junction between the somites[13] (epimere), and the lateral plate[14] (Balfour and Parker, 1882; Kunz, 2004). Anteriorly the duct does not become shut off from the body cavity, and bends inwards, towards the middle line. This inflected part of the duct is the first rudimentary pronephros (Fig. 8). Shortly afterwards it becomes considerably dilated in comparison to the posterior part of the duct (Balfour and Parker, 1882). At the stage when the lumen in the optic primordia appears (stage 7th day) the cephalic extremity of the pronephric duct becomes slightly convoluted (Balfour, 1885). The pronephric segments are modified into pronephric chambers, which become filled by vascular glomeruli formed along a branch of the aorta. After this process is accomplished, the excretory part of the kidney is essentially equipped with all the parts typically present in an opisthonephros (Balfour and Parker, 1882). The differentiated kidney of the gar, like that of the sturgeon and the teleostei, consists of the anterior pronephric part which is a lymphoid tissue, and the middle-posterior opisthonephros (Balfour and Parker, 1882; Jaroszewska and Dabrowski, 2006). In *A. tropicus* no changes from anterior pronephros to lymphatic tissue were observed from the stage of 56°D (6 hph) (Fig. 9) until 112°D (54 hph) stage. In the larvae of *L. osseus* the presence of anterior lymphatic tissue, followed by the excretory opisthonephros, was observed prior to the first feeding at the total body length (BL) of 23.5 mm (Jaroszewska and Dabrowski, 2006). For instance, in the case of the teleost *Danio rerio*, only at the age of 6 months the pronephric tubules have degenerated into solid cords of cells which have

---

[11]In the original text named 'archinephric duct' or 'segmental duct' (Balfour and Parker, 1882).

[12]According to Neidhart (1985) in non-teleost ray-finned fish, bichir, sturgeon, garfish and bowfin, the first pronephric ducts appear at the stage of 3 somites.

[13]In the original version 'mesoblastic somites' (Balfour and Parker, 1882).

[14]In the original text of Balfour and Parker (1882) named as 'hypomere'. According to Kunz (2004): 'mesoblast'.

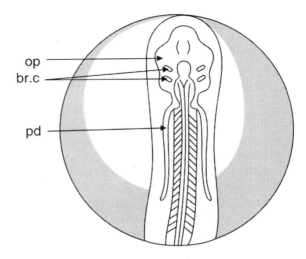

Fig. 8. Morphology of the embryo of *L. osseus* on the 6th day after fertilization (Balfour and Parker, 1882 with modified labeling); br.c—branchial clefts; op—optical lobe; pd—pronephric duct.

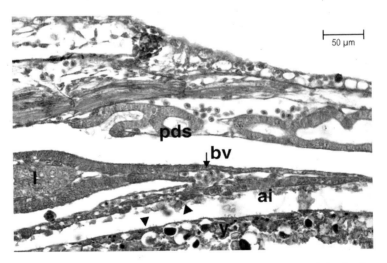

Fig. 9. Longitudinal parasagittal section of the anterior part of the body cavity in *A. tropicus* larva at stage of 56°D (6 hph). The simple structure of the anterior pronephros, composed of the pronephric ducts (pds), is visible. Below: the liver (l) is shown with the blood vessel (bv) which drains the anterior intestine (ai) and supplies the liver. The luminal epithelium of the presumptive anterior intestine forms an extensive intestine-yolk contact area (arrowheads); y—yolk.

the appearance of mammalian lymph nodes (Kunz, 2004).

Balfour and Parker (1882) described two of the most distinct stages during the development of the posterior part of the pronephric ducts in *L. osseus*. The first begins at the opening of the posterior part of this duct into the cloacal section of the alimentary tract. The second stage, in the specimens with the total body length above 11 mm, consists of the process of hypoblastic cloaca division into two sections. The posterior part of the cloaca receives the coalesced primary ureter[15] (Wolffian ducts), and the anterior remains connected with the alimentary tract. The opening of the posterior portion forms the urogenital opening, and the anterior the anus. In the larvae of *L. osseus* at 24.5 mm BL (11 dph) the distinct posterior connection of both Wolffian ducts (primary ureters), which form the urinary bladder, was evident in our preliminary observation. The primary ureter performing the function of the common outlet of the urinary and sperm ducts empties into a urinary bladder with trabecular inner structure (Lukáš, 1989; *Atractosteus tristoechus*).

GONADS

The development of the gonad in *L. osseus* has been described by Allen (1911). He mentions cells reminiscent of primordial germ cells PGC(s) (precursor germ cells) dispersed on the edges of the body cavity which then migrate towards the location of the gonadal primordium in 4 mm embryos. When based only on histological routine, the identification of primordial germ cells offered by Allen (1911) is somewhat questionable. The most convincing evidence of the presence of a gonadal crest in *Amia calva* was illustrated in fish of much larger size (16 mm). It is worth to emphasize that the phenomenon of PGC(s) migration, described by Allen (1911), from the part of the embryo where they are formed to their final location, was confirmed in the most recent studies conducted in a teleost, the zebrafish (Kunwar and Lehmann, 2003). However, the specific results related to the fate map of the migration of PGC(s) are completely different from those obtained in *L. osseus* by Allen (1911). This author suggested the endoderm as the source of PGC(s). In the case of the zebrafish it was proven that PGC(s) originate at random positions with respect to the body axis and that 14 days after fertilization they aligned with the presomatic mesoderm resulting in one cluster on either side of the embryo (Knaut *et al.*, 2003).

The gonadal crests with PGC(s) appeared in *L. osseus* at 31 mm BL, after the completion of yolk absorption and an exogenous period of feeding (Fig. 10) (Dabrowski *et al.*, 2006). They are attached to the abdominal wall

---

[15]In the original version still called 'segmental duct' (Balfour, 1882).

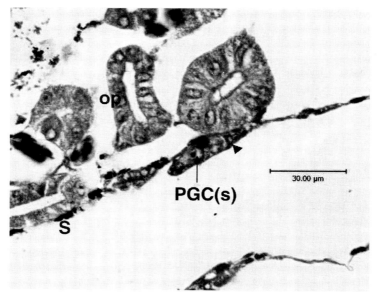

Fig. 10. Transverse sections of the middle part of the body cavity of *L. osseus* at 31 mm body length (BL) with gonadal crest (arrowhead) with primordial germ cells (PGC(s)) without epithelia, and opisthonephros (op), i.e., excretory part of the kidney; S—splachnopleura.

on the dorsal side, near the inner border of the kidneys and project into the body cavity. These structures are very distinct in the middle part of the body cavity. At this stage of the gar development no epithelia can be distinguished in the gonadal crests.

In the gars, the separation of the genital duct from the urinary duct, with the exception of the urogenital sinus region at the posterior end, is a fundamental characteristic of most vertebrate male reproductive systems. In female fish, as in the higher vertebrates, the reproductive duct is always distinct from the urinary duct. In the case of Semionotiformes and telostei there is no evidence of an exact homology of the reproductive duct with the Müllerian duct (Nelsen, 1953).

INTESTINE-YOLKSAC CONTACT

When the lepisosteids hatch, they have an external yolksac (Fig. 11). This feature constitutes a difference between the *Acipenseriformes* with internal yolk in the gut and *Semionotiformes* and is indicative of the closer affinity to teleostei. During the embryonic and early larval development of the lepisosteid fish there is a connection between the anterior part of the

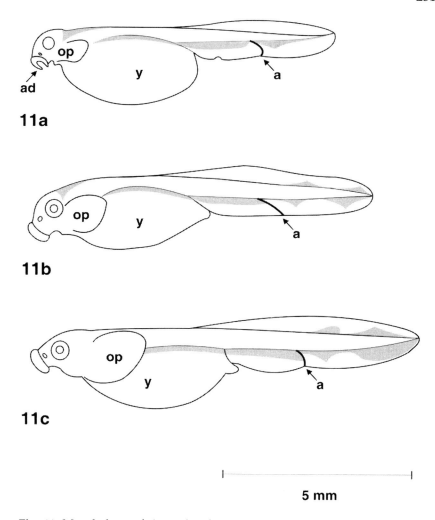

Fig. 11. Morphology of *A. tropicus* larva (a) at stage of 70°D (18 hph; 8.04 mm BL). At this stage the buds of filaments are present on the surface of the gill arches, lying under the elongated operculum; the eye lenses and the nasal pits are barely recognizable; pectoral fin buds and pigment aggregates in the ventral fin fold are visible; larva (b) at stage of 84°D (30 hph; 8.5 mm body length) and (c) at stage of 112°D (54 hph; 9.04 mm body length) with straightened body and lengthened snout; at both stages the eye lenses are visible; the adhesive organ (ad) is in the form of a disc. a—anus; ad—adhesive organ; grey shade—pigment aggregates indicating the location of future anal and caudal fins; op—operculum; y—yolksac, diminishing in size with development.

Fig. 12. Transverse section of the *A. tropicus* embryo at stage of 45.5°D (39 hpf; just before hatching). Presumptive anterior intestine (ai) is open to the yolk (y).

presumptive intestine and the yolk. This morphological connection was observed in *A. tropicus* at the stage before and after hatching (Figs. 9, 12, 13, 14) as well in *L. osseus* (specimens at 9 mm BL) (Beard, 1896). The location of this fragment of the intestine lies behind the differentiating liver. It is ventrally open to the yolk and forms the wide passage from the yolksac to the lumen of the embryonic alimentary tract (Figs. 9, 12, 13, 14). Balfour and Parker (1882) described the part of the intestine present behind the opening of the bile duct and open to the yolk as a 'vitelline duct' which is analogous with the 'umbilical cord' in sharks (Fishelson and Baranes, 1998). In our opinion, in the case of lepisosteids, where the intestine provides an extensive and flat area of contact with the yolk, 'vitelline duct' would be a mistaken term. However, it seems that in the gar larvae the course of the closing of the alimentary tract is very similar to that of the bowfin, where at the end of this process the presence of a *ductus omphaloentericus* was described (Piper, 1902; Ballard, 1986a) (see below in relation to *Amia* [Fig. 20]). In the gar a similar structure, in the form of a narrowed area of the connection between intestine and yolksac, was observed. It could be assumed that the observations by Balfour and Parker (1882) were made at the latest stages of the "vitelline duct", or the intestine-yolksac junction.

During the development of the alimentary tract its anterior and posterior ends are first formed in the form of a solid tube. Balfour and Parker (1882) stated that in *Lepidosteus* these two parts gradually elongate, so as to approach each other at a very short distance behind the opening of the bile duct into the intestine. The anus arises as a solid outgrowth of the formative epiblast in the place where the blastopore once existed (Beard, 1889). It was concluded that in *L. osseus* the anterior and posterior parts of the alimentary tract are still closed shortly before hatching of the embryos (Balfour, 1885). However, our studies in *A. tropicus* (stage of 45.5°D and 39 hpf) have shown that just before hatching the posterior intestine is provided with a lumen, i.e., earlier than the oesophagus. After hatching of *A. tropicus*, i.e., beginning at the stage of 49°D, the presumptive oesophagus, stomach and the developing respiratory gas bladder (RGB) are visible. On the transverse sections of 84°D, 30 hph these parts of the alimentary tract are identifiable in parallel position with the differentiated liver (Fig. 15). During early ontogenesis the posterior intestine forms a very long, straight section of the alimentary tract in embryos of both species, *A. tropicus* as well as *L. osseus* (Balfour and Parker, 1882). There is no spiral valve observed in the posterior intestine in larvae of *A. tropicus*. In *L. osseus* the spiral fold is not formed until about three weeks after hatching, when the remnants of the yolk are still present in the body cavity (Balfour and Parker, 1882; Jaroszewska and Dabrowski, 2006).

At the stage of 84°D and 112°D of tropical gar development, the process of anterior intestine closure begins (Figs. 13, 14). It is reflected by the limitation of the intestine-yolk contact area. The anterior part on both sides begins to roll up to the middle of the intestine, gradually separating it from the *ysl*. It is suggested that this structure, where the intestine forms restricted special connection with the yolksac, is homologous to the *ductus omphaloentericus* in the bowfin, as described below (Fig. 20a,b). In the larvae of the gar at 11 dph and 23.5 mm BL, the remnants of the yolk are present exclusively in the body cavity. The whole alimentary tract is in the form of a closed tube and there is no intestine-yolk connection (Fig. 16) (Jaroszewska and Dabrowski, 2006). Balfour and Parker (1882) stated that in *L. osseus* the stomach grows ventralwards and forwards shortly after hatching. In *A. tropicus* at the stage 84°D (30 hph) and 112°D (54 hph), the curvature of the presumptive pylorus is visible (Fig. 17), closely located to the posterior end of the liver. The most characteristic feature of the gar alimentary tract is the absence of the anatomical distinction between oesophagus and stomach during the larval development (described in *A. tropicus*), and in the adult specimens of *L. osseus* (Balfour and Parker, 1882;

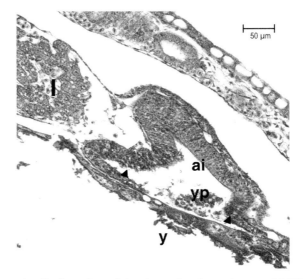

Fig. 13. Longitudinal section of the *A. tropicus* larva (stage of 84°D; 30 hph). The anterior intestine (ai), located behind the liver (l), is still open to the yolk (y) but the process of closing the lumen begins (arrowheads indicate places where the wall of the anterior intestine begins to roll up on both of its sides to the middle of the organ). Yolk platelets (yp) are found inside the intestinal lumen.

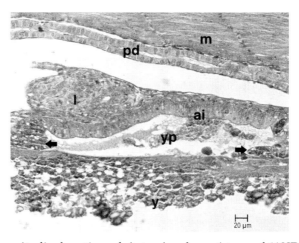

Fig. 14. Longitudinal section of *A. tropicus* larva (stage of 112°D; 54 hph) showing the closing of the anterior intestine. ai—anterior intestine; l—liver; m—muscles; pd—pronephric duct; y—yolk; yp—yolk platelets inside the intestinal lumen. Arrows indicate places where the wall of the anterior intestine started to close in.

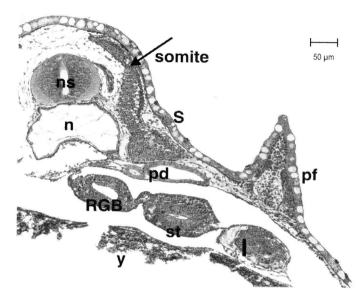

Fig. 15. Transverse section of the *A. tropicus* larva (stage of 84°D; 30 hph). The topography of the organs of the alimentary system shows: l—distal part of the liver adjacent to anterior intestine; n—notochord; ns—neural system; pd—pronephric duct; pf—pectoral fin; RGB—respiratory gas bladder; s—skin; st—stomach; y—yolk.

Macallum, 1886). It is believed that the oesophagus will not be morphologically separated from the stomach until a glandular region of the stomach has developed (Balfour and Parker, 1882). This stage becomes established in *L. osseus* at the commencement of the exogenous feeding (23.5 mm BL), although the remnants of the yolk are still present in the body cavity (Jaroszewska and Dabrowski, 2006). The pyloric caeca arise as outgrowths of the anterior intestine in the larvae aged about 3 weeks when they become rapidly elongated and more prominent (Balfour and Parker, 1882; Jaroszewska and Dabrowski, 2006).

In *A. tropicus* until the stage of 112°D (54 hph) there is no lumen in the anterior oesophagus, which immediately follows the pharynx. The air bladder arises in the tropical gar as a dorsal unpaired diverticulum from the anterior part of the oesophagus (Fig. 18). Present results confirmed observations by Balfour and Parker (1882) on *L. osseus*. At the same time the arising longitudinal slit and differentiating glottis above the oesophagus become visible. The wall of the longitudinal slit (Jaroszewska and Dabrowski, 2008a), and the glottal ridges are composed of connective tissue under the epithelium. The primordium of the air bladder is noticeable at the stage immediately after hatching. Balfour and Parker (1882) described

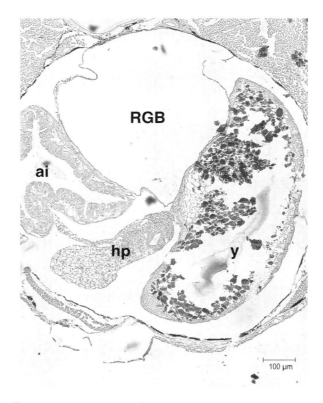

Fig. 16. Transverse sections of the larva of *L. osseus* at 23.5 mm BL, 11 dph: ai—anterior intestine; hp—hepatopancreas; RGB—respiratory gas bladder; y—yolk remnants in the body cavity.

it, as at first a very short and narrow, and then as growing in succeeding stages longer and wider, making its way backwards into the mesentery of the alimentary tract. In *A. tropicus*, at stage of 112°D, the structure of the gas bladder is still undifferentiated but visible at 84°D (Fig. 15). In the larva of *L. osseus* (23.5 mm) at the age of 11 dph it has still a fairly simple form, and is without traces of alveoli on the inner surface (Fig. 16) (Jaroszewska and Dabrowski, 2006). However, at this stage fish continuously take air at the surface and release air bubbles afterwards. This is suggestive of a respiratory function.

In all stages of examined larvae of *A. tropicus* the enterocytes adjoined to the yolk were observed to contain yolk platelets (Figs. 12, 14, 17). It confirmed the description by Allen (1911) that in *L. osseus* at 4 mm BL large quantities of yolk are found in the one-layered endoderm of the presumptive intestine. The yolk platelets present in the intestinal endoderm cells

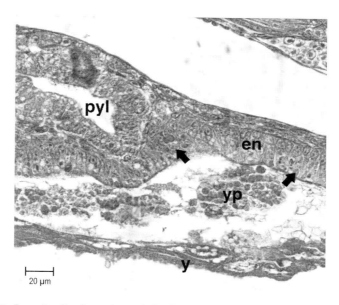

Fig. 17. Longitudinal section of the larva of *A. tropicus* (stage of 112°D; 54 hph). en—enterocytes with the endocytosed yolk platelets (arrows); pyl—pylorus, distal part of the stomach, is present above the anterior intestine; y—yolk; yp—yolk platelets.

were also described in the embryo of the frog *Eleutherodactylus cogui*, in which there is an intestine-yolksac connection (Buchholz et al., 2007). It is believed that similarly the cells of the intestinal wall of lepisosteids participate in yolk digestion through some kind of membrane invagination and endocytosis. A process with bearing similarity was also described in the primordial hepatocytes of cichlid fishes (Fishelson, 1995) and avian embryos. In larvae of *A. tropicus* the platelets of yolk material were also found in the space between the enterocytes of the anterior intestine and the yolk (Figs. 13, 14, 17). This fact confirmed a discovery by Beard (1896) who described "the yolk-emulsion (small yolk platelets) in the embryonic gut" in *Lepisosteus*. He stated that this emulsion (a 'mass of a very few large platelets, a great number of very small ones, and a large quantity of finely divided yolk') had come from the *ysl*[16] at the place where the intestine is open to the yolk.

In the case of *L. osseus* the *ysl* was observed the first time at the stage of a very late blastula in its central and peripheral regions of the yolk (Dean, 1895a). As Balfour and Parker (1882) emphasized, in the later stages of

---

[16]In the original text 'merocytes' (Beard, 1896).

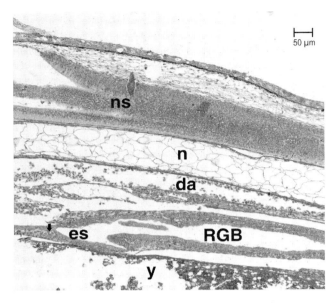

Fig. 18. Longitudinal section of the larva of *A. tropicus* (stage of 112°D; 54 hph) showing the simple structure of the gas bladder (RGB) wall; da—dorsal aorta; es—part of the oesophagus without a lumen in the anterior part (arrow); n—notochord; ns—neural system; y—yolk.

development of *L. osseus*, when the yolksac cannot be observed as an external appendage, yolk is still present within the abdominal cavity and numerous yolk nuclei in *ysl* are still to be found. At this stage, in the larvae of the gar at 11 dph (23.5 mm BL), with the remnants of the yolk in the body cavity, there is no "functional" connection of intestine and yolk (Fig. 16). However, the *ysl* with very large and numerous nuclei is still very distinctly visible under the light microscopy (Jaroszewska and Dabrowski, 2006). Balfour and Parker (1882) have summarized the presence of the nuclei in *ysl* at the late stages of the larval gar development as a feature similar to teleostei and elasmobranchii. In the larvae of *A. tropicus* the yolksac is surrounded by vitelline circulation which consists of many capillaries and large blood collecting vessels (Fig. 19) (for *L. osseus* data are presented in Jaroszewska and Dabrowski, 2006). It can be presumed that in the advanced larvae of the gar the absorption of the mobilized yolk nutrients is entirely released into the vitelline blood circulation system and then transported to the growing tissue. The degradation and transport of yolk-derived material in the developing and energy-depending embryo has been described in selachian (chondrichtyan) fish, in which the yolksac, from the onset of embryogenesis, is in contact with the primordial endo-

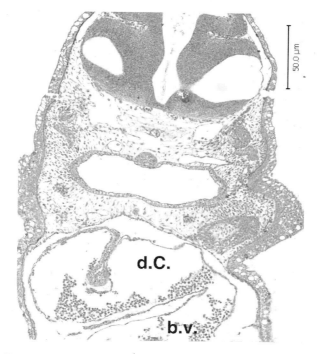

Fig. 19. Transverse section of the larva of *A. tropicus* (stage of 70°D; 18 hph). Vessels collecting blood (b.v.) carry the blood from the yolk directly to the left *ductus Cuvieri* (d.C.) and then to the heart.

dermic gut epithelium (Fishelson, 1995). There is no contact in teleosts fish, between the presumptive gut and yolk reserves during the whole embryonic and early larval development (Kunz, 2004) while, for example, in amphibia, the epithelial cells of the embryo that border the yolk contain yolk particles in their cytoplasm (Fishelson, 1995). In birds the hypoblast of the embryo becomes the origin of the gut endodermis (Yoshizaki et al., 2004). It lies over the yolk and serves the major role in yolk absorption.

Fishelson (1995) stated that the cells, similar to the endodermic cells in birds, are also found in the extra-embryonic *ysl* of the teleost cichlid embryos, where they are engaged in yolk absorption. However, this conclusion is questionable in the light of the present findings on silver arowana, *Osteoglossum bicirrhosum*, which clearly demonstrate "exocytosis" on the surface of the yolk cell which is a part of the *ysl* (Jaroszewska and Dabrowski, 2008b). Ho *et al.* (1999) alluded to the fact that *ysl* in teleosts is the "functional equivalent" of mammalian visceral endoderm which is also difficult to reconcile with our present discovery (Jaroszewska and Dabrowski, 2008b). Therefore, we recommend further studies in order to recognize trophic functions of *ysl* at larval stages of the yolk absorption in garfish.

Free-swimming larvae of lepisosteids begin to feed on zooplankton before completion of absorption of yolk occurred (Jaroszewska and Dabrowski, 2006). The existence of a mixed nutrition phase (lecito-endotrophic) was found in *A. spatula* between 5 and 8 dph in water temperature of 28°C. At this stage, the stomach and pancreatic tissue were formed and pepsin-like acidic proteolytic activity was detected in the stomach before the complete absorption of the yolk. Moreover, between 2 and 9 dph, the gradually increased activity of other digestive enzymes, like trypsin, chymotrypsin and aminopeptidase was found in the alimentary tract of this species. The increased activity of alkaline phosphatase at 8 dph coincided with the maturation of enterocytes in the intestine of *A. spatula* (Mendoza *et al.*, 2002). Histological studies on *L. osseus* (Balfour and Parker, 1882; Jaroszewska and Dabrowski, 2006) and *A. spatula* (Mendoza *et al.*, 2002) indicated that the alimentary tract of lepisosteids is completely functionally formed at the beginning of exogenous feeding. Digestive enzyme activity studied in the Cuban gar (*A. tristoechus*) also indicated the presence of a functional stomach in the early stages of larval development (5-18 dph) (Comabella *et al.*, 2006).

Liver is the first formed alimentary gland and is already a compact structure before the larva hatches (Balfour and Parker, 1882). The studies on *A. tropicus* confirmed that the liver exhibits the same cytological structure in the embryo as it does in the adult. In the differentiated digestive system of the gar the exocrine pancreas was located within the mesentery, connecting the bile ducts, along the abdominal blood vessels, the outer surface of the gastrointestinal tract, the gall bladder, and in the interstitial space between pyloric caeca (Groff and Youson, 1997; Jaroszewska and Dabrowski, 2006). The exocrine pancreas may also enter the liver tissue along the branches of the portal vein (Groff and Youson, 1997). In the gar the islet tissue (endocrine part) is always found closely associated with exocrine pancreatic tissue, i.e., near the portal vein, common bile duct, accumulated near the papilla of the extra-hepatic common bile duct. However, small islets are also scattered along the gall bladder, and in the mesentery and connective tissue on the surface of the anterior intestine (Groff and Youson, 1997). The diffused structure of the pancreas of the gar could be the reason why Balfour and Parker (1882) did not describe its development.

During the embryonic development of the bowfin, there is, as in the garfish, the connection between intestinal endoderm and yolksac (Ballard, 1986a; Eycleshymer and Wilson, 1906). The stomach and the anterior intestine are open posteriorly to the subgerminal cavity above the giant

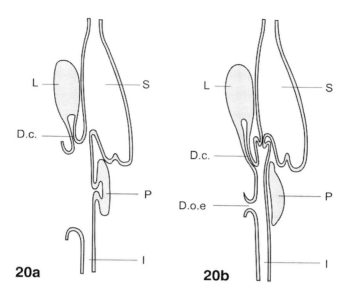

Fig. 20 Development of the liver and *ductus omphalo-entericus* in bowfin individuals (stages of development according to Piper, 1902). (a) at stage 5 the endodermal sheet is open to the giant blastomeres. (b) at stage 6 the endodermal sheet is closed from the sides and from both ends forms the duodenal tube, *ductus omphalo-entericus*, which remains open to the yolky blastomeres until yolk is exhausted. D.c.—ductus choledochus; D.o.e—ductus omphalo-entericus; L—liver; I—intestine; P—pancreas; S—stomach. (Drawing after Piper, 1902).

blastomeres. These stages of differentiation of the alimentary tract in *Amia* correspond to stage 21 described by Ballard (1986a). At this time the presumptive anterior and middle intestine are in the form of an unclosed flat endodermal sheet, extended to the posterior intestine. The oesophageal section and posterior intestine are in the form of the tubular structure, while the oesophagus lacks a lumen. Moreover, according to Ballard (1986a) and Piper (1902), the liver, which arises as a dorsal evagination of the anterior end of the flat endodermal tissue, along with the stomach, share a broad opening downward into the subgerminal cavity (stage 22; stadium I, see Piper, 1902). On the contrary, Dean (1896) stated that in *Amia* the liver has no connection with the yolk, and during its growth it is supplied with nutrients by the vitelline circulation and the intestinal veins. This discrepancy with regard to the liver as an organ connected with the yolksac is further addressed under 'Discussion'. The first indication of the formation of a spiral valve in the posterior intestine of the bowfin is visible at stage 25 (10 dpf, Ballard, 1986a; stadium V according to Piper, 1902). At this time, the early differentiation of the pancreas, spleen and gas bladder, is also evident. As it was described earlier, from stage 26 (Ballard, 1986a; stadium V, Piper, 1902) the endodermal sheet is no longer open to the yolk but it closes from the sides and from both ends to form the duodenal tube, leaving only a prominent round foramen midway of its ventral line (stadium VI, Piper, 1902). This becomes the *ductus omphalo-entericus*, an internal umbilicus (Fig. 20a,b) which opens downward to the yolky giant blastomeres and persists until stage 31, when the yolk has been completely exhausted. The function of the *ductus omphalo-entericus* in the bowfin is not elucidated. It may be assumed as becoming involved in the uptake of nutrients but its function was not yet defined. No blood vessels were noticed between giant blastomeres, but the presence of capillary beds lying close to the coelom lining was evident (Ballard, 1986a).

ADHESIVE ORGAN

The adhesive organ of the gars and bowfin is a structure used by the larvae to attach themselves to aquatic objects. It allows the fish to remain motionless for several days while absorbing yolk reserves. The adhesive organ is located on the tip of the larval snout (Fig. 11a,b,c).

The adhesive organ in the gar is a product of the epiblast and is developed very early (third day after fertilization) with the commencement of differentiation (Beard, 1889)[17]. However, description[18] of the adhesive organ in *L. osseus* was mostly restricted to the hatched individuals (Agassiz, 1879a; Balfour and Parker, 1882; Balfour, 1885; Long and Ballard, 2001). After hatching this organ is completely formed in *L. osseus* and *A. tropicus* (Fig. 11a-c). Our initial results of the studies on *A. tropicus* larvae have shown that the adhesive organ is mainly made up of high columnar cells with a very strong PAS-positive reaction, which is evidence of secretion of polysaccharides and neutral mucopolysaccharides (Fig. 21). In the literature there is no information available about the histological structure of the adhesive gland in embryos of lepisosteid fish. The epiblastic origin of the adhesive organ in the *L. osseus*, described by Beard (1889), contradicts the suggestion expressed by Eycleshymer and Wilson (1908) regarding *Amia*. According to these authors, who studied in detail the development

---

[17] In the original term: 'larval suckers' (Beard, 1889); according to A. Agassiz (1879a) 'sucking disc'.

[18] Beard (1889) described the structure of the 'functional suckers' as composed of two types of cells: 1) long glandular cells with slightly granular hyaline contents and a nucleus lying near the inner end of the cell. Between the glandular cells are present supporting cells (2), with the nucleus in the middle of the cell.

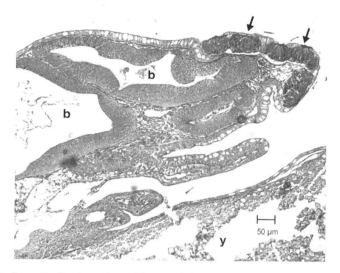

Fig. 21. Longitudinal section of larva of *A. tropicus* (stage of 49°D; 42 hpf). Adhesive organ (arrows) is composed of cells with strong PAS-positive reaction; b—brain; y—yolk.

of the adhesive organ in *Amia*, the primordium of the adhesive organ appears as paired diverticula of the anterior end of the presumptive alimentary tract, below the pre-cerebral mass, already at 70 hpf, when the embryo extends 160 degrees of the egg circumference. During the development of the adhesive organ, the structure of this part of the endoderm changes; first it is composed of flat or cuboidal and then of columnar cells (Eycleshymer and Wilson, 1908). Morphologically, the first indication of the adhesive organ appears at the stage when the entire head region of the embryo enlarges in the form of two bulges anteriorly to the optic primordia, on either side of the medial line. Shortly afterwards the epithelium of the adhesive organ is in contact with the ectoderm, but there is an open connection to the alimentary tract (Ballard, 1986). When the bowfin larva reaches 8-9 mm of BL, the ectoderm no longer extends over the adhesive organ, which now consists of pseudo-stratified columnar cells secreting mucous substances. After this stage these two cell masses of the organ transform from the rectangular into the broad C-shape, and each of them concave medially (Ballard, 1986; Eycleshymer and Wilson, 1908) (Fig. 22). In the free-swimming bowfin larvae (13-14 mm BL) the regression of the adhesive organ is observed. It simply sinks under the ectoderm, but is still in contact with the exterior through a funnel-shape opening in the epithelium (Ecleshymer and Wilson, 1908).

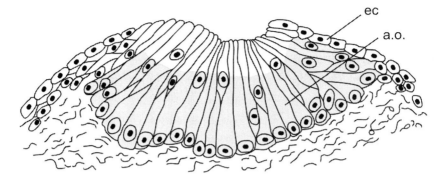

Fig. 22. Sagittal section of one of the adhesive organs in the bowfin embryo (about 130 hpf); a.o.—adhesive organ; ec—ectoderm (redrawn from Eycleshymer and Wilson, 1908).

ARCHES

At the stage when the length of the gar embryo axis exceeds 180° of the egg circumference the thickened protrusion in the subgerminal cavity of the roof marks the beginning of the pharyngeal arches (Long and Ballard, 2001: the number of somites is from 10 to 20). The hyomandibular and hyobranchial pharyngeal pouches are visible externally at the stage when 25-30 somites are present (Long and Ballard, 2001). The oral region is bound posteriorly by a well-marked mandibular arch. It is separated by a shallow depression from a still more prominent hyoid arch (Balfour, 1885). Between the hyoid and mandibular arches a double lamella of hypoblast, which represents the hyomandibular pouch, is continued from the throat to the skin but does not posses a lumen (Balfour, 1885). The hyomandibular pouch appears shortly before hatching (embryo of 10 mm BL) and disappears shortly after the embryo is hatched, without leaving an opening to the exterior (Balfour and Parker, 1882). Prior to hatching, the embryo has the hyoid arch prolongated backwards into a considerable opercular fold, which to a great extent overshadows presumptive gill filaments (Fig. 11c). The opercular edge begins its extension at the stage of 30-40 somites (Long and Ballard, 2001). The second, hyobranchial cleft is open before the larva is hatched. Behind the hyobranchial cleft there are four pouches on each side, they become perforated and converted into branchial clefts shortly before hatching (Balfour and Parker, 1882). They are the rudiments of the four branchial clefts of the adults (Balfour, 1885). Shortly after hatching of *Lepisosteus* the mandibular arch is placed on the posterior border of the mouth and is separated by a deep groove from the hyoid arch (Balfour, 1885). The groove is connected with the hyomandibular cleft. The posterior border of the hyoid arch is extended into an opercular fold. The free edge of operculum begins to extend, but does not yet cover the branchial arches (Balfour, 1885; Long and Ballard, 2001). At the 30th stage of Ballard when the dorsal fin fold has been in part absorbed, and the diameter of operculum is twice as big as the diameter of the eye, gill filaments appear on the third branchial arch in *L. osseus* (Long and Ballard, 2001). In *A. tropicus*, this process was observed at stage of 70°D (Fig. 11a). In the larvae at the stage 84°D and 112°D the operculum flap covers the gills and partially the pectoral fin. In both stages the gills have well developed filaments on the arches (Fig. 11b,c).

THYMUS GLAND

In all vertebrates the primordium of the thymus is formed by interaction between derivatives of the three germ layers: the endoderm of the pharyngeal pouches, the paraxial mesoderm and the neural crest cells, originating from the embryonic midbrain and hindbrain (Zapata *et al.*, 2006). The fish thymus is an intraepithelial bilateral organ in the dorso-lateral region of the gill chamber intimately associated with the pharyngeal cavity (Boehm *et al.*, 2003). It was described in *Amia* that the endodermal, syncytial primordial of the thymus are present on the dorso-lateral ends of the second, third and fourth visceral pouches (Hill, 1935a). They appear in the embryos at 8-9 mm BL (5-7 dpf) and during development they fuse by growth and migration. Occasionally, the third one does not fuse with the other primordia. The bowfin thymus remains a single gland without lobules, surrounded by connective tissue, in some places attached to the skin. Hill (1935a) described the path of the lymphopoiesis; however, the medulla of the bowfin thymus was not studied.

THYROID GLAND

The thyroid development in the bowfin was described by Hill (1935b). The thrust of this work leads us to the conclusion that in this species the thyroid primordium appears as a slid structure from the medial part of the pharyngeal floor (embryo at 6.5 mm BL) and during the development becomes a scattered gland, composed of numerous follicles. The thyroid follicles are present in the mesh of the venous plexus which surrounds the ventral aorta. As it was summarized by Kunz (2004), the thyroid in the teleosts is also a scattered organ, which spreads in its main part along the ventral aorta and into the basis of branchial arches. Additionally, in the teleosts the thyroid follicles are dispersed as an ectopic thyroid in the pronephros, eye, brain, and spleen (Leatherland, 1994). Hill's work (1935b) does not provide any evidence that in the bowfin any part of the thyroid follicles can be found in other organs. It was stated that in the bowfin the colloid is well pronounced in the follicles before hatching (embryo at 9-10 mm BL). In addition, in teleosts prior to the completion of development of the larval thyroid gland, fish eggs and then the yolksac contain THs (thyroid hormones) of maternal origin (Leatherland *et al.*, 1989; Einarsdóttir *et al.*, 2006). There is no information on THs concentrations in tissues of embryos or larvae during early development of bowfin or garfishes. Mendoza *et al.* (2002) reported that a dietary treatment with THs did not affect growth, however, it allowed juveniles to feed on live prey and prevented cannibalism. Larvae exposed to 3,3', 5-triiodo-1-thyronine (T3) had a faster development. This potentially advantageous characteristic of the function of thyroid hormones in early development of non-teleost bowfin and garfish is worth exploring further to elucidate the mechanism and the importance in embryonic development and larval metamorphosis.

Notochord

The development of the notochord takes its origin at the very early stage of the garfish embryo (Fig. 6). According to Dean (1895a) this is the stage when the head eminence is the widest part of the body of the embryo, with lanceolate outline, flat dorsal surface, but with the presence of marked cords of cells in its median line and feebly visible tail. A description of the development of the vertebral column of the gar has been given by Balfour and Parker (1882). They studied the vertebral column as early as at the stage when the notochord is in the form of a large circular rod, composed of vacuolated cells, forming a delicate wide-mesh reticulum. The notochord then becomes surrounded by the thin cuticular sheath (Balfour and Parker, 1882). In fact, they described the first indication of the future vertebral column, found in the formation of a distinct mesoblastic investment of the notochord, in "the advanced larva". Neural processes start to develop at the time when an elastic membrane, the *membrana elastica externa,* is formed around the cuticular sheath of the notochord. This was more precisely described in specimens at 21 mm in BL (Balfour and Parker, 1882).

The hypochord[19] is a transient rod-like structure situated between the notochord and dorsal aorta in embryos of lamprey, fish and amphibia (Löfberg and Colazzo, 1997; Cleaver *et al.*, 2000). It was also observed during the embryonic development in *Amia* (Eycleshymer and Wilson, 1906; Ballard, 1986a). In the gar *L. osseus* the hypochord was described by Balfour (1885) who came to the general conclusion that it originates from the

---

[19]Also named 'subnotochordal rod' by Balfour (1885) and Cleaver *et al.* (2000); 'subchorda' by Löfberg and Colazzo (1997).

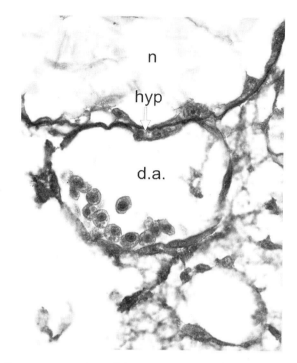

Fig. 23. Transverse section of the tail of *A. tropicus* larva (stage of 49°D; 42 hpf). d.a.—dorsal aorta; hyp—hypochord; n—notochord.

hypoblast. In *Amia* this structure arises from the hypoblast and forms the dorsal wall of the alimentary tract (Eycleshymer and Wilson, 1906). The most recent literature indicated explicitly that it is derived from the endoderm (Löfberg and Colazzo, 1997). A drawing by Balfour (1885) and also an illustration by Balfour and Parker (1882) of the *Lepisosteus* hypochord show a rod-like structure. It is consistent with the literature where it is emphasized that the wall of the hypochord is made up by a one row of cells (Cleaver et al., 2000; Kunz, 2004) (Fig. 23). Balfour and Parker (1882) and Balfour (1885) do not speculate upon the function of hypochord in the gar. Löfberg and Colazzo (1997) suggest that the hypochord could be involved in the positioning of the dorsal aorta during embryonic development. According to Cleaver et al. (2000) and Cleaver and Krieg (2001) notochord signals are required for the formation of the hypochord and then the hypochord is involved in patterning of the dorsal aorta development. It is believed that the functions of the hypochord, mentioned above (in *Xenopus*), are also characteristic for lepisosteids.

DISCUSSION AND SUMMARY

Despite the extensive literature on the embryonic development of the amiids and lepisosteids there is still the diversity of characteristics and multitude of data which require an explanation and do not allow to make comparisons without a risk of overinterpretation. For this reason, the authors of the present contribution discuss the differences and similarities in embryonic and larval development between *Semionotiformes* and *Amiiformes*, and other Actinopterygians to the best of their knowledge and understanding of the state-of-the-art. Simultaneously, they hope that their doubts and omitted matters persuade others to undertake studies on lepisosteids and amiids which will provide answers to the embryological questions in an evolutionary context. The embryonic development of the bowfin and gars possess transitional characteristics between the chondrostean and teleost fish. However, as Cooper and Virta (2007) summarized it, despite the intermediate position of the gars and bowfin between the ancestral and the most modern fishes, in both cases the comparison leads us to teleostei.

However, even in the most recent literature on the early development of *Amia* and the gars, the process of cleavage, even though different in both systematic groups, is still described as an intermediate mode between holoblastic and meroblastic division (Colazzo et al., 1994; Cooper and Virta, 2007). As it was concluded, the similarity of the egg division to the holoblastic cleavage model is better pronounced in *Amia*. The mode of meroblastic cleavage with the presence of undivided yolk and the yolk syncytial layer (YSL) is the most significant feature of the gar embryonic development. This is indicative of close resemblance to teleostei. The similarity of the gars to teleosts also stems from the division of epiblast into two layers: the outer one—the enveloping layer (EVL), inside of which is the second one—the inner layer of precursor cells of the ectoderm of the embryo. It is thought that the inner part of the epiblast in gars represents the future ectoderm: epidermis and neuro-ectoderm. Following Ballard (1986b) who cited Balfour (1881), we can state that in *Amia* embryos the epiblast is also divided into outer and inner layer with the latter one becoming the neuro-ectoderm. However, molecular data are necessary to establish the homology of the outermost epithelium in lepisosteids, amiids and teleosts. Also, the aspect of homology of the gastrulation of gars with that of *Amia* and teleostei, and the role of the gar *ysl* and bowfin hypoblast in the process of epiboly need to be further examined. Another feature, which is common for the gars, *Amia* and the teleosts, is the formation of the neural system as a solid keel of the neuro-ectoderm (Lowery and Sive, 2004). It becomes a hollow neural tube by cavitation. It is known that in gars as well as in *Amia*, and teleosts alike, the optic primordia arise from the forebrain as a pair of solid structures.

The gastrulation of *Amia* was studied in more detail and it is possible to emphasize that its embryonic development has no major resemblance to the developmental features of amphibians, that are, however observed in sturgeon (Salensky, 1881). *Amia*'s late blastula arises in a similar manner to teleosts (Ballard, 1986b). As it was described above, in contrast to sturgeon gastrulation, there is no involution in the superficial marginal zone of the epithelium in the bowfin (Ballard, 1986b; Cooper and Virta, 2007). Among the differences that are worth to mention is that the *Amia* blastoderm spreads during epiboly over the surface of the cleavaged yolk composed of giant blastomeres, whereas blastoderm of teleosts spreads over a single yolky cell filled with a clear granular content. However, similarly to teleostei, at the initiation of the development of *Amia*, the material for the definitive epithelial endoderm lies deep in the upper part of the egg (Ballard, 1986b). Also, in the bowfin, as in the gar and the teleosts, the outermost epithelium of the embryo does not largely contribute to development of internal organs, which is in contrast to amphibia. Both, *Amia* and nearly all teleosts, simultaneously form a germ ring and an anterio-ventral "evacuation zone" during the period of epiboly. Some of the cells migrating from this zone are relocating to the hypoblast of the ring.

However, the opening of the presumptive anterior intestine to the yolk during embryonic development of the gar and *Amia* is their common feature which is strikingly different from Acipenseriformes and teleosts. There are some ambiguities between observations in *A. tropicus* and the data in literature on the bowfin. As it was mentioned above, Ballard (1986a) and Piper (1902) stated that in the bowfin, the stomach, anterior intestine, as well as the liver are open to the giant blastomeres. However, the comparison of figures in the publication on *A. calva* by Piper (1902) and results of the present studies on *A. tropicus* confirmed the opinion of Dean (1896) that in the bowfin only the anterior part of the alimentary tract forms a junction between the intestine and yolk and that the liver is not directly connected to the yolk cell. It can be concluded that the *ductus omphalo-entericus* present in the gar as in the bowfin is a unique structure probably common just for these orders and not present in other fishes.

In addition to the listed gaps regarding early gastrulation in *Amia* as well as in the gars, the development of some inner organs still requires advanced and complex studies. The growth of the reproductive system and blood circulation system seems to be poorly described in Semionotiformes and Amiiformes.

Acknowledgment

Financial support for these studies came from USAID Grant No. LAG-G-00-96-90015-00 through the Aquaculture Collaborative Research Support Program (ACRSP). The Aquaculture CRSP accession number is 1332. The senior author was also supported by the International Studies Scholarship for Young Scientists founded in 2006 by Vice-Rector Prof. dr hab. Andrzej Tretyn for Research and International Relationship of Nicolaus Copernicus University in Toruń, Poland. The main part of the histological documentation was made with the microscope Nikon Eclipse 80i and photo camera Spot RTKe by courtesy of Dr. M. Boehm from Department of Plant Pathology, The Ohio State University, Columbus, Ohio, USA.

We are grateful to Wilfrido Contreras and Arlette Hernandez-Franyutti from the Department of Biology, Universidad Juárez Autónoma de Tabasco, Mexico. We thank Yvette Kunz for discussions and comments on the final version of the manuscript.

References

Afzelius, B.A. 1978. Fine structure of the garfish spermatozoon. *Journal of Ultrastructure Research* 64: 309-314.

Afzelius, B.A. and S.D. Mims. 1995. Sperm structure of the bowfin, *Amia calva* L. *Journal of Submicroscopic Cytology and Pathology* 27: 291-294.

Agassiz, A. 1879a. The development of *Lepidosteus*. *Proceedings of the American Academy of Arts and Sciences* 14: 65-75.

Agassiz, A. 1879b. Embryology of the garpike (*Lepidosteus*). *Science News* 1: 19-20.

Allen, B.M. 1911. The origin of the sex-cells of *Amia* and *Lepidosteus*. *Journal of Morphology* 22: 1-35.

Balfour, F.M. 1885. Ganoidei. In: *A Treatise on Comparative Embryology*, F.M. Balfour (Ed.). Macmillan and Co., London, pp. 102-119, 2nd Edition.

Balfour, F.M. and W.N. Parker. 1882. On the structure and development of *Lepidosteus*. *Philosophical Transactions of the Royal Society Part II (London)* 173: 359-442.

Ballard, W.W. 1984: Morphogenetic movements in embryos of the holostean fish, *Amia calva*: a progress report. *American Zoologist* 24: 539-543.

Ballard, W.W. 1986a. Stages and rates of normal development in the holeostean fish, *Amia calva*. *Journal of Experimental Biology* 238: 337-354.

Ballard, W.W. 1986b. Morphogenetic movements and a provisional fate map of development in the holeostean fish, *Amia calva*. *Journal of Experimental Biology* 238: 355-372.

Beard, J. 1889. On the early development of *Lepidosteus osseus*—preliminary notice. *Proceedings of the Royal Society of London* 46: 108-118.

Beard, J. 1896. The yolk-sac, yolk and merocytes in *Scyllium* and *Lepidosteus*.

Anatomischer Anzeiger 12: 334-347.
Betchaku, T. and J.P. Trinkaus. 1978. Contact relations, surface activity, and cortical microfilaments of marginal cells of the enveloping layer and of the yolk syncytial and yolk cytoplasmic layers of *Fundulus* before and during epiboly. *Journal of Experimental Zoology* 206: 381-426.
Boehm, T., C.C. Bleul and M. Schorpp. 2003. Genetic dissection of thymus development in mouse and zebrafish. *Immunological Review* 196: 15-27.
Buchholz, D.R., S. Singamsetty, U. Karadge, S. Williamson, C.E. Langer and R.P. Elinson. 2007. Nutritional endoderm in a direct developing frog: a potential parallel to the evolution of the amniote egg. *Developmental Dynamics* 236: 1259-1272.
Cleaver, O., D.W. Seufert and P.A. Krieg. 2000. Endoderm patterning by the notochord: development of the notochord in *Xenopus*. *Development* 127: 869-879.
Cleaver, O. and P.A. Krieg. 2001. Review. Notochord patterning of the endoderm. *Developmental Biology* 234: 1-12.
Colazzo, A., J.A. Bolker and R. Keller. 1994. A phylogenetic perspective on teleost gastrulation. *American Naturalist* 144: 133-151.
Comabella, Y., R. Mendoza, C. Aguilera, O. Carrillo. A. Hurtado and T. García-Galano. 2006. Digestive enzyme activity during early larval development of the Cuban gar *Atractosteus tristoechus*. *Fish Physiology and Biochemistry* 32: 147-157.
Cooper, M.S. and L.A. D'Amico. 1996. A cluster of noninvoluting endocytic cells at the margin of the zebrafish blastoderm marks the site of embryonic shield formation. *Developmental Biology* 180: 184-198.
Cooper, M.S. and V.C. Virta. 2007. Evolution of gastrulation in the ray-finned (Actinopterygian) fishes. *Journal of Experimental Zoology* 308: 591-608.
Dabrowski K., M. Jaroszewska, G. Rodríguez and J. Rinchard. 2006. Gonadal differentiation in longnose gar *Lepisosteus osseus*, *Primera Reunión Internacional para la Investigación de Lepisostéidos: Red International para la Investigación de Lepiosteidos*. Villahermosa, Tabasco, Mexico, 3.
Dean, B. 1895a. The early development of gar-pike and sturgeon. *Journal of Morphology* 11: 1-62.
Dean, B. 1895b. The early development of *Amia*. *Quarterly Journal of Microscopical Science* 38: 413-444.
Dean, B. 1896. On the larval development of *Amia calva*, *Zoologische Jahrbuecher Abteilung fuer Systematik. Oekologie und Geographie der Tiere* 9: 639-672.
De Pinna, M.C.C. 1996. Teleostean Monophyly. In: *Interrelationship of Fishes*, M.L.J., Stiassny, L.R. Pareti and G.D. Johnson (Eds.) Academic Press, San Diego. pp. 147-162.
Einarsdóttir, I.E., N. Silva, D.M. Power, H. Smáradóttir and B.T. Björnsson. 2006. Thyroid and pituitary gland development from hatching through metamorphosis of a teleost flatfish, the Atlantic halibut. *Anatomy and Embryology* 211: 47-60.
Essner, J.J., J.D. Amack, M.K. Nyholm, E.B. Harris and H.J. Yost. 2005. Kupffer's vesicle is a ciliated organ of asymmetry in the zebrafish embryo that initiates left-right development of the brain, heart and gut. *Development* 132: 1247-1260.

Eycleshymer, A.C. 1899. The cleavage of the egg of *Lepidosteus osseus*. *Anatomischer Anzeiger* 16: 529-536.

Eycleshymer, A.C. 1903. The early development of *Lepidosteus osseus*. *University of Chicago, Decennial Publication* 10: 259-275.

Eycleshymer, A.C. and J.M. Wilson. 1906. The gastrulation and embryo formation in *Amia calva*. *American Journal of Anatomy* 5: 133-162.

Eycleshymer, A.C. and J.M. Wilson. 1908. The adhesive organ of *Amia*. *Biological Bulletin*. 14: 134-149.

Fishelson, L. 1995. Ontogenesis of cytological structures around the yolk sac during embryologic and early larval development of some cichlid fishes. *Journal of Fish Biology* 47: 479-491.

Fishelson, L. and A. Baranes. 1998. Observations on the Oman shark *Iago omanensis* (Triakidae), with emphasis on the morphological and cytological changes of the oviduct and yolk sac during gestation. *Journal of Morphology* 236: 151-165.

Groff, K.E. and J.H. Youson. 1997. An immunohistochemical study of the endocrine cells within the pancreas, intestine, and stomach of the gar (*Lepisosteus osseus* L.). *General and Comparative Endocrinology* 106: 1-16.

Hamlett, W.C., F. Schwartz, and L.J.A. DiDio. 1987: Subcellular organization of the yolk syncytial-endoderm complex in the preimplantation yolk sac of the shark, *Rhizoprionodon terraenovae*. *Cell and Tissue Research* 247: 275-285.

Hall, B.K. 1999. Bony and Cartilaginous Fishes. In: *The Neural Crest in Development and Evolution*, Springer pp. 77-87.

Hill, B.H. 1935a. The early development of the thymus glands in *Amia calva*. *Journal of Morphology* 57: 61-89.

Hill, B.H. 1935b. The early development of the thyroid gland in *Amia calva*. *Journal of Morphology* 57: 533-545.

Ho, C.Y., C. Houart, S.W. Wilson and Y.R. Stainier. 1999. A role for the extra-embryonic yolk syncytial layer in patterning the zebrafish embryo suggested by properties of the *hex* gene. *Current Biology* 9: 1131-1134.

Hurley, I.A, R.L. Mueller, K.A. Dunn, E.J. Schmidt, M. Friedman, R.K. Ho, V.E. Prince, Z. Yang, M.G. Thomas and M.I. Coates. 2007. A new time-scale for ray-finned fish evolution. *Proceeding of the Royal Society B* 274: 489-498.

Jamieson, M.B. 1991. *Fish Evolution and Systematics: Evidence from Spermatozoa*. Cambridge University Press, Cambridge.

Jaroszewska, M. and K. Dabrowski. 2006. Morphological features of digestive tract development in the longnose gar *Lepisosteus osseus*, *Primera Reunión International para la Investigación de Lepisostéidos: Red International para la Investigación de Lepiosteidos*. Villahermosa, Tabasco, Mexico, 8.

Jaroszewska, M. and K. Dabrowski. 2008a. Morphological analysis of the functional design of the connection between alimentary tract and the gas bladder in air-breathing lepisosteid fish. *Annals of Anatomy* 190: 383-390.

Jaroszewska, M. and K. Dabrowski. 2008b. The yolk syncytial layer in ancient osteoglossid fish silver arowana—the first report. *World Aquaculture 2008: Aquaculture for Human Wellbeing—the Asian Perspectives,* Busan, Korea, 150.

Kimmel, C.B. and R.D. Law. 1985. Cell lineage of zebrafish blastomeres. 2. Formation of the yolk syncytial layer. *Developmental Biology* 108: 86-93.

Knaut, H., C. Werz, R. Geisler, T. Tübingen, S. Consorium and C. Nüsslein-

Volhard. 2003. A zebrafish homologue of the chemokine receptor *Cxcr4* a germ-cell guidance receptor. *Nature* (London) 421: 279-282.

Kunwar, P.S. and R. Lehmann. 2003. Germ-cell attraction. *Nature* (London) 421: 226-227.

Kunz, Y.W. 2004. *Development Biology of Teleost Fishes*. Springer. The Netherlands.

Lanzi, L. 1909. Richerche sui primi momenti di sviluppo degli *Oleostei (od Euganoidi) Amia calva* Bonap. e *Lepidosteus osseus* L. Con speciale riguardo al cosi detto ispessimento prostomale. *Arch. Ital. Anat. Embriol* 8: 292-306.

Leatherland, J.F. 1994. Reflection on the thyroidology of fishes: from molecules to humankind. *Guelph Ichthyology Reviews* 2: 1-67.

Leatherland, J.F., L. Lin, N.E. Down and E.M. Donaldson. 1989. Thyroid hormone content of egg and early development stages of three stocks of goitred coho salmon (*Onconrhynchus kisutch*) from the Great Lakes of North America, and a comparison with a stock from British Colombia. *Canadian Journal of Fisheries and Aquatic Sciences* 46: 2146-2152.

Löfberg, J. and A. Colazzo. 1997. Hypochord: an enigmatic embryonic structure. Study of the axolotl embryo. *Journal of Morphology* 232: 57-66.

Long, W.L. and W.W. Ballard. 2001. Normal embryonic stages of the longnose gar, *Lepisosteus osseus*. *BMC Development Biology* 1:6, http://www.biomed central.com/1471-213X/1/6

Lowery, L.A. and H. Sive. 2004. Strategies of vertebrate neurulation and a re-evaluation of teleost neural tube formation. *Mechanisms of Development* 121: 1189-1197.

Lukáš, J. 1989. Visceral anatomy of the garpike, *Atractosteus tristoechus*. *Folia Zoologica* 38: 265-274.

Macallum, A.B. 1886. The alimentary canal and pancreas of *Acipenser, Amia* and *Lepidosteus*. *Journal of Anatomy and Physiology* 21: 604-636.

Melby, A.E., R.M. Warga and C.B. Kimmel. 1996. Specification of cell fates at the dorsal margin of the zebrafish gastrula. *Development* 122: 2225-2237.

Mendoza R., C. Aguilera, G. Rodríguez, M. Gonzáles and R. Castro. 2002. Morphophysiological studies on alligator gar (*Atractosteus spatula*) larval development as a basis for their culture and repopulation of their natural habitats. *Reviews in Fish Biology and Fisheries* 12: 133-142.

Neidhart, I. 1985: Pro- and opistonephros relation in *Acipenser ruthenus* L. Fortschritte der Zoologie, 30: 469-471. Duncker/Fleischer (Eds.). Vertebrate Morphology. Gustav Fischer Verlag. Stuttgart.

Nelsen, O.E. 1953. Comparative Embryology of the Vertebrates. Blakiston. New York.

Nelson, J.S. 1994. *Fishes of the World*. John Wiley and Sons, New York. 3rd Edition.

Nelson, J.S. 2006. *Fishes of the World*. John Wiley and Sons, Hoboken, New Jersey. 4th Edition.

Nilsson, E.E. and J.G. Cloud. 1992. Rainbow trout chimeras produced by injection of blastomeres into recipient blastulae. *Proceedings of the National Academy of Sciences of the United States of America* 89: 9425-9428.

Ninhaus-Silveira, A., F. Foresti and A. de Azevedo. 2006. Structural and ultra-structural analysis of embryonic development of *Prochilodus lineatus* (Valenciennes, 1836) (Characiforme; Prochilodontidae). *Zygote* 14: 217-229.

Piper, H. 1902. Die Entwickelung von Magen, Duodenum, Schwimmblase, Leber,

Pankreas and Milz bei *Amia calva*. *Archiv für Anatomie und Entwickelungsgeschichte*: (*Anat. Abth*). 1-78.

Salensky, W. 1881. Recherches sur le dévelopment du sterlet. *Archive de Biologie* 2: 233-341.

Solnica-Krezel, L. 2005. Conserved patterns of cell movements during vertebrate gastrulation. Review. *Current Biology* 15: R213-R228.

Trinkaus, J.P. 1984. Mechanism of *Fundulus* epiboly—a current review. *American Zoologist* 24: 673-688.

Whitman, C.O. and A.C. Eycleshymer. 1897. The egg of *Amia* and its cleavage. *Journal of Morphology* 12: 309-354.

Yoshizaki, N., M. Soga, Y. Ito, K.M. Mao, F. Sultana and S. Yonezawa. 2004. Two-step consumption of yolk granules during the development of quail embryos. *Development Growth and Differentiation* 46: 229-238.

Zapata, A., B. Diez, T. Cejalvo, C. Gutiérrez-de Frías and A. Cortés. 2006. Ontogeny of immune system of fish. *Fish and Shellfish Immunology* 20: 126-136.

5 Early Development in Sarcopterygian Fishes

Felisa Kershaw, Gregory H. Joss and Jean M.P. Joss*

Biological Sciences, Macquarie University, Sydney, NSW 2109, Australia.
E-mail: jjoss@rna.bio.mq.edu.au

INTRODUCTION

There is a dearth of recent knowledge about early development for the eight living species of sarcopterygian fish. These living sarcopterygians belong to the coelacanths, two species, and the lungfish, six species. Of these eight species, only the lungfish, *Neoceratodus forsteri*, has provided recent information about sarcopterygian early development. Coelacanths, *Latimeria chalumnae* and *L. mendosa*, are ovoviviparous (Smith et al., 1975) and live at considerable ocean depths, rendering the early developmental stages virtually inaccessible. Most of what is known about coelacanth biology is based on the few (~100) hand-line-captured individuals that survived for a short time after capture and a very limited number of formalin-fixed or frozen specimens (Balon et al., 1988). For an excellent account of the maternal-embryonic relationship in the coelacanth obtained from a fixed specimen of a gravid female held by the American Museum of Natural History, I refer you to Wourms et al. (1991) and also Balon's chapter in the same volume (1991). The following chapter, although entitled "sarcopterygian fishes" will primarily discuss recent studies on the early development of *Neoceratodus forsteri*, with reference to older descriptions of all three genera of lungfish, *Neoceratodus*, *Protopterus* and *Lepidosiren*, especially those of Kerr (1909, 1919). The stages of development for *Neoceratodus* used in this chapter are based on previous staging tables of Semon (1893) as reproduced in Olsson et al. (2004) and those of Kemp (1982) and our own website: http://mac-0170.bio.mq.edu.au/~gjoss/lungfish_development/lungfishSQL.php

Apart from reproductions of Semon's illustrations, the images on this website are all of live embryos and hatchlings of Australian lungfish, *Neoceratodus forsteri* (Krefft), developing from eggs collected from the Macquarie University lungfish breeding ponds during the 2004 spawning season, from 14/9/2004 to 19/11/2004. They were imaged from 14/9/2004 to March 2005. As far as possible, individual fish were followed as they grew.

The imaging and website preparation was undertaken in an effort to make it easier for researchers and students to identify stages of interest to them and the morphological changes leading to these stages. This was necessary as even though the illustrations of Semon (1893) are magnificent, they are only one instant in time per stage and they don't always correspond with the simple line drawings of Kemp (1982), together with the fact that photographs are more representative depictions than illustrations and, therefore, more helpful when attempting to stage live embryos for study.

EARLY DEVELOPMENT OF *NEOCERATODUS FORSTERI*, WITH COMMENTS ON OTHER LIVING DIPNOANS

In his conclusion to his normal tables for the development of *Lepidosiren* and *Protopterus*, Kerr (1909) states how closely they resemble one another in their development and how strikingly they differ from that of *Neoceratodus*, as described by Semon (1893). The early cleavages, gastrulation and neurulation are more amphibian-like in the latter, and the yolk is distributed more evenly throughout the embryonic gut so that the development of the head is more rapid in *Neoceratodus* than in the lepidosirenid lungfishes. However, he then concedes that the early development of all three genera have sufficient similarities for them to be considered merely as subdivisions of the same group—the Dipnoi or lungfishes. In the following accounts we will consider first the general developmental features, shared in common by all developing lungfish, followed by brief comments on their differences.

The Egg

Lungfish are oviparous. Their eggs more closely resemble those of amphibians than they do those of teleost fishes (see website mentioned above). The large size of their eggs is at least in part due to their very large genome (Rock *et al.*, 1996), requiring a large nucleus to contain the DNA, but is also due to the large amount of yolk contained therein. The eggs of *Lepidosiren* appear to be the largest (6.5-7 mm), followed by those of *Protopterus* (3.5-4 mm) (Kerr, 1909) and *Neoceratodus* (3-5 mm). The yolk content is not so great as to cause cleavage divisions, following fertilization,

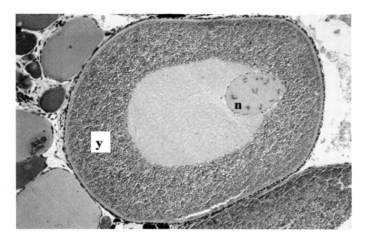

Fig.1. Section through portion of the ovary of a female *Neoceratodus*, showing a growing oocyte containing yolk around the periphery. The nucleus clearly contains a number of nucleoli. Haematoxylin and eosin stain, photographed using the 20X objective lens. n—nucleus; y—yolk.

to be restricted to the embryonic cell mass as it is in telolecithal eggs such as those found in reptiles and birds, but it is much greater than in most teleost fish eggs. Lange (1983) reported that the crystal structure of yolk found in the eggs of all mesolecithal vertebrates appears to be very similar. His study included eggs from coelacanth, lungfish, several ray-finned fish and both anuran and urodele amphibians.

While still in the ovary prior to ovulation, recrudesced eggs of *Neoceratodus* are characterized by several large nucleoli in the nucleus (Fig. 1). Similar nucleoli also occur in the nuclei of eggs of some amphibians (Lofts, 1974). Once the eggs are ovulated, they are picked up by the oviducts which convey them to the outside environment, during quite elaborate courtship rituals (Grigg, 1965). At oviposition, the yolk-filled ovum is surrounded by several non-cellular egg envelopes: an outer and an inner albumin egg coat and between the embryo and the inner albumin coat there is a vitelline membrane. At oviposition, the outermost egg envelope is able to receive sperm but soon afterwards this envelope becomes very sticky and attaches to the water plants over which it has been released. The stickiness also attaches any passing detritus, so that the inner egg/embryo is rapidly obscured from view (Kemp, 1982). Demersal eggs sink when oviposited, as opposed to the pelagic eggs of many seawater fish. The demersal eggs of lungfish may be further classified as phytophilic because of their preference of attaching to plants.

During the courtship of a pair of lungfish (*Neoceratodus forsteri*) considerable interest seems to be taken in choosing the exact site for oviposition. A general principle appears to be finding a substrate for development that will provide the most protection, the warmest temperature, and the most access to sufficient oxygen. The most favoured spawning sites are those that are shallow, with good plant cover over the substrate and slow water movement. Spawning between a pair of lungfish may proceed for several hours, usually during the early evening. Most commonly the female deposits just one or two eggs at a time which the male then fertilizes. Following this, the pair will swim off to choose another area for the next oviposition. Thus they appear to take great care in placing the eggs where they will experience the most favourable conditions for development and survival (personal observations). But following each oviposition and subsequent fertilization there is no further care of the young. The lepidosirenid lungfish appear to offer much more care of the young in that they construct nests into which the eggs are deposited and fertilised and which they may guard and aerate throughout early development and even beyond hatching (Greenwood, 1957, 1986).

Sperm

Lungfish sperm are also very large, again due to their large nuclei/genomes and in contrast to the small size of teleost sperm, and indeed the sperm of most other vertebrates. The sperm of *Neoceratodus forsteri* have been described in detail by Jesperson (1971), and Boisson *et al.* (1967) provide similar descriptions of the sperm of the African lungfish, *Protopterus annectens.* The two descriptions share similarities but also have some striking differences. The sperm heads of both species are almost equally large even though the DNA content of the nucleus in *Protopterus* is almost twice that of the already large genome of *Neoceratodus* (Ohno and Atkin, 1966; Pedersen, 1971). *Neoceratodus* has two rods which extend from the acrosome through either side of the nucleus for about 4/5 of its length where they emerge to terminate in the cytoplasm. These rods are not present in the sperm of *P. annectens,* but similar structures have been described in the sperm of the primitive ray-finned fish, *Polypterus senegalis* (Mattei, 1969) and a lamprey, *Lampetra planeri* (Follenius, 1965). The tail of the sperm in *P. annectens* is paired with the two centrioles lying parallel to each other. The tail of *Neoceratodus* sperm is not biflagellate.

Fertilisation and Cleavage

Fertilised eggs of *Neoceratodus* attach to plants in the spawning area several hours before cleavage begins. Early cleavages follow a regular deuterostomic pattern for the first two cleavages, after which the yolk distribution between cells formed from each cleavage starts to interfere with the regularity of the size and shape of some cells as they become distributed into an animal pole and a vegetal pole (Plates 1A,B). This closely resembles the situation in *Lepidosiren* and most amphibians where it is referred to as holoblastic. As cleavages proceed, the pigmented animal pole becomes increasingly differentiated from the yolk-filled larger fewer cells of the vegetal pole. However, all cells, even those in the animal pole, contain at least some yolk platelets in their cytoplasm. Ultrastructural examination of the developing CNS in *Neoceratodus* clearly shows yolk platelets in all cells, even as late as stage 33 (Fig. 2).

The blastula forms gradually as cleavages continue from stages 9 to 11. By stage 12, the dorsal lip of the blastopore becomes visible and signals the start of gastrulation (Plate 1C). Through all these early stages the lungfish embryo bears a remarkable similarity to the developing amphibian embryo (cf. Huettner, 1941 with the website).

Gastrulation

At the very beginning of gastrulation, a small groove becomes apparent in the vegetal pole (stage 12, Plate 1C). This groove becomes deeper and more crescent-shaped as gastrulation proceeds (stage 13, 14). The striking resemblance to amphibian gastrulation cannot be missed. The early work of Kerr (1919) on *Protopterus* and *Lepidosiren* includes a more detailed description of gastrulation than is available for *Neoceratodus*. He describes early gastrulation as being governed by the three processes of 1) involution, 2) overgrowth by the dorsal lip and 3) delamination. From his descriptions, it is quite clear that Kerr is referring to what is now more commonly termed 1) involution, which is the rolling in of the ectoderm to become later mesoderm, 2) epiboly, which is the spreading of the ectoderm across the outer surface of the embryo, as the cells of the future entoderm are moving inside the embryo, and 3) convergent extension, which describes the extension of the mesoderm cells between the outer ectoderm and the inner endoderm.

During the later stages of gastrulation, the animal pole which is developing into the embryonic lungfish begins to elongate, extending out over the yolk mass (Plate 1D). As it does so, the eversion of the neural plate (medullary plate of Kerr, 1909) begins leading to the formation of the neural folds and as they meet, the neural tube of the central nervous system, becomes more obvious (Plate 1E).

Thus the process of gastrulation in lungfish is governed by the intermediate amount of yolk present in the fertilised egg and most closely resembles gastrulation in the amphibian *Xenopus*, which contains a similar amount of yolk in its fertilised egg.

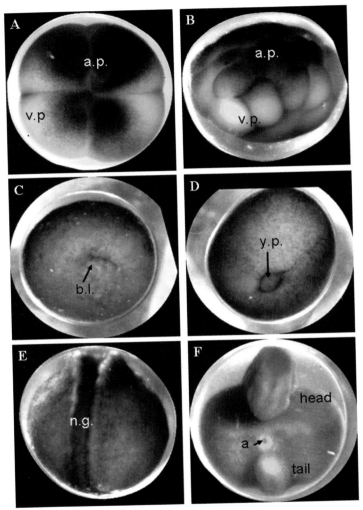

Plates 1 and 2 show selected stages of development in *Neoceratodus forsteri* as shown on the website: http://mac-0170.bio.mq.edu.au/~gjoss/lungfish_development/lungfishSQL.php

Plate 1. A—Stage 3: 4-cell, after equal second cleavage. B—Stage 6: morula, cleavages no longer synchronised, clear difference between macromeres of vegetal pole and micromeres of animal pole. C—Stage 12: early gastrula, showing the dorsal lip of the blastopore. D—Stage 15: late gastrula, showing yolk plug in the blastopore. E—Stage 22: neurula, mid stage, neural folds not yet closed to form neural tube. F—Stage 31: phylotypic (tail bud) stage, blastopore now clearly visible as anus. a—anus, a.p.—animal pole, b.l.—blastopore lip, n.g.—neural groove, v.p.—vegetal pole, y.p.—yolk plug.

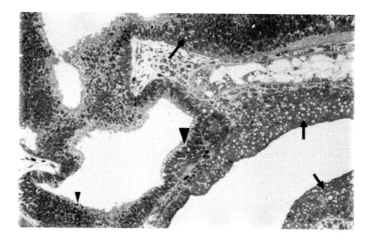

Fig. 2. Semi-thin section (methylene blue stained) through the ventral forebrain of a stage 33 embryonic *Neoceratodus*. Note the yolk platelets (lipovitellin—white [arrows]; phosvitin—black [arrowheads]) distributed throughout the brain cell bodies (arrows/arrowheads indicate just some of these). Photographed using the 20X objective lens.

Neurulation, including Neural Crest

As the blastopore lips begin to close in late gastrulation, the ectoderm of the animal pole adjacent to the blastopore begins to evert in the manner characteristic of amphibian neurulation (stage 16-19) in response to signals from the notochord which is already forming from the chorda-mesoderm during late gastrulation. As the neural folds produced by the eversion of the neural plate ectoderm start to close, beginning anteriorly and gradually progressing posteriorly, the neural crest starts to become apparent (Plate 1E). This is close to the end of neurulation in *Neoceratodus*.

Scanning electron micrographs prepared from *Neoceratodus* embryos at the end of neurulation show that soon after the neural crest forms, their cells start to migrate laterally and ventrally in distinct streams (Fig. 3) to contribute to a variety of head structures. As the neural crest continues to form alongside the neural tube beyond the most anterior head end of the embryo, its cells migrate to form many other features, such as melanophores and ganglia of the sympathetic nervous system (Fig 4, Plates 2A,B). This information is derived largely by inference from fate maps made for neural crest migrating cells in amphibians, chicks and mice but fate mapping is also underway for *Neoceratodus* (Ericsson et al., 2007; Kundrat et al., 2008 and unpublished data) which is finding that cranial neural crest cells appear to contribute to all the same features that they do in tetrapods.

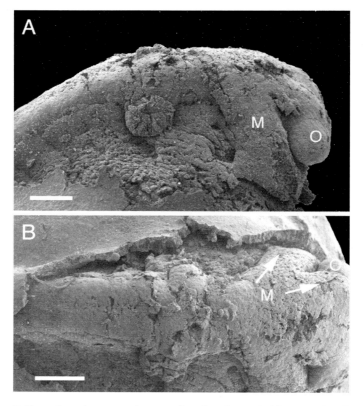

Fig. 3. Scanning electron micrographs of embryonic *Neoceratodus* showing the cranial neural crest from a lateral view in A and from a dorsal view in B, anterior to the right. Epidermis overlying the neural tube has been removed. Edges of unremoved epidermis are at top of figure. A—Stage 27 embryo, showing a conspicuously migrating mandibular crest stream (M). B—Same embryo in dorsal view, showing the mandibular stream dividing and migrating along two different routes. One stream goes rostrally between the epidermis and the neural tube dorsal to the optic vesicle (O), while the other is a rostroventral route (arrows). Scale bars are 0.2 mm.

Cranial neural crest cells have been shown to contribute to the characteristic head of all vertebrates (Hanken and Hall, 1993). In most vertebrates studied, the cranial neural crest, in contrast to the trunk, contributes to most skeletal elements in the head.

There has been some controversy over whether lungfish have cranial neural crest (Kemp, 1995, 1999, 2000). The scanning electron microscopic study that clearly showed cranial neural crest streams in *Neoceratodus*

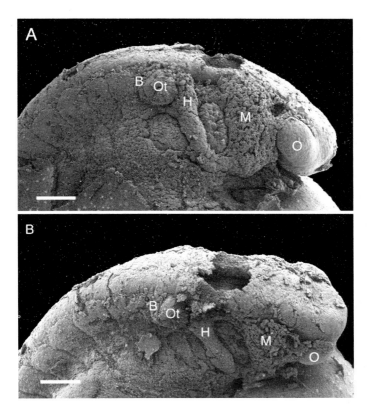

Fig. 4. Scanning electron micrographs of a stage 28 embryonic *Neoceratodus*, showing the cranial neural crest streams after removal of the overlying epidermis. A—Embryo showing the mandibular (M), hyoid (H) and branchial (B) neural crest streams, the developing optic vesicle (O) and the otic placode (Ot). B—The other side of the same animal (flipped 180 degrees so that anterior is to the right) showing the mandibular (M), hyoid (H) and branchial (B) streams. Scale bars are 0.2 mm.

forsteri was published as recently as 2000 by Falck and colleagues. Since then, this latter study has been used to identify the location and timing of the streams for fate mapping the neural crest cells by injecting them with the vital dye, DiI, prior to, or during early migration in living embryonic lungfish (*Neoceratodus*) and then tracing the fate of the DiI-labeled cells (Ericsson *et al.*, 2007; Kundrat *et al.*, 2008, 2009).

Hatching

Hatching is not a morphological stage in the life of *Neoceratodus* as it is for most other fish. Young lungfish are staged according to morphological changes during development and hatching can occur over a range of these in *Neoceratodus*, the earliest being stage 39 (Plate 2B), which is an animal without paired fins, and considerable yolk still in the intestine, through stages 42 (Plate 2C) and stage 45 (Plate 2D) to stage 46 with pectoral fins developing, the lateral line complete and yolk reserves almost depleted. Having said this, the majority of lungfish hatch at stage 45, in which the hatchling still has yolk reserves and does not feed, the lower and upper lips have met and the pectoral fins are more than just buds (Plate 2D) (Joss *et al.*, 1997; see web images). The external gills, which are a characteristic feature of the lepidosirenid lungfishes, are not present in hatchling *Neoceratodus*, or at any other stage. Their internal gills are much better developed than in *Protopterus* or *Lepidosiren* and appear to provide adequate gaseous exchange at this stage, prior to the development of a working lung.

The cement organs of *Lepidosiren* and *Protopterus* are present at hatching in the ventral pharyngeal region. Kerr (1909) describes these hatchlings as being more tadpole-shaped than *Neoceratodus*, and *Protopterus* more so than *Lepidosiren*, due to the accumulation of yolk at the anterior end of the gut of the little hatchlings. Not long after hatching the external gills start to appear. In *Lepidosiren* these grow rapidly, persist for some time and then equally rapidly degenerate in what Kerr (1909) considered a metamorphosis, characterised by the rapid atrophy of the external gills, the loss of the cement organ, and a change in coloration as they become much darker and more active. In *Protopterus*, such a rapid change does not occur. The external gills atrophy much more slowly and may even be present, albeit as vestigial fragments well into maturity.

Hatchling *Neoceratodus* remain close to the egg capsule, from which they have escaped, for the next few days and well beyond continue to stay close by their hatching site among the water plants in the warm shallow water where they are well camouflaged against predators. At about stage 47, they will have depleted their yolk and start to feed on small invertebrates living amongst the same plants (Plate 2E). They have quite well-developed teeth of the conical kind at this stage which enable them to restrain worms, rotifers, etc. as they proceed to swallow them. Their teeth are not of the characteristic crushing plate type of juvenile and adult lungfish yet, so it is not uncommon to see a small worm departing a lungfish via the operculum rather than being crushed and continuing into the spirals of its intestine. Early feeding stages of 48 and 50 are illustrated in Plates 2F and 2G, G')

285

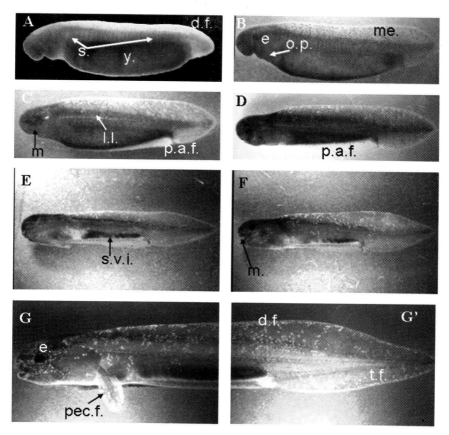

Plates 1 and 2 show selected stages of development in *Neoceratodus forsteri* as shown on the website: http://mac-0170.bio.mq.edu.au/~gjoss/ lungfish_development/lungfishSQL.php

Plate 2. A—Stage 36: beginning to resemble a lungfish, melanophores developing, somites still visible, dorsal fin and tiny post-anal fin, yolk-filled gut. B—Stage 39: more melanophores, lateral eye becoming more obvious, mouth ventral. C—Stage 42: Mouth moving from ventral to anterior, pectoral fins appearing as buds, lateral line extending past half way along body. D—Stage 45: yolk in gut rapidly diminishing, upper and lower lips of mouth meet, melanophores more dense, lateral line extends to tail. E—Stage 47: spiral valve of intestine clearly visible as yolk reserves almost completed and feeding has begun, mouth now completely anterior. F—Stage 48: well developed pectoral fin and pelvic fin primordium appearing. G, G'—Stage 50: fish now growing more rapidly, mouth terminal, pectoral fins very large and pelvic fins beginning to show more clearly. d.f.—dorsal fin, e.—eye, l.l.—lateral line, m.—mouth, me.—melanophores, o.p.—operculum, p.a.f.—preanal fin, pec. f.—pectoral fin, s.—somites, s.v.i.—spiral valve intestine, t.f.—tail fin, y.—yolk.

Neoteny

Neoteny is the condition wherein a normally metamorphosing animal fails to metamorphose before becoming a reproductively mature larva. It is a common life strategy among several families of urodele amphibians, where it is also related to the possession of a large genome. As a general rule, the longer a family of urodeles has been neotenic, the larger the genome they have and the more neotenic they are (Martin and Gordon, 1995). Neoteny can occur by different mechanisms. It may only occur if the water habitat remains beyond the normal larval period. For this group, if the water dries up, the larval form can metamorphose into a more terrestrial salamander. This is said to be facultative neoteny. In the second form of neoteny metamorphosis never occurs in wild populations but can be induced experimentally by administration of thyroid hormones. This is called inducible obligatory neoteny. In the last type of neoteny, metamorphosis never occurs in wild populations and it cannot be induced experimentally —non-inducible obligate neoteny. In a continuum, the first type of neoteny is generally characteristic of recent families and the last type of non-inducible obligatory neoteny correlates with the oldest families, eg. Amphiumidae. Lungfish, like the Amphiumidae, have very large genomes. Larval or paedomorphic characters in adult forms have been recognized for a number of years (e.g., Bemis, 1984), and recently the thyroid axis of *Neoceratodus forsteri* has been shown to be very similar to that of neotenic amphibians (Joss, 2006). These features of adult lungfish considered together provide strong support for proposing that nowadays lungfish are non-inducible neotenes. However, the developmental patterns of the lepidosirenid lungfish need to be considered more closely in the light of this hypothesis, especially as Kerr (1909) considered *Lepidosiren* to undergo metamorphosis fairly soon after hatching.

General Discussion

The most striking feature of lungfish early development is the strong similarities with equivalent stages in amphibian development. While we were following the external features of *Neoceratodus* embryos as they underwent early development and collecting the images now placed on the website, we were continually struck with these similarities, starting with the bottom heavy morula (stage 6-7, Plate 1B) and then the 'grey crescent' of the dorsal lip of the blastula as gastrulation began (stage 12, Plate 1C) and at a later stage of gastrulation, the clearly recognisable yolk plug stage (stage 15, Plate 1D), followed by the distinctive amphibian stages of neurulation (stages 17-23, Plate 1E) and finally the phylotypic tail bud stage (stage 31, Plate 1F). Beyond this stage, the little embryos still resemble those of amphibians but begin to show more distinctly lungfish characteristics (gill arches, median fins, lateral line, cranial ribs, etc.). As development proceeds beyond the early stages, they not only retain some resemblance with larval amphibians, but continue with this appearance into adulthood so that a display tank with adult *Neoceratodus* alongside one containing Chinese or Japanese Giant salamanders, on casual viewing appear to differ only in the paired lobe fins of the one and the paired limbs of the other.

There is, of course, a whole group of tetrapodomorph fossil fishes separating lungfish from the ancestral tetrapods, but the similarities, we argue are sufficient to warrant the study of lungfish early development in the hope of shedding light on the genetic mechanisms that underlay this major transformation during the mid-late Devonian some 350-375 mya.

Acknowledgements

We would like to acknowledge the expert assistance of Libby Eyre in collecting lungfish eggs from our breeding ponds and in the subsequent care of the embryos and hatchlings. JMJ is also very grateful to Lennart Olsson and Zerina Johanson for retrieving some of the old literature on *Lepidosiren* and *Protopterus* and Lennart Olsson for providing Figs 3 and 4. Support was provided over many years by several grants from the Australian Research Council and Macquarie University Research Committee to JMJ.

References

Balon, E.K. 1991. Probable evolution of the coelacanth's reproductive style: lecithotrophy and orally feeding embryos in cichlid fishes and in *Latimeria chalumnae*. *Environmental Biology of Fishes* 32: 249-265.

Balon, E.K., M.N. Bruton and H. Fricke. 1988. A fiftieth anniversary reflection on the living coelacanth, *Latimeria chalumnae*: some new interpretations of its natural history and conservation status. *Environmental Biology of Fishes* 23: 241-280.

Bemis, W.E. 1984. Paedomorphosis and the evolution of the Dipnoi. *Paleobiology* 10: 293-307.

Boisson, C., C. Mattei and X. Mattei. 1967. Troisième note sur la spermiogenenèse de *Protopterus annectens* [Dipneuste] du Sénégal. *Bulletin de l'Institute Fondamental Afrique Noire*, A29, 1097

Ericsson, R., J. Joss and L. Olsson. 2007. The fate of cranial neural crest cells in the Australian lungfish, *Neoceratodus forsteri*. *Journal of Experimental Zoology* B 308: 757-768.

Falck, P., J. Joss and L. Olsson. 2000. Cranial neural crest cell migration in the Australian lungfish, *Neoceratodus forsteri*. *Evolution and Development* 2: 179-185.

Follenius, E.J. 1965. Particularités de structure des spermatozoïdes de *lampetra planeri*: Etude au microscope électronique. *Journal of Ultrastucture Research* 13, 459-468.

Greenwood, P.H. 1957. Reproduction in the East African lung-fish *Protopterus aethiopicus* Heckel. *Royal Zoological Society of London* 130: 547-567.

Greenwood, P.H. 1986. on p. 5 of your ms the year mentioned is 1987. Please check. The natural history of African lungfishes. *Journal of Morphology* (Supplement) 1: 163-179.

Grigg, G. 1965. Spawning behaviour in the Queensland lungfish, *Neoceratodus forsteri*. *Australian Natural History* 15: 75.

Hanken, J. and B.K. Hall. 1993. *The Skull*, Volumes I-III. University of Chicago Press, Chicago.

Huettner, A.F. 1941. *Fundamentals of Comparative Embryology of the Vertebrates*. The Macmillan Company, New York.

Jespersen, A. 1971. Fine structure of the spermatozoan of the Australian lungfish *Neoceratodus forsteri* (Krefft). *Journal of Ultrastructure Research* 37: 178-185.

Joss J.M.P. 2006. Lungfish evolution and development. *General and Comparative Endocrinology* 148: 285-289.

Joss, J.M.P., P.S. Rajasekar, R.A. Raj-Prasad and K. Ruitenberg. 1997. Developmental endocrinology of the dipnoan, *Neoceratodus forsteri*. *American Zoologist* 37: 461-469.

Kemp, A. 1982. The embryological development of the Queensland lungfish, *Neoceratodus forsteri* (Krefft). *Memoirs of the Queensland Museum* 20: 553-597.

Kemp, A. 1995. On the neural crest cells of the Australian lungfish. *Bulletin du Museum National d' Histoire Naturelle Section C* 17: 343-357.

Kemp, A. 1999. Ontogeny of the skull of the Australian lungfish *Neoceratodus forsteri* (Osteichthyes: Dipnoi). *Journal of Zoology* (London) 248: 97-137.

Kemp, 2000. Early development of neural tissues and mesenchyme in the Australian lungfish *Neoceratodus forsteri* (Osteichthyes: Dipnoi). *Journal of Zoology* (London) 250: 347-372.

Kerr, J.G. 1909. Normal plates of the development of *Lepidosiren paradoxa* and *Protopterus annectens*. In: *Normentafeln zur Entwicklungsgeschichte der Wirbeltiere*. F. Keibel (Ed.). Gustav Fischer, Jena, pp. 1-27.

Kerr, J.G. 1919. *Textbook of Embryology*, Volume II. Vertebrata. MacMillan and Co. Ltd., London.

Kundrát, M., J.M.P. Joss and M.M. Smith. 2008. Fate mapping in embryos of *Neoceratodus forsteri* reveals cranial neural crest participation in tooth developmental as conserved from lungfish to tetrapods. *Evolution and Development* 10(5): 531-536.

Kundrát, M., J.M.P. Joss and L. Olsson. 2009. Prosencephalic neural folds in Australian lungfish. *Journal of Experimental Zoology B (Molecular and Developmental Evolution)* 312B: 83-94.

Lange, R.H. 1983. Les cristaux de lipovitelline-phosvitine dans l'ovocyte de *Latimeria chalumnae* Smith 1939 (Coelacanthidae, Pisces). Etude comparative. *C R Acad Sc Paris* Series III 297: 393-396.

Lofts, B. 1974. *Physiology of the Amphibia*, Volume II. Academic Press, New York.

Martin, C.C. and R. Gordon. 1995. Differentiation trees, junk DNA molecular clock, and the evolution of neoteny in salamanders. *Journal of Evolutionary Biology* 8: 339-354.

Mattei, X. 1969. *Contribution a l'étude de la spermiogenèse et des spermatozoïdes de poissons par les méthodes de la microscopie électronique*. Faculty of Science, University of Montpellier (Thèse).

Ohno, S and N.B. Atkin. 1966. Comparative DNA values and chromosome complements of eight species of fishes. *Chromosoma* 18: 455-466.

Olsson, L., U. Hossfeld, R. Bindl and J.M.P. Joss. 2004. The development of the Australian lungfish *Neoceratodus forsteri* (Osteichthyes, Dipnoi, Neoceratodontidae): From Richard Semon's pioneering work to contempoary approaches. *Rudolstadter naturhistorische Schriften* 121: 51-128.

Pedersen, R.A. 1971. DNA content, ribosomal gene multiplicity, and cell size in fish. *Journal of Experimental Zoology* 177: 65-78.

Rock, J., M. Eldridge, A. Champion, P. Johnston and J.M.P. Joss. 1996. Karyotype and nuclear DNA content of the Australian lungfish, *Neoceratodus forsteri* (Ceratodidae: Dipnoi). *Cytogenetics and Cellular Genetics* 73: 187-189.

Semon, R. 1893. Die äussere Entwicklung von *Ceratodus forsteri*. *Denkschr. Med.-Naturwiss. Ges. Jena* 4: 113-135.

Smith, C.L., C.S. Rand, B. Schaeffer and J.W. Atz. 1975. *Latimeria*, the living coelacanth, is ovoviviparous. *Science* 190: 1105-1106.

Wourms, J.P., J.W. Atz and M.D. Stribling. 1991. Viviparity and the maternal-embryonic relationship in the coelacanth *Latimeria chalumnae*. *Environmental Biology of Fishes* 32: 225-248.

Subject Index

Acrodin 159
Acrosome 174, 175, 176, 232, 278
Acrosome reaction 176
Actinopterygii 104, 105, 127, 151, 154, 155, 158, 159, 160, 161, 162, 163, 166, 167, 168, 169
Actinopterygii: Ganoin 159
Actinotrichs 143
Adelphophagous 36
Adelphophagy 6, 25, 31, 36
Adhesive organs 251, 262, 263, 264
Alimentary canal 216
Alimentary tract 208, 210, 249, 252, 253, 256, 260, 261, 263, 267, 269
Amphibian development 286
Amphibian gastrula 152
Amphibian gastrulation 279
Amphibian neurulation 281
Amphibians 120, 150, 151, 159, 161, 166, 277, 279, 281, 286, 287
Anal fin 150
Anal opening 118
Anchor 173, 183
Animal 178
Animal cap 184, 221
Animal hemisphere 178, 182, 184, 186, 221
Animal pole 109, 111, 113, 114, 115, 116, 117, 150, 152, 172, 173, 177, 178, 182, 183, 184, 221, 233, 234, 235, 242
Animal pole micromere 113

Anovular egg cases 5
Anterior pit line 144
Aorta dorsalis 137
Appendicula 49
Appendiculae 41, 43, 45
Apterolarvae 155
Archenteron 78, 119, 120, 121, 184, 185, 186, 188, 221, 237, 243
Artificial insemination 13
Atrium 143
Attachment glands 106, 121, 122, 123, 125, 126, 127, 128, 129, 130, 132, 133, 134, 135, 136, 137, 138, 139, 140, 142, 144, 148, 153, 154, 157, 159, 160
Auditory organ 246
Auditory placodes 160
Auditory primordia 246

Batoids 3, 67, 69, 74, 75, 76, 78
Blastocoel 115, 116, 118, 152, 183, 184, 221, 236, 237, 238
Blastoderm 221, 234, 236, 237, 238, 239, 240, 241, 242, 244, 269
Blastodiscs 13, 47, 52, 54, 55, 57, 60, 61, 62, 64, 72, 233, 236, 237, 238, 241, 242
Blastomeres 110, 113, 114, 115, 152, 178, 182, 183, 234, 235, 236, 237, 242, 261, 269
Blastopores 45, 69, 78, 79, 117, 118, 120, 122, 152, 182, 183, 184, 185, 186, 188, 239, 240, 241, 247, 253, 279, 281

Blastula 110, 115, 116, 183, 184, 186, 221, 234, 235, 236, 237, 241, 257, 269, 286
Blastula cavity 183
Blood cells 133, 134, 136
Bottle cells 116, 117, 152, 183, 221, 242, 243
Brain 120, 122, 125, 126, 127, 128, 130, 132, 138, 162, 166, 168, 182, 186, 189, 190
Bulbus cordis 143
Byssal fibers 8, 11

Candle 26
Capillary networks 140, 142, 143
Carbohydrates 15, 16, 31
Caudal fin 147, 148, 155, 158
Caudal fin fold 142
Caudal finrays 149
Cells 206
Cement organs 284
Central nervous system 118, 125, 127
Chalazae 15
Chalaziferous chamber 15, 16, 17
Chiasma opticum 127
Chondromesoderm 187
Chondrostei 105, 147, 151, 154, 159, 160
Chorion 173
Ciliated epidermal cells 120, 125, 133, 135, 138, 169
Cladistia 104, 158, 162, 163, 164, 167, 168, 169
Cladistian 163, 169
Clear crescent 177, 178, 183, 187, 220
Cleavage furrows 113, 114, 115
Cleavages 2, 47, 52, 54, 59, 61, 79, 178, 182, 210, 220, 221, 222, 231, 232, 233, 234, 235, 236, 238, 242, 268, 276, 278, 279
Cloacal aperture 129, 130
Cloacal tube 132, 135, 145
Coelacanths 275, 277
Coelom 145
Colloid 15, 16, 17
Colloid jelly 14, 15, 16, 17
Convergent extension 279
Cornea 145
Coronoid fangs 144

Coronoid teeth 143
Cortical 173
Cortical granules 177, 178
Cortical reaction 177, 220
Courtship 277
Cushion 241, 243
Cytoplasm 129, 178, 216, 218

Demersal eggs 277
Dental arcades 143
Dermohyal 159, 162
Deuterencephalon 120, 122, 125, 126, 127
Deuterostomic 278
Developing amphibian 279
Development 276, 278, 284, 286
Developmental 276, 286
Diapause 45, 47, 49, 51
Diapausing 50
Dipnoans 151, 154, 169
Dipnoi 154, 159, 165
Dorsal 184, 238
Dorsal aorta 267
Dorsal fin 148, 155, 168
Dorsal forerunner cells 243
Dorsal lip 183, 184, 239, 240, 243
Dorsal scales 25
Dorsolateral denticles 21
Dorsolateral line 147
Dorsolateral scales 21
Ducts of Cuvier 205
Ductus endolymphatici 152
Ductus omphaloentericus 252, 253, 261, 262, 269
Ductus vitellointestinalis 79, 83

Eclosion 6, 13, 16, 17, 18, 21, 23, 69, 83
Eclosion gland 16, 17
Ectoderm 184, 186, 187, 188, 190, 192, 239, 240, 243, 244, 246, 263, 264, 268
Egg blastula 238
Egg candles 26
Egg case eclosion 8
Egg case hatching 11
Egg cases 2, 6, 8, 9, 10, 11, 13, 14, 15, 16, 17, 18, 19, 21, 23, 25, 28, 30, 33, 35, 36, 38, 41, 45, 47, 52, 54, 72

Egg cytoplasm 173, 177
Egg envelopes 108, 110, 111, 132, 133, 134, 145, 150, 157, 158, 159, 160, 165, 168, 172, 233, 277
Egg jelly 2, 5, 13, 14, 15, 16, 26, 30, 36, 41, 43
Egg laying rates 21
Egg membrane 153, 154, 164
Eggs 166, 172, 173, 176, 177, 178, 183, 186, 188, 210, 222, 233, 235, 263, 264, 268, 269, 276, 277, 278
Embryonic 79
Embryonic axis 45, 60, 62, 64, 83
Embryonic blastopore 69, 72
Embryonic fin-fold 127
Embryonic shield 238, 242
Endoderm 184, 186, 188, 222, 236, 242, 243, 249, 256, 259, 260, 263, 265, 267, 269
Enterocytes 216
Entoderm 116, 122
Entodermal 116, 120, 121, 129, 130, 137, 138, 159
Envelope 177, 233
Enveloping layer (EVL) 236, 239, 240, 245, 268
Epiblast 239, 242, 245, 246, 253, 262, 268
Epiboly 13, 26, 45, 57, 58, 60, 61, 69, 72, 79, 184, 236, 238, 242, 243, 268, 269, 279
Epidermis 120, 122, 125, 126, 127, 129, 130, 132, 135, 136, 139, 154, 156, 169
Epiphysis 127, 130, 132
Erythrocytes 136
Ethmoidal commissure 140, 144, 146
Ethmoidal line 146
Ethmoidal neuromast organ 146
EVL 236, 239, 240, 245, 268
Excretory 205
Excretory organs 202
Extended oviparity 9, 25
External gill (arteria hyoidea) 135
External gills 120, 125, 126, 127, 128, 129, 130, 132, 133, 134, 135, 136, 137, 139, 140, 142, 143, 144, 148, 150, 153, 156, 157, 159, 284

External gills 125
External yolk sac (EYS) 3, 28, 30, 36, 38, 40, 72, 75, 79, 83
Extra-embryonic coelom 79
Eye vesicle 127
EYS 3, 28, 30, 36, 38, 40, 72, 75, 79, 83

Fate mapping 281, 283
Fate maps 281
Fertilization 176, 177
Fin 76
Fin folds 130, 133, 136, 137, 140, 142, 143
Flagellum 174, 175, 176
Follicle 108, 110, 151
Foregut 123, 130, 132, 153, 158
Free embryonic phase 107

Gametes 176
Gas bladder 253, 255, 256, 258
Gastric stomach 210
Gastrula 117, 118, 184, 185, 187, 237, 239, 240
Gastrulation 47, 54, 60, 62, 64, 78, 79, 84, 116, 117, 118, 120, 151, 152, 158, 163, 165, 166, 168, 169, 183, 184, 185, 186, 187, 188, 220, 221, 232, 238, 241, 242, 243, 268, 269, 276, 279, 281, 286
Genomes 276, 278, 286
Germ cells PGC 249
Germ layers 183
Germ ring 238, 242, 269
Germinal vesicle 172
Germinative discs 54, 55, 59
Gill 171, 195, 231, 251, 264, 265
Gill filaments 2, 3, 18, 19, 31, 33, 38, 43, 64, 67, 69, 70, 74, 75, 76, 77, 78, 81, 84
Glandular 220
Glandular lumen 130, 132, 133
Glia cells 127
Gonadal crests 205, 206, 249, 250
Guanophores 135, 137, 142
Gular pit line 144
Gular regions 143
Gut 120, 122, 129, 130, 158, 184, 186, 188, 210, 216

Hatching 3, 6, 9, 10, 11, 13, 15, 17, 18, 19, 21, 25, 26, 28, 35, 38, 41, 67, 69, 70, 75, 78, 83, 84, 278, 284, 286
Hatchlings 284
Head enlargement 62, 64
Heart 49, 67, 69, 80, 81, 205, 207, 208, 209, 222
Hemisphere 183
Hepatocyte cytoplasm 220
Hepatocytes 257
Hepatopancreas 256
Histotroph 26, 30, 31, 33, 38, 39, 74
Histotrophic 6, 38, 76, 84
Histotrophy 6, 25, 26, 30, 33
Holoblastic 279
Horizontal pit line 144
Hyoid arch 67, 69
Hyoid artery 143
Hyomandibular cleft 67, 68, 69, 70
Hypoblast 236, 239, 240, 242, 244, 245, 259, 264, 267, 268, 269
Hypochord 266, 267
Hypochorda 127

Infraorbital lateral line 140, 144, 146
Infraorbital neuromast organ 146
Insipient histotrophy 26
Internal gills 284
Internal yolk sac (IYS) 39, 84
Intersegmental vessels 136, 137, 139
Intestinal spiral valve 171
Intestine 122, 132, 136, 137, 139, 142, 145, 158, 186, 187, 188, 210, 213, 214, 216, 217, 219, 220, 221, 248, 252, 253, 254, 255, 256, 257, 258, 260, 261, 269
Intrauterine cannibalism 6
Invagination 184, 195, 202
Involution 183, 184, 186, 221, 232, 239, 242, 269, 279
Iridiophores 137
Iris 145
IYS 69, 79, 83

Jaws 137, 139, 142, 144, 157
Juvenile infraorbital line 146
Juveniles 107, 138, 146, 147, 148, 149, 150, 153, 156, 161, 163

Kidneys 202, 204, 247, 250
Kupffer's vesicle 243

Labial folds 139, 140, 142, 143
Labial furrows 143
Larval phase 138, 148, 155, 156
Lateral line 125, 140, 142, 143, 144, 146, 147, 150, 153
Lateral line neuromast organs 140, 144
Lateral lips 186
Lecithotrophic 6, 8, 30, 39, 67
Lecithotrophy 26
Lens 125, 135, 137, 138
Lepidotrichia 148, 150
Lip 184, 238
Lipophores 135, 137
Lipovitellin 5
Liquefaction 16, 17, 41
Liver 184, 208, 216, 222, 252, 253, 254, 260, 261, 269
Lung 284
Lungfish 276, 277, 278, 279, 284, 286, 287
Lungfish early development 287

Macromeres 114, 115
Mandibular arch 67, 68, 69, 70, 74
Mandibular pit lines 144, 147
Marginal zone 183, 184, 186, 221
Matrotrophic 26, 30
Matrotrophy 6
Medullary plate 118
Megalecithal 5
Meiotic division 177
Melanin plug 220
Melanocytes 152
Melanophores 120, 123, 125, 129, 132, 137, 139, 142, 145
Mermaid purses 8, 30
Meroblastic 52
Meroblastic cleavage 5, 84
Mesenchyme 130
Mesentoderm 117
Mesentodermal 116
Mesoderm 184, 186, 187, 190, 195, 198, 202, 205, 239, 240, 243, 247, 249, 265
Mesolecithal 277

Mesonephros 158
Metamorphosing 286
Metamorphosis 284, 286
Micromeres 114, 115
Micropyle 173, 177
Microplicae 118
Micropylar region 172
Micropyles 108, 109, 110, 150, 151, 159, 160, 163, 169, 173, 176, 233
Midpiece 174, 175, 176, 232, 233
Mitochondria 128, 129, 176
Mitochondrial 105
Morula 115
Mucus secretion 106
Müllerian ducts 4, 158, 205, 250
Multioviparity 13
Multiple oviparity 6, 9, 10, 21, 23
Multiple paternity 4
Myomeres 132, 133, 135, 137, 139, 142, 145, 146, 153, 155, 159

Nasal openings 140, 142, 143, 145, 150
Nasal pit line 147
Nasal sac 139, 152, 153
Nasal tentacles 146, 150
Neopterygii 153, 160
Neotenic 286
Neoteny 286
Nephrotome 122
Nervus olfactorius 138
Nests 278
Neural 188
Neural arches 145
Neural crest 120, 122, 123, 125, 126, 130, 138, 152, 159, 187, 281, 282
Neural folds 47, 63, 64, 67, 118, 119, 120, 122, 123, 125, 126, 152
Neural groove 119, 122, 186
Neural plate 118, 119, 120, 122, 153, 182, 184, 186, 187, 188, 279, 281
Neural system 186, 187, 244, 258
Neural tube 122, 123, 125, 126, 127, 133, 152
Neuroectoderm 118, 120
Neuroectodermal 118, 119
Neurohypophysis 127
Neuromast organs 142, 143, 146, 147, 148, 150
Neurula 120, 123, 126, 187
Neurulation 118, 119, 123, 125, 152, 184, 186, 187, 188, 189, 202, 244, 276, 281, 286
Non-glandular stomach 213, 220
Normal tables 276
Notochord 120, 122, 123, 126, 127, 128, 130, 133, 140, 145, 171, 186, 187, 198, 200, 202, 205, 243, 244, 245, 255, 258, 266, 267, 281
Notochordal 132, 138
Notochordal triangle 62, 64, 83
Nucleoli 277
Nucleus 175, 176

Oesophagus 209, 210, 211, 212, 213, 214, 220, 253, 255, 258, 261
Olfactory bulbs 246
Olfactory nerve 246
Olfactory organ 189, 190, 191
Olfactory placodes 138, 139, 189, 190
Oocytes 172, 177, 182
Oolemma 173, 176
Oophagous 6, 35, 40
Oophagy 6, 25, 31, 36, 38
Ooplasm 177, 178
Opercular folds 140, 143, 144, 145
Operculum 79
Opisthonephric placodes 202
Opisthonephros 247
Optic 192, 193, 246
Optic lenses 189, 190
Optic placodes 189
Optic primordia 246, 247, 268
Optic vesicles 190
Oral cavity 127, 137
Organizer 241
Otic placodes 195
Otic vesicle 125, 195
Ovarian eggs 108, 110
Ovarian pore 35
Oviducal gland 45
Oviparity 6, 8, 9, 25
Oviparous 276
Oviposition 4, 6, 8, 9, 11, 16, 23, 25, 52, 62, 71, 278

Oviposition rate 19
Ovoviviparous 275

Paedomorphic 286
Pancreas 184, 216, 219, 260, 261, 262
Parthenogenesis 51
Parturition 6, 8, 28, 30, 33, 36, 38, 41, 47, 49, 51, 67, 76, 78, 84
Pectoral 76
Pectoral fin fold 136, 137, 139
Pectoral fins 31, 49, 68, 70, 71, 74, 75, 76,78, 81, 135, 139, 140, 143, 144, 145, 150, 153, 157, 158
Pectoral ridges 69
Pelagic eggs 277
Pelvic fins 149, 150, 155
Periblast 58, 232
Pericardium 122
Perivitelline fluid 173
Perivitelline space 111, 134, 151, 165, 177, 178
PGC 249
Pharyngeal pouches 64, 67, 68
Pharynx 135, 136, 137
Phosvitin 5
Phylotypic tail bud stage 287
Pigment granules 178, 210, 213
Pineal organ 134
Pit lines 140, 143, 144
Placental viviparity 6, 38
Placental viviparous 6, 41
Polyandry 4
Polyovular egg cases 9
Polypteridae 104, 108, 127, 138, 154, 155, 162, 164, 167, 168, 169
Polypterus 220
Polyspermy 2, 52, 55
Postembryonic 133, 135, 145, 153, 156, 157, 139
Postembryos 106, 107, 133, 134, 135, 136, 137, 139, 140, 142, 144, 145, 153, 155, 156, 159
Posterior crescent 59, 60, 61
Posterior lobes 64, 72, 83
Post-temporal lateral line 144
Preanal fin fold 145
Preopercular lateral line 140, 144

Preopercular line 146
Primary nasal pits 135, 136, 139, 140, 152
Primary nostril 140, 145
Primordial germ cells (PGCs) 204, 205, 250
Pronephric ducts 202, 203, 246, 247, 248, 249, 254, 255
Pronephros 118, 125, 128, 158, 182, 188, 202, 247, 248, 265
Pronephros body 126
Prosencephalon 120, 123, 125, 126, 127, 128, 130
Protease 16
Pterolarva 149
Pterolarval 157, 158
Pterygiophores 158, 161
Pyloric caeca 216

Quadratojugal pit line 144, 147
Quinine tanning 11

Rathke's pouch 189
Rectal occlusion 83
Rectal tube 133
Respiratory canals 2, 15, 16, 17, 18, 23, 28
Retained oviparity 9
Retina 145
Retinal 145
Rhombencephalon 127, 130, 132
Rostral bulb 78, 80, 81, 84
Rostrum 31, 69, 71, 74, 75, 76, 79, 146

Sarcopterygii 105, 127, 138, 151, 155, 158
Sclerotization 11
Secretory system 187, 188
Segmentation 52, 54, 55
Segmentation cavity 50, 54, 55, 59, 60, 61
Sexual organs 205
Shell gland 5, 8, 10, 11, 15, 36, 54
Somatopleura 145
Somites 119, 120, 122, 127, 129, 130, 132, 133, 134, 135, 137, 144, 148, 159, 187, 198, 200, 202

Spawning 278
Sperm 4, 176, 277, 278
Sperm cells 174, 175, 176
Sperm pronuclei 54
Sperm storage 4, 36, 51
Sperm tail 232
Spermatozoa 174, 176, 177, 232
Spermatozoon 174, 177
Sphincter 213
Spines 21, 25, 74, 76, 78
Spiracles 31, 33, 43, 49, 67, 69, 75, 77, 78
Spiral 210, 220
Spiral intestine 77, 79, 83, 216
Spiral valve 205
Splanchnopleura 145
Splanchnopleure 79
Spleen 262, 265
Staging tables 275
Stomach 188, 205, 210, 212, 213, 214, 215, 216, 220, 253, 255, 257, 260, 261, 269
Stomodaeum 135, 136, 139, 140
Store sperm 4, 49
Subepidermis 120, 138
Subgerminal cavity 237, 238, 239, 243, 244, 260, 264
Supraorbital lateral line 140, 144, 146, 147
Syncytium 52, 54, 55, 56, 58

Tail filament 19, 23
Tail-bud 122
Teeth 284
Telencephalon 127, 138
Teleost 276
Teleostei 151, 153, 154, 155, 160, 162, 167
Telolecithal 5, 84, 277
Temporal lateral line 144, 147
Tendrils 8, 9, 11, 13, 30, 59
Tertiary egg envelope 5, 6, 8, 26, 30, 41, 43, 45
Tetrapodomorph fossil fishes 287
Thymus 265
Thyroid 265
Trophonemata 30, 31, 33, 41, 74, 76, 78

Umbilical artery 39
Umbilical stalk 39, 45, 49
Umbilical vein 39
Urmund-lippe 116
Urostyle 140, 144, 146, 147, 150
Uterine compartments 25, 26, 45
Uterine partitions 49

Vegetal 183
Vegetal hemisphere 178, 182, 186, 221
Vegetal pole 111, 113, 117, 172, 173, 177, 178, 182, 183, 184, 221
Vein 83
Vena caudalis 133, 136, 137
Vena subintestinalis 133, 137
Venae cardinales 137
Venae vitellinae 133, 134, 136
Ventricle 143
Vertical pit lines 140, 144, 147
Vitelline artery 39, 72, 83
Vitelline blood vessels 47
Vitelline circulation 258, 261
Vitelline membrane 111, 113, 153, 154
Vitelline scar 28, 30, 79
Vitelline vein 39, 68, 72, 74
Vitelline vessels 72, 80
Vitellogenin 5
Vomeral 143

Wolffian duct 202, 205, 207

Xenopus 222

YCL 60, 79, 80, 83
Yolk 178, 182, 183, 184, 186, 188, 198, 210, 213, 216, 220, 221, 276, 277, 278, 279, 284, 286
Yolk blastopore 72
Yolk cytoplasmic layer (YCL) 58
Yolk endoderm 205
Yolk globules 120, 122, 132, 138, 140
Yolk granules 54, 56
Yolk merocytes (yolk nuclei) 52
Yolk nuclei 52, 54, 55, 56, 57, 58, 80, 83
Yolk platelets 5, 54, 56, 79, 80, 279
Yolk sac viviparity 6, 8, 25, 28
Yolk sac viviparous 6, 13, 25, 26, 28

Yolk stalks 33, 39, 74, 76, 77, 80, 83
Yolk syncytial layer 79, 210, 221, 232, 239, 268
Yolk-plug 116, 117, 118, 128
Yolksac 107, 118, 120, 122, 128, 152, 195, 205, 213, 220, 234, 250, 251, 253, 257, 258, 260, 266
Yolksac endoderm 210
Yolky 185
YSL 234, 237, 238, 240, 253, 257

Zona radiata 108, 109, 110, 111, 150, 151, 153, 173, 233
Zona radiata interna 173
Zygote 110, 111, 177

Species Index

Acipenser 202, 205, 221, 231
Acipenser baerii 189, 195, 200, 202, 210, 211, 213, 214, 216, 217
Acipenser brevirostrum 174, 175, 176
Acipenser fulvescens 174, 175
Acipenser gueldenstaedti 176, 178, 190, 202, 210, 212, 213, 215, 218, 219, 220
Acipenser gueldenstaedti colchicus 171, 172, 175
Acipenser medirostris 210, 213, 216, 220
Acipenser naccarii 205, 208, 210, 213, 216, 220
Acipenser oxyrhynchus 172, 175, 176
Acipenser oxyrhynchus oxyrhynchus 171, 174
Acipenser ruthenus 150, 168, 171, 172, 173, 182, 186, 187, 188, 189, 190, 191, 192, 193, 194, 195, 196, 197, 198, 199, 200, 201, 202, 203, 204, 205, 206, 207, 209, 211, 213, 214, 215, 216, 217, 218, 220
Acipenser sinensis 174, 175
Acipenser stellatus 171, 172, 173, 174, 175, 176, 178, 183
Acipenser transmontanus 163, 170, 171, 172, 174, 175, 176, 178, 183, 184, 185, 187, 195, 213
Acipenseridae 170, 173, 178
Acipenseriformes 170, 172, 176, 198
Actinopterygii 170
Aetobatus flagellum 47

Aetobatus narinari 84
Agonus cataphractus 164
Alopias pelagicus 35
Alopias superciliosus 35
Alosa sapidissima 169
Amblyraja radiata 19, 83
Amblystoma punctatum 165
Amia 210, 232, 234, 236, 237, 241, 242, 244, 246, 252, 261, 262, 265, 266, 268, 269
Amia calva 152, 153, 154, 161, 163, 230, 231, 232, 249
Anoxypristis cuspidata 71
Atractosteus 230, 231, 234
Atractosteus spatula 260
Atractosteus tristoechus 249, 260
Atractosteus tropicus 231, 247, 248, 251, 252, 253, 254, 255, 256, 257, 258, 259, 260, 262, 263, 265, 267, 269

Barbus schuberti 169
Bathyraja parmifera 2, 19, 23
Betta splendens 154
Bombinator igneus 164
Brachydanio rerio 164, 169

Calamoichthys calabaricus 166, 167, 168, 169
Callorhinchus milii 8, 10, 21, 80, 81, 83
Carcharhinus acronotus 79
Carcharhinus altimus 4

Carcharhinus isodon 84
Carcharhinus limbatus 79, 84
Carcharhinus melanopterus 3, 67
Carcharhinus plumbeus 4, 83
Carcharias taurus 35, 36, 38, 84
Carcharodon carcharias 35
Centrophorus granulosus 5
Centrophorus uyato 1
Centroscymnus coelolepis 28
Centroscymnus owstoni 28
Cephaloscyllium ventriosum 10, 19, 25
Ceratodus forsteri 168
Cetorhinus maximus 35
Chiloscyllium griseum 18, 21
Chiloscyllium plagiosum 21, 51
Chiloscyllium punctatum 3, 18, 19, 71
Chlamydoselachus anguineus 1, 5, 28, 51, 54
Chondrostei 170
Cichlasoma dimerus 166
Cichlasoma urophthalmus 166
Conger myriaster 165
Cynolebias viarius 160

Danio malabaricus 169
Danio rerio 164, 244
Dasyatis akajei 33, 75
Dasyatis americana 4
Dasyatis brevis 47
Dasyatis laevigata 33, 75
Dasyatis sabina 75
Dasyatis sayi 47, 49, 50, 52

Eleutherodactylus cogui 257
Eptatretus burgeri 166
Erpetoichthys calabaricus 162
Esox lucius 154, 162, 168
Etmopterus perryi 1
Etroplus maculatus 165

Galeocerdo 2
Galeocerdo cuvier 2
Galeorhinus cuvier 2
Galeorhinus galeus 2, 23
Galeus atlanticus 23
Galeus melastomus 16, 23, 52
Ginglymostoma cirratum 2, 4, 5, 6, 28, 30, 71, 85
Gobio gobio 167
Gollum attenuatus 38, 40
Gymnura micrura 31, 33

Halaelurus boesemani 23
Halaelurus buergeri 13, 23
Halaelurus lineatus 23
Halaelurus natalensis 23
Halaelurus quagga 23
Haploblepharus edwardsii 54, 84
Hemiscyllium ocellatum 3, 18
Herotilapia multispinosa 154
Heterodontus francisci 10, 21
Heterodontus galeatus 21
Heterodontus japonicus 51, 52, 54, 71
Heterodontus portusjacksoni 3, 13, 16, 18, 21
Hexanchus griseus 2
Huso dauricus 172
Huso huso 171, 172, 173, 178, 186
Hydrolagus colliei 78, 79

Iago omanensis 41
Isurus oxyrinchus 35
Isurus paucus 35

Lamna nasus 35
Lamnids 36, 38
Lampetra planeri 278
Latimeria mendosa 275
Latimeria chalumnae 275
Lepidosiren 275, 276, 279, 284, 286
Lepidosiren paradoxa 165
Lepidosteus 253
Lepisosteus 210, 230, 231, 232, 234, 236, 238, 244, 264
Lepisosteus osseus 166, 235, 239, 240, 245, 246, 247, 248, 249, 250, 252, 253, 255, 256, 257, 260, 262, 266
Lethrinus nebulosus 161
Leucoraja erinacea 11, 13, 16, 19, 23, 67, 75
Lota lota 167

Megachasma pelagios 35
Micrura 31, 33

Mitsukurina owstoni 31, 35
Moenkhausia oligolepis 169
Mustelus canis 3, 39, 41, 43, 45, 47
Mustelus manazo 26

Negaprion brevirostris 4, 79
Neoceratodus 275, 276, 277, 278, 279, 281, 283, 284, 286, 287
Neoceratodus forsteri 165, 275, 276, 277, 278, 282, 286
Noemacheilus barbatulus 167

Odontaspis ferox 35
Odontaspis noronhai 31
Oncorhynchus mykiss 167
Oryzias latipes 169
Osteoglossum bicirrhosum 259

Parascyllium variolatum 21
Perca fluviatilis 168
Petromyzon 233
Petromyzon planeri 244
Polyodon 170
Polyodon spathula 150, 154, 161, 162, 171, 177
Polyodontidae 170, 171, 178
Polypterus 231
Polypterus bichir 104, 160
Polypterus delhezi 168
Polypterus endlicheri 161
Polypterus lapradei 163
Polypterus mokelembembe 168
Polypterus ornatipinnis 106, 108, 109, 110, 111, 112, 113, 114, 115, 117, 124, 125, 126, 128, 129, 130, 132, 133, 134, 135, 136, 137, 139, 140, 142, 145, 148, 149, 150, 154, 155, 161, 162, 163, 169
Polypterus senegalis 278
Polypterus senegalus 105, 106, 108, 109, 110, 111, 112, 113, 114, 115, 119, 121, 123, 124, 125, 126, 128, 129, 130, 132, 133, 135, 137, 139, 140, 142, 143, 144, 145, 148, 149, 150, 154, 155, 161, 162, 163, 165, 166, 167, 169
Polypterus teugelsi 162
Potamotrygon circularis 74
Prionace glauca 2, 41

Pristis 69, 78
Proscyllium habereri 13, 23
Proscyllium magnificum 23
Protopterus 275, 276, 278, 279, 284
Protopterus annectens 162, 165, 278
Protopterus dolloi 162
Psephurus 170
Pseudotriakis microdon 38
Psuedocarcharias kamoharai 35
Pterophyllum eimekei 166
Pterophyllum scalare 154
Pteroplatytrygon violacea 47, 49

Raja asterias 16
Raja binoculata 9, 10, 13
Raja clavata 4, 21
Raja eglanteria 3, 13, 18, 19, 21, 52, 62, 67, 71, 72, 84
Raja pulchra 9, 10
Rhincodon typus 1, 6, 28, 30
Rhinobatos cemiculus 47
Rhinobatos halavi 3, 67
Rhinobatos horkelii 47, 51
Rhinobatos hynnicephalus 47
Rhinobatos productus 47, 51
Rhinobatos rhinobatos 45, 47
Rhinoptera bonasus 76
Rhizoprionodon taylori 47, 49
Rhizoprionodon terraenovae 3, 43, 79
Rhynchobatus djiddensis 3, 67

Salmo gairdneri 164, 169
Salmo salar 164
Scaphirhynchinae 170
Scaphirhynchus albus 174, 175
Scoliodon laticaudus 5, 41, 43, 45, 49, 84
Scyliorhinus canicula 3, 13, 16, 18, 19, 52, 54, 59, 60, 64, 67, 68, 69, 70, 72, 83
Scyliorhinus retifer 18, 21
Scyliorhinus stellaris 16
Scyliorhinus torazame 13, 14, 15, 16, 17, 18, 21, 23, 85
Scymnodalatias albicauda 2
Serenoichthys kemkemensis 163
Serranus atrarius 244
Silurus glanis 151
Sphyrna tiburo 4, 41, 84

Sphyrna tudes 51
Squalus acanthias 2, 13, 26, 51, 79
Stegostoma fasciatum 19, 21, 23, 25
Stizostedion vitreum 166
Sympterygia acuta 8
Sympterygia bonapartei 21

Taxiphyllum barbieri 106
Torpedo marmorata 2, 16, 52, 60
Torpedo ocellata 16, 52
Triaenodon obesus 84
Triakis megalopterus 28
Trygonoptera personalis 47, 51
Trygonorrhina fasciata 47, 49

Urolophus halleri 52, 76, 77
Urolophus lobatus 76

Xenopus 183, 184, 185, 208, 221, 279
Xenopus laevis 167, 221